ELECTRONIC PROPERTIES
OF POLYMERS

ELECTRONIC PROPERTIES OF POLYMERS

Edited by

J. MORT
Xerox Corporation
Webster, New York

G. PFISTER
Cerberus Corporation
Männedorf, Switzerland

A WILEY-INTERSCIENCE PUBLICATION

JOHN WILEY AND SONS
New York • Chichester • Brisbane • Toronto • Singapore

Copyright © 1982 by John Wiley & Sons, Inc.

All rights reserved. Published simultaneously in Canada.

Reproduction or translation of any part of this work beyond that permitted by Section 107 or 108 of the 1976 United States Copyright Act without the permission of the copyright owner is unlawful. Requests for permission or further information should be addressed to the Permissions Department, John Wiley & Sons, Inc.

Library of Congress Cataloging in Publication Data:

Main entry under title:

Electronic properties of polymers.

   "A Wiley-Interscience publication."
   Bibliography: p.
   Includes index.
   1. Polymers and polymerization—Electric properties. I. Mort, J. II. Pfister, G. (Gustav)

QD381.9.E38E38	547.7	82-2814
ISBN 0-471-07696-1		AACR2

Printed in the United States of America

10 9 8 7 6 5 4 3 2 1

CONTRIBUTORS

D. Baeriswyl   *RCA Laboratories, Zurich, Switzerland*

G. Harbeke   *RCA Laboratories, Zurich, Switzerland*

H. Kiess   *RCA Laboratories, Zurich, Switzerland*

W. Klöpffer   *Battelle Institut, Frankfurt am Main, West Germany*

W. Meyer   *RCA Laboratories, Zurich, Switzerland*

J. Mort   *Xerox Corporation, Webster, New York*

G. Pfister   *Cerberus Corporation, Männedorf, Switzerland*

J. J. Ritsko   *Xerox Corporation, Webster, New York*

G. M. Sessler   *Technische Hochschule, Darmstadt, West Germany*

Y. Wada   *University of Tokyo, Tokyo, Japan*

PREFACE

The study of the electronic properties of polymers and polymeric systems extends back in time over several decades. Nonetheless, this area of solid state science is still perhaps largely unfamiliar to many involved in materials science research or to those for whom the combination of electronic function and mechanical features of polymers might suggest practical applications. This book, which consists of contributions by leading research workers in various areas of the electronic properties of polymers, is intended to acquaint this broader audience with the most recent developments and the current understanding and use of this aspect of polymers.

J. Mort
G. Pfister

Webster, New York
Männedorf, Switzerland
January 1982

CONTENTS

1 Introduction   **1**
  by J. Mort and G. Pfister

2 Electronic States and Triboelectricity   **13**
  by J. J. Ritsko

   2.1 Introduction, 14
   2.2 Extended States—Band Structure, 15
   2.3 Localized States, 24
   2.4 Spectroscopic Techniques, 29
   2.5 Results, 35
   2.6 Triboelectricity, 46
   2.7 Summary and Acknowledgments, 52
   References, 54

3 Charge Storage   **59**
  by G. M. Sessler

   3.1 Introduction, 59
   3.2 Charging Techniques, 61
   3.3 Charge Measuring Techniques, 64
   3.4 Experimental Results and Interpretation of Real-Charge Storage, 77
   3.5 Experimental Results and Interpretation of Dipole Polarization, 90
   3.6 Applications, 96
   Acknowledgments, 100
   References, 100

4 Piezoelectricity and Pyroelectricity   **109**
  by Y. Wada

   4.1 Introduction and Basic Definitions, 111

- **4.2** General Equations of Piezoelectricity and Pyroelectricity in Polymers, 114
- **4.3** Piezoelectricity of Polypeptides, 117
- **4.4** Formation and Structure of Poly(vinylidene fluoride) Electrets, 121
- **4.5** Piezoelectricity and Pyroelectricity of Polar Crystals, 127
- **4.6** Piezoelectricity and Pyroelectricity of Electrets of Poly(vinylidene fluoride) and Other Polymers, 132
- **4.7** Piezoelectricity and Pyroelectricity of Polymer–Ceramic Composites, 140
- **4.8** Piezoelectric Relaxation, 142
- **4.9** Measuring Techniques, 145
- **4.10** Applications, 147
  Acknowledgments, 150
  References, 150

5 Energy Transfer 161
by W. Klöpffer

- **5.1** Introduction, 161
- **5.2** Basic Concepts of Electronic Energy Transfer, 164
- **5.3** Measuring Techniques, 180
- **5.4** Results and Interpretation, 187
- **5.5** Applications, 205
- **5.6** Summary, 206
  References, 208

6 Photoelectronic Properties of Photoconducting Polymers 215
by J. Mort and G. Pfister

- **6.1** Introduction, 215
- **6.2** Experimental Techniques, 217
- **6.3** Charge Transport, 220
- **6.4** Photogeneration and Photosensitization, 246
- **6.5** Chemical Control of Conductivity, 256
- **6.6** Summary, 262
  References, 263

7 Conducting Polymers: Polyacetylene 267
by D. Baeriswyl, G. Harbeke, H. Kiess, and W. Meyer

- **7.1** Introduction, 268

CONTENTS

- 7.2 Preparation and Characterization of Polyacetylene, 270
- 7.3 Theoretical Approach, 276
- 7.4 Pristine and Lightly Doped Polyacetylene: A One-Dimensional Semiconductor, 295
- 7.5 The Semiconductor–Metal Transition and Heavily Doped Polyacetylene, 305
- 7.6 Other Conducting Polymers, 313
- 7.7 Concluding Remarks, 316
    Acknowledgements, 317
    References, 317

Index **327**

ELECTRONIC PROPERTIES
OF POLYMERS

Chapter 1

INTRODUCTION

J. Mort
Xerox Corporation
Webster, New York

G. Pfister
Cerberus Corporation
Männedorf, Switzerland

Polymers are one of the best examples of the tangible transfer of laboratory research to the marketplace. They play a pervasive role in our everyday life, ranging from highly specialized applications in high technology contexts to more mundane mass-produced products found in almost every home. Until very recently, the applications have capitalized on the mechanical and chemical properties. In the hands of synthetic chemists these can be tailored to an amazing degree. Thus the ability to mold lightweight materials of controllable mechanical properties coupled with chemical inertness and durability accounts for what may truly be called the Polymer Age. These aspects of polymeric solids initially became the single focus of this fascinating class of materials, in large part because of the onset of World War II, which dictated the need for synthetic rubber and natural product substitutes. In fact, the most valued electrical property of polymers was their ability to inhibit conductivity, that is, to act as insulators.

By contrast with this development was the discovery and exploitation of the electronic properties of solids centered almost exclusively on elemental inorganic materials such as silicon and germanium. This resulted in the birth of the semiconductor industry. As a consequence the study of the electronic properties of organic solids, and in particular polymers, received only sporadic attention of solid state and materials scientists as potential electronic materials. For an excellent overall survey of this early work the reader is directed to *Organic Semiconductors* by F. Gutmann and L. E. Lyons (John Wiley & Sons, New York, 1967) and *Organic Semiconducting Polymers,* edited by J. E. Katon (Marcel Dekker, New York, 1968). It should be stressed that the progress made was constrained, as it is today, by the complexity of the problems, including the synthesis, purification, and characterization of materials that are inherently more complex than their inorganic counterparts. In addition, the physical nature and character of the electronic processes and phenomena are and can be expected to be qualitatively different than in inorganic solids. Indeed these questions are at the heart of current research and constitute one of the themes of this book.

During the 1970s a considerable renaissance of effort occurred in the study of polymers as electronic materials. This effort is characterized by a multidisciplinary approach involving physicists, chemists, and materials and device scientists. The renewed interest has been stimulated by the mutual interplay of both scientific and technological motivations. For the scientist, these polymer materials provide a challenge for new and often unconventional interpretations of the experimental results. Questions relating to interesting phenomena such as the origin of ferroelectric behavior in poly (vinylidene fluoride), soliton transport in polyacetylene, dispersive transport of charge and excitons in a hopping system, the origin and nature of localized states, as well as the microscopic processes involved in carrier and exciton motion will be a central theme of this book. Although substantial progress has been made in both theoretical interpretation and experimental capabilities, intriguing and challenging scientific questions remain.

From the technological point of view, it was an explicit goal of the early pioneers in the field to marry the materials properties of polymers with useful electronic characteristics. Significant economic advantage and technologic flexibility were perceived in the ability to fabricate polymers as low-cost, large-area elements or lightweight fibers coupled with electronic function. In this respect, although all the hopes or expectations have not yet been realized, the commercial use of polymeric solids as electronic materials is already an established fact and is perhaps more extensive than is widely perceived.

Perhaps the most dramatic example is the use of polymers in the multibillion dollar electrophotography business. Here, the phenomena of electronic transport, photosensitization, and triboelectricity are utilized in highly sophisticated synergy to yield a technology which is, in its own way, as impressive as the more familiar inorganic electronic applications. Another technological application of different phenomena found in polymers is the exploitation of piezoelectricity and pyroelectricity in polymers such as poly (vinylidene fluoride). As transducers that can convert thermal or sonic energy into electrical impulses, and vice versa, they have found or are finding applications in both commercial and military spheres as broad-band thermal or sonar detectors. Bolometers or stereophonic headphones or loudspeakers of high fidelity are commercially available. The good acoustic impedance match to water has led to significant activity in the area of underwater detection and in medical diagnostics of the body. The objective of a semiconducting or metallic polymer of demonstrated technologic value is a yet unattained goal. Nevertheless, these phenomena have been realized in polymers such as polyacetylene, including the ability to produce p- or n-type conductivity, and constitute one of the most active areas of research at the present time. These materials and phenomena clearly reside in the research phase and their full potential has yet to be defined.

Therefore, we have invited leading research workers and experts in the various areas of research into polymers as electronic materials to review in this book the current understanding and current problems in the field. Given limitations on length, the chapters are not intended to be all-embracing reviews of the field, but rather to present a detailed discussion of key ideas and directions with the objec-

INTRODUCTION

tive of delineating the most pressing and exciting problems. Specifically excluded is any detailed discussion of vibrational, morphological, mechanical, or thermal properties or synthesis except as these are specifically relevant to the electronic phenomena. Our objective has been to collect in one book the most up-to-date assessment of our understanding of the electronic properties of polymers and their actual or potential technologic applications. It is our hope that it will further highlight this rapidly developing and important field and stimulate current and new researchers and ideas in the field.

An essential precursor to understanding electronic properties of polymers is the knowledge of their electronic states. Chapter 2 discusses the progress made recently in the electronic spectroscopy of polymers and its interpretation in terms of the molecular structure of the polymers. High precision electron energy-loss spectroscopy, ultraviolet absorption, ultraviolet photoemission, and x-ray photoemission have all recently been applied to the spectroscopic investigation of a wide array of polymers. Electron excitations range from localized molecular-ion states in pendant-group polymers to essentially conventional delocalized band states in polymers such as polyacetylene. Such information, particularly that involving molecular ion states, has given the first quantitative understanding of important technological phenomena such as triboelectricity.

The phenomenon of charge storage, particularly in electrets, is described in Chapter 3. This probably constitutes one of the oldest and yet most difficult phenomenon involving, as it does, deeply bound electronic charges in polymers. Formidable experimental difficulties are encountered in acquiring and interpreting the appropriate experimental parameters, and much effort and ingenuity have been used in improving these techniques. Here again this phenomenon constitutes the basis of a significant technological application, namely, electret microphones with an annual production of ~ 100 million units. Thus a significant technological base requires a continuing effort to understand more fully the nature of electrets.

Piezoelectricity and pyroelectricity have their origin in volume charges and induced or permanent dipoles aligned in an external field at elevated temperatures and frozen in upon cooling. A prime objective in this field of research is to unravel the various contributions to the resulting permanent electrical (metastable) polarization and to understand the correlation between the phenomenon and the underlying molecular structure of the polymers. This is clearly complicated by the different structural forms in which the polymers can exist. The role of injected charge from contacts during the necessary poling process is not yet fully understood. These types of polymers are also an interesting example of attempts to develop composite materials that possess improved properties over either of the constituents. These questions are discussed in Chapter 4.

This is followed in Chapter 5 by a discussion of energy-transfer processes in polymer systems. Such phenomena are of great importance in molecular systems, and their study can reveal much about the underlying microscopic processes. The various mechanisms of energy dissipation following the absorption of light in polymers are described. The discussion includes the photophysical phenomena of fluorescence, phosphorescence, and excimer and exciplex emission. These mechanisms in turn necessitate a description of energy transfer in polymers, for instance, exciton migration and dipole–dipole interactions. Many of these processes are controlled by the same (localized) states which determine electronic transport and photogeneration. It has therefore been an important aim of research in this area to establish or suggest correlations between experimental results on energy transfer and photoelectronic processes in certain polymeric solids. Recent developments in this field involve the use of time-resolved spectroscopy in the picosecond and subpicosecond domain which allows one to actually freeze the molecular motion in time.

Chapter 6, on photoconductivity and transport, deals with the remarkable progress made in the last decade in understanding the broad features of electronic transport in polymeric systems based on direct measurement of transport phenomena. This progress has been greatly aided by the exploitation of molecularly doped polymers and by advances in the understanding of such processes in related disordered inorganic solids. A discussion of recent developments in the study of the photoelectronic properties of disordered organic solids of which polymers are a particular class is given. Topics cover electronic mobilities, lifetimes, photogeneration and photosensitization, and chemically controlled conductivity. Materials systems range from intrinsic photoconducting polymers to extrinsic systems such as molecularly doped polymers. The commonality between solid-state photoelectronic properties and general intermolecular charge/energy transfer involved in solid-state chemical reactions is stressed. New results on magnetic interactions in a molecularly doped polymer are presented. Much of the progress in these phenomena as well as the exciton and charge transport properties discussed in Chapters 5 and 6, respectively, draws heavily on the unique flexibility polymers offer in preparation and doping.

In the final chapter, recent work on conjugated organic polymers is presented which shows that it is possible by chemical doping to control electrical conductivities over many orders of magnitude and indeed achieve metallic-like conductivities. This most current and exciting study of the electronic properties of polymers became possible after major problems in materials preparation, notably of polyacetylene, were overcome and polymer films suitable for electrical measurements could be produced. This field is still in its infancy, and many of the current experimental results and interpretations are contradictory. The chapter describes the most commonly proposed theoretical models and gives an account of the essential experimental results of conductivity, magnetic susceptibility, and optical measurements. Although the most widely studied material is polyacetylene, other interesting materials with similar or related properties have recently been reported and are briefly discussed.

It is common to abbreviate long and frequently used polymer names. For quick reference, we list in Tables 1.1 and 1.2 respectively, the chemical formula and acronym of the polymers and small molecules most often referred to in this book.

Table 1.1 Polymers

Name	Repeat Unit	Acronym	Trade Name
Polyacetylene	$-(CH=CH)_x-$	$(CH)_x$	
Polyethylene	$-(CH_2-CH_2)_n-$	PE	
Poly(tetrafluoroethylene)	$-(CF_2-CF_2)_n-$	PTFE	Teflon
Poly(vinyl chloride)	$-(CH_2CHCl)_n-$	PVC	
Poly(vinyl fluoride)	$-(CH_2CF_2)_n-$	PVF	Tedlar
Poly(vinylidene fluoride)	$-(CH_2CF_2)_n-$	PVDF	Kynar
Polypropylene	$-(CH_2CH)_n-$ $\quad\quad\mid$ $\quad\ CH_3$	PP	
Poly(trifluoroethylene)	$-(CHF-CF_2)_n-$	PTrFE	
Polystyrene	$-(CH-CH_2)_n-$ $\quad\mid$ $\ \ C_6H_5$	PS	

Table 1.1 (*Continued*)

Name	Repeat Unit	Acronym	Trade Name
Poly(2-vinylnaphthalene)	$-(\text{CH}-\text{CH}_2)_n-$ with 2-naphthyl	P2VN	
Poly(1-vinylnaphthalene)	$-(\text{CH}-\text{CH}_2)_n-$ with 1-naphthyl	P1VN	
Poly(acenaphthylene)	$-(\text{CH}-\text{CH})_n-$ with acenaphthylene	PACN	
Poly(vinyl phenyl ketone)	$-(\text{CH}-\text{CH}_2)_n-$ with C=O–phenyl	PVPK	

Poly(vinylbenzophenone) — PVB

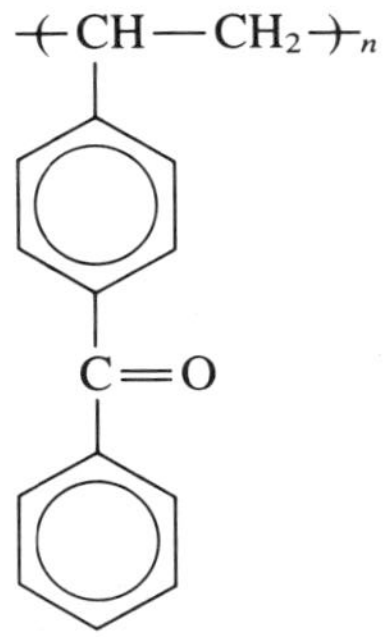

$\{CH-CH_2\}_n$ — C6H4—CO—C6H5

Poly(N-vinylphthalamide) — PVPI

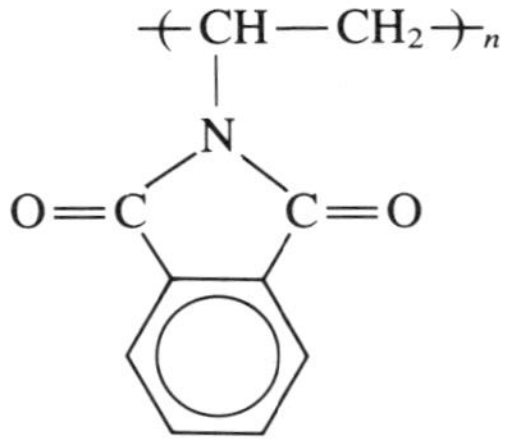

$\{CH-CH_2\}_n$ —N(CO)2C6H4

Poly(ethyl methacrylate) — PEMA

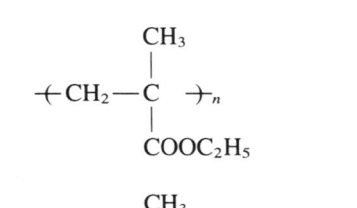

$\{CH_2-C(CH_3)(COOC_2H_5)\}_n$

Poly(methyl methacrylate) — PMMA

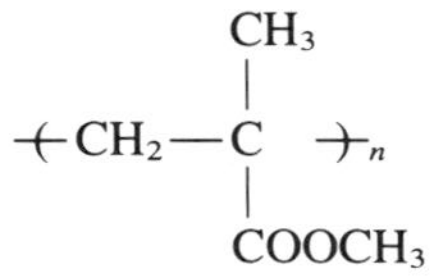

$\{CH_2-C(CH_3)(COOCH_3)\}_n$

Table 1.1 (*Continued*)

Name	Repeat Unit	Acronym	Trade Name
Poly(ethylene terephthalate)	$-(OC-C_6H_4-COOCH_2CH_2O)_n-$	PET	Mylar
Poly(*N*-vinylcarbazole)	$-(CH-CH_2)_n-$ with N-carbazole substituent	PVCA or PVK	
Poly(*N*-vinyl-2-bromocarbazole)	$-(CH-CH_2)_n-$ with N-(2-bromocarbazole) substituent	2Br-PVCA or 2Br-PVK	
Poly(*N*-vinyl-3-bromocarbazole)	$-(CH-CH_2)_n-$ with N-(3-bromocarbazole) substituent	3Br-PVCA or 3Br-PVK	

Poly(2-*N*-carbazolylethyl vinyl ether)   PCEVE

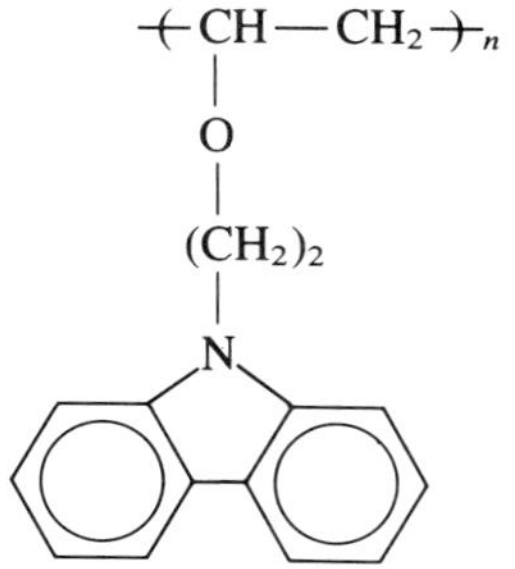

Bisphenol A polycarbonate   PC   Lexan / Makrolon

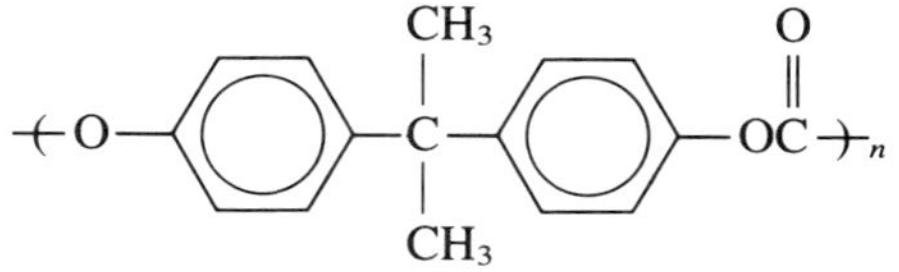

Poly(γ-benzyl-L-glutamate)   PBLG

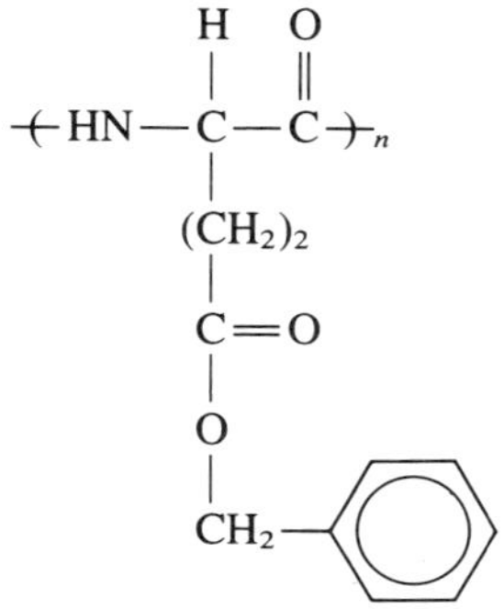

Table 1.1 (*Continued*)

Name	Repeat Unit	Acronym	Trade Name
Poly(γ-methyl-L-glutamate)	$-[\mathrm{HN-CH(CH_2)_2{-}C(=O){-}OCH_3}{-}C(=O)]_n-$	PMLG	

Poly(γ-methyl-L-glutamate) repeat unit:

$$-\!\!\left[\mathrm{HN}-\underset{\underset{\underset{\mathrm{OCH_3}}{|}}{\underset{\mathrm{C=O}}{|}}}{\underset{(\mathrm{CH_2})_2}{\overset{\mathrm{H}}{|}}}\mathrm{C}-\overset{\mathrm{O}}{\overset{\|}{\mathrm{C}}}\right]_n\!\!-$$

Table 1.2 Small Molecules

Name	Structure	Acronym
N-Isopropylcarbazole		NIPCA
Trinitrofluorenone		TNF
Triphenylamine		TPA
Tri-p-tolylamine		TTA
7,7,8,8-Tetracyanoquinodimethane		TCNQ
2,6-Diphenyl-4-(p-diethylaminophenyl) thiapyrylium tetrafluoroborate		BF_4^-

Table 1.2 (*Continued*)

Name	Structure	Acronym
Rhodamine B	$(C_2H_5)_2N$—[xanthene ring with O bridge]—$\overset{+}{N}(C_2H_5)_2$, with pendant phenyl-COOH group	Cl^-
Dimethyl terephthalate	$CH_3-O-\overset{O}{\underset{\|}{C}}$—[C$_6H_4$]—$\overset{O}{\underset{\|}{C}}-O-CH_3$	DMTP
Hexachloro-*p*-xylene	Benzene ring with CH_2Cl groups para, and four Cl substituents	HCX

Chapter 2

ELECTRONIC STATES AND TRIBOELECTRICITY

John J. Ritsko
Xerox Corporation
Webster, New York

2.1	Introduction	14
2.2	Extended States—Band Structure	15
	2.2.1 Tight-Binding Theory, 15	
	2.2.2 Detailed Calculations, 18	
	2.2.3 Limits of Band Theory, 23	
2.3	Localized States	24
	2.3.1 Molecular-Ion Model, 24	
	2.3.2 Dielectric Response, 25	
	2.3.3 Relaxation Energy, 26	
2.4	Spectroscopic Techniques	29
	2.4.1 Photoemission, 29	
	2.4.2 Optical Spectroscopy, 32	
	2.4.3 Inelastic Electron Scattering, 32	
2.5	Results	35
	2.5.1 Polyethylene, 35	
	2.5.2 Polyacetylene, 38	
	2.5.3 Polydiacetylenes, 40	
	2.5.4 Pendant-Group Polymers, 43	
2.6	Triboelectricity	46
	2.6.1 Extrinsic Surface States, 47	
	2.6.2 Intrinsic Bulk States, 49	
2.7	Summary and Acknowledgments	52
	References	54

This chapter reviews the current understanding of electronic states in polymers. Theoretical concepts and calculations are outlined and compared to experimental results obtained by photoemission, optical absorption, and inelastic electron scattering spectroscopy. The significance of these states for electrical conductivity, charge storage, and energy transfer is the subject of subsequent chapters, while the present understanding of electronic states involved in "triboelectricity," or contact electrification, is briefly described here.

2.1 INTRODUCTION

What are the states of electrons in polymers? A polymeric solid consists of an assembly of very long molecules such that within the molecular chain there are strong covalent bonds but that between chains only weak bonding occurs, usually of a van der Waals nature. The chains are made up of a very large number (sometimes 10^5 or more) of small identical units or a small number of units which are repeatedly bonded together. Each unit can be thought of as a separate molecule with electronic states consisting of the molecular orbitals of the molecule. In building the polymer model, the molecular orbitals which are degenerate on each unit overlap in space and lift their degeneracy by forming a series of extended electronic states, that is, energy bands (1). Thus bonding and antibonding molecular unit orbitals lead to polymer valence and conduction bands, respectively. Hence polymers can be considered as organic semiconductors, and the concepts of energy band theory can be used to characterize their electronic states and properties (2).

While the energy band picture of extended states has important consequences for many polymers and band structure calculations on a wide variety of polymers have been performed (3, 4), the electronic properties of narrow band systems such as saturated backbone pendant-group polymers and molecularly doped polymers are determined by localized electronic states characteristic of the small molecule or molecular unit. The photoconductive properties of such systems are dealt with in Chapter 6. In these systems, charge carriers do not exist in extended energy band states but rather are localized by polarization and relaxation effects that create a distribution of states in energy which exceeds the bandwidth due to overlap of the originally degenerate molecular orbitals (5). How disorder in energy localizes states is discussed in this chapter.

Structural disorder, that is, defects, is also very important for the electronic properties of polymers. Extrinsic defects such as chain end groups, impurities left over from the fabrication process, deliberate commercial additives, and the ever-present oxygen (in molecular form or bound in oxidized products) can be crucial to electronic transport and contact electrification in low conductivity polymers.

The ground, or lowest, energy state of polymers involves individual electrons in bonds, bands, or defects. The neutral excited electronic states are two-particle states involving electrons and holes which occur in the conduction and valence bands or form neutral molecular bound states called excitons. When generated by an external probe, the excitation spectrum provides the information needed to reveal the nature of the electronic states. In addition, as is described in Chapter 5, energy absorbed by molecular units can migrate through the polymers by energy transfer between molecular units. A bound state of an excited molecule and an unexcited molecule, or exciplex, may even occur.

There are also many-electron states in polymers which are due to correlated electronic motion. Plasmons are collective oscillations of the valence electrons.

EXTENDED STATES-BAND STRUCTURE

They are of practical importance because they are the main energy loss mechanism of fast charged particles in polymers. Plasmon excitation determines the mean free path and ultimately the range of fast charged particles used, for example, in making electrets (Chapter 3). Spectroscopically, the momentum dependence of plasmon excitations can give important information about the nature of electronic states—are they extended or localized?

This chapter describes the nature of electronic states in polymers. The present understanding of these states in terms of energy band theory and localization effects will be sketched. The theoretical predictions are compared with some recent experimental measurements of electronic states and their excitation spectra. The present discussion is not intended as a comprehensive review but rather as a description of some of the key ideas in our present understanding and of what further work needs to be done. The reader is referred to other works for reviews of the literature (2–5).

2.2 EXTENDED STATES—BAND STRUCTURE

2.2.1 Tight-Binding Theory

Many of the concepts associated with extended electronic states in polymers can be illustrated with a simple tight-binding band structure model originally due to Bloch (1, 6, 7). The polymer is considered as an infinite periodic array of sites with which are associated atomic wave functions φ that overlap slightly from site to site. The wave function $\psi(\mathbf{r})$ of one electron in the polymer must satisfy the Schrödinger equation,

$$H\psi = E\psi \qquad (2.1)$$

where E is the eigenvalue of the energy operator and H is the Hamiltonian,

$$H = \frac{-\hbar^2 \nabla^2}{2m} + V(\mathbf{r}) \qquad (2.2)$$

$V(\mathbf{r})$ is the potential energy. In this treatment, the many-electron wave function is a product of the single-particle solutions $\varphi(\mathbf{r})$, and $V(\mathbf{r})$ is a function of the resulting charge density. The tight-binding wave function made up of a linear combination of atomic orbitals which satisfies the Bloch condition for translational invariance is then

$$\psi(\mathbf{r}) = \sum_{ij} C_i\, e^{i\mathbf{k}\cdot\mathbf{R}_j}\, \varphi(\mathbf{r} - \boldsymbol{\xi}_i - \mathbf{R}_j) \qquad (2.3)$$

where $\mathbf{R}_j$ is the position of the unit cell, $\boldsymbol{\xi}_i$ is the position of atoms within the unit cell, and C_i is the coefficient for the ith atom. Putting $\psi(\mathbf{r})$ from Eq. (2.3) into Eq. (2.1) and taking the outer product with $\varphi(\mathbf{r} - \boldsymbol{\xi}_n - \mathbf{R}_m)$, we obtain

$$\langle \varphi(\mathbf{r} - \boldsymbol{\xi}_n - \mathbf{R}_m)|H| \sum_{ij} C_i e^{i\mathbf{k} \cdot \mathbf{R}_j} \varphi(\mathbf{r} - \boldsymbol{\xi}_i - \mathbf{R}_j)\rangle = EC_n e^{i\mathbf{k} \cdot \mathbf{R}_m} \quad (2.4)$$

where we used the approximation $\langle \varphi(\mathbf{r} - \boldsymbol{\xi}_n - \mathbf{R}_m)|\varphi(\mathbf{r} - \boldsymbol{\xi}_i - \mathbf{R}_j)\rangle = \delta_{in}\delta_{mj}$ since overlap between sites is small. In general, basis functions may be chosen to rigorously satisfy this condition. Considering nearest-neighbor interactions only, two types of integrals can arise, namely, on-site energies

$$\langle \varphi(\mathbf{r} - \boldsymbol{\xi}_i - \mathbf{R}_j)|H|\varphi(\mathbf{r} - \boldsymbol{\xi}_i - \mathbf{R}_j)\rangle = H_{ij} \quad (2.5)$$

and interaction energies, sometimes called resonance or transfer integrals,

$$\langle \varphi(\mathbf{r} - \boldsymbol{\xi}_n - \mathbf{R}_j)|H|\varphi(\mathbf{r} - \boldsymbol{\xi}_i - \mathbf{R}_j)\rangle = H_{ni} \quad (i \neq n) \quad (2.6)$$

At this point, Eq. (2.4) must be applied to a specific structure. As an example, consider the structure given in Fig. 2.1, which is a simple one-dimensional model for polyacetylene but which illustrates many of the consequences of tight-binding bands in general. A single atomic orbital (carbon p_z in the case of polyacetylene) is associated with each site. We consider the case where there are two atoms per unit cell of length a to allow the possibility of different bond lengths b_1 and b_2. Then let $H_{11} = \alpha = H_{22}$, $H_{12} = \beta_1$, and $H_{21} = \beta_2$. Writing out Eq. (2.4) for each atom (1 and 2) in the unit cell, a simple 2×2 matrix equation follows for the determination of the coefficients C_i. A nontrivial solution is possible only if the determinant of the resulting matrix is zero:

$$\det \begin{vmatrix} \alpha - E & \beta_1 + \beta_2 e^{-ika} \\ \beta_1 - \beta_2 e^{ika} & \alpha - E \end{vmatrix} = 0 \quad (2.7)$$

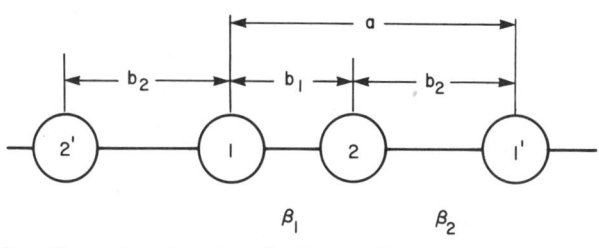

Figure 2.1 One-dimensional model of polyacetylene: carbon $2p_z$ orbitals on atoms 1 and 2 are separated by distances b_1 and b_2 resulting in transfer integrals β_1 and β_2.

EXTENDED STATES-BAND STRUCTURE

This matrix always has a characteristic form with on-site energies α along the diagonal and the transfer integrals β as off-diagonal elements. The solution to Eq. (2.7) is

$$E = \alpha \pm (\beta_1^2 + \beta_2^2 + 2\beta_1\beta_2 \cos ka)^{1/2} \qquad (2.8)$$

These energy bands are drawn schematically in Fig. 2.2a. Since there are two atoms per unit cell and one electron per atom, the lower band is completely filled, yielding a semiconductor with a band gap $\Delta = 2|\beta_1 - \beta_2|$ and with a total bandwidth $W = 2|\beta_1 + \beta_2|$. Electrons in these band states move along the chain as if they had an effective mass (1)

$$m^* = \frac{\hbar^2}{d^2E/d^2k} \qquad (2.9)$$

At the bottom of the valence band ($k = 0$), $m^* = 8\hbar^2/a^2W$; and at the bottom of the conduction band ($k = \pi/a$), $m^* = \hbar^2|\beta_1 - \beta_2|/a^2\beta_1\beta_2$. These expressions are important in understanding transport properties in a band model since the electrical conductivity σ is (1)

$$\sigma = ne\mu \qquad (2.10)$$

where n is the carrier density, e is the electrical charge, and μ is the mobility defined by

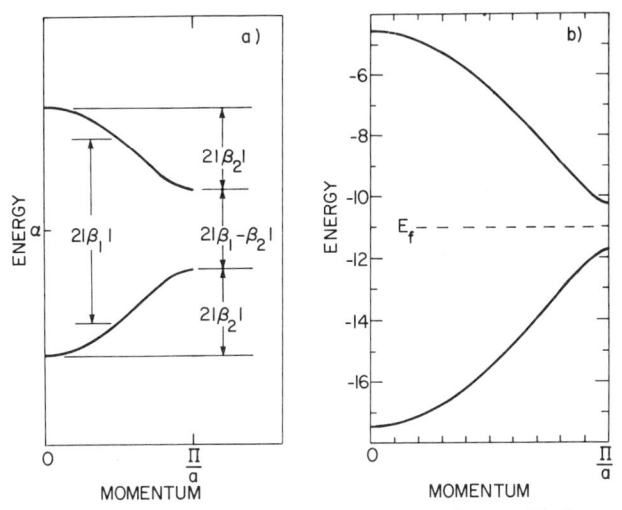

Figure 2.2 Tight-binding energy bands: (*a*) general form. (*b*) for polyacetylene modeled in Fig. 2.1 with parameters taken from *ab initio* calculation for graphite.

$$\mu = \frac{e\tau}{m^*} \qquad (2.11)$$

where τ is the electron-lattice scattering time which depends on the strength of the coupling between the charge carrier and the polymer backbone. Thus, once τ and the transfer integrals β are known, the electronic structure and measurable electrical properties can be calculated.

2.2.2 Detailed Calculations

At the cost of much computation time, the one-electron Schrödinger equation can be solved from first principles. Such *ab initio* techniques use special orthogonalized basis wave functions and realistic atomic potentials to calculate interaction integrals. Interactions beyond nearest-neighbor interactions, electron correlation, and exchange may be included. A less expensive alternative approach is to fix some interaction integrals empirically as is done in CNDO and other semiempirical formulations. For details of these calculations, the reader is referred to original papers. In this section, the tight-binding model is used to calculate bands in model polymers and a brief comparison is made with more complicated theoretical calculations. While some experimental results are discussed, a fuller comparison is made in Section 2.5.

In our simple tight-bonding model, the interaction energies β cannot be accurately calculated since the atomic orbitals on different sites are not orthogonal. With some effort, orthogonalized basis functions can be constructed and β computed using an appropriate superposition of atomic potentials. However, the method can also be made accurate and useful if the transfer integrals are simply determined empirically by fitting to first-principle calculations or accurate experimental measurements (8). This can be done easily for conjugated polymers because of the similarity of the bonding in these materials to that in graphite. All contain sp^2 hybridized carbon atoms with differing bond length x; and by fitting to an *ab initio* calculation for graphite (9), we obtain $\beta = -3.0$ eV for $x = 1.42$ Å. However, the differing bond lengths in conjugated systems require that we know $\beta(x)$. This can be computed for bond lengths near 1.4 Å from Eq. (2.6) if we assume that in the region between sites the atomic wave functions fall off exponentially as expected for Slater orbitals, that is, $\varphi(r) \sim e^{-\mu r}$. Then it is not hard to show that $d\beta = \mu\beta \, dx$, which leads to

$$\beta = \beta_0 \, e^{-\mu x} \qquad (2.12)$$

For carbon $2p$ orbitals, $\mu = 2.65$ Å^{-1} (9), and we find $\beta_0 = -128.59$ eV. Eq. (2.12) provides a surprisingly good estimate of π-electron resonance integrals as given in a recent theoretical review (10).

Recent calculations have shown that polyacetylene, $(CH)_x$, contains bonds of alternating lengths (3, 11). Although this point has been the subject of some controversy (12), *ab initio* calculations show this to be the lowest energy state

(11), and such behavior is expected from chemical intuition applied to short polyenes and by the necessary Peierls distortion in infinite one-dimensional systems (13). Using typical bond lengths of 1.35 and 1.44 Å (14), the π-band structure for $(CH)_x$ shown in Fig. 2.2b is calculated using the simple tight-binding model with transfer integrals β computed from Eq. (2.12).

This calculation can be compared with more detailed extended tight-binding (ETB) calculation by Grant and Batra (15) shown in Fig. 2.3, where band structures for different bond lengths are shown and both π and σ bands are calculated. The ETB method uses a linear combination of Bloch-adapted Gaussian orbitals as a basis set and constructs the crystal potential from a superposition of atomic potentials generated from self-consistent Hartree–Fock charge densities which include exchange effects. In our simple theory, the π bandgap Δ at $k = \pi/a$ goes to zero when the transfer integrals (i.e., the bond lengths) are equal as expected from the more detailed calculation, namely, $\Delta = 2|\beta_1 - \beta_2|$. The present calculation predicts $\Delta = 1.52$ eV and a total bandwidth $W = 2|\beta_1 + \beta_2| = 12.8$ eV, in good agreement with both detailed calculations (15) and optical absorption experiments (14–16). The effective mass m^* is $0.6m_0$ at $k = 0$, but m^* is only $0.1m_0$ at $k = \pi/a$. Assuming that $\tau \sim 10^{-14}$ s, which is a typical value (17), the free-carrier mobility along the chain, μ, should be very large, ~ 200 cm^2/V-s. Attempts to measure this quantity in polyacetylene have not yet been successful since the experimental drift mobility μ_D is strongly influenced by trapping of carriers and the necessity of hopping between polymer molecules that have finite lengths.

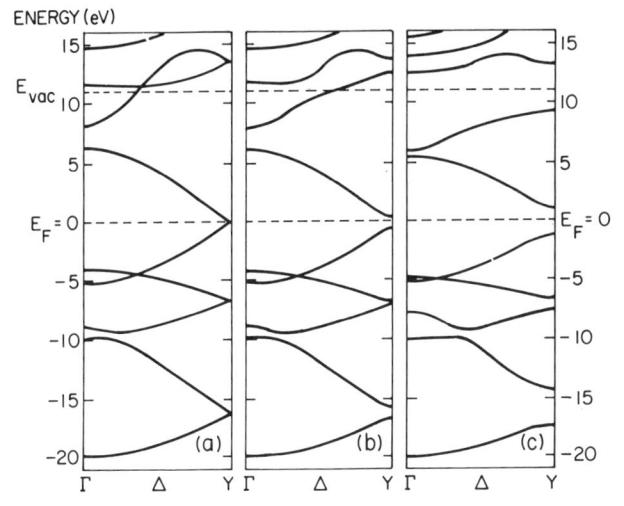

Figure 2.3 Band structure of trans-$(CH)_x$ for different carbon–carbon bond lengths: (a) uniform (1.39 Å); (b) weakly alternating (C—C 1.36 Å, C—C 1.43 Å); and (c) strongly alternating (C—C 1.34 Å, C—C 1.54 Å). Note the lifting of the degeneracy at Y as bond alternation occurs. (From Grant and Batra [15].)

The main point to the calculations of Grant and Batra (15) is that a one-electron model can provide an adequate description of the electronic structure of polyacetylene although their results are critically dependent on the bond lengths which are assumed. The important effect left out of these calculations is the correlated motion of electrons which is expected to be particularly important in one-dimensional systems (12). Ovchinnikov et al. (12) found that the optical bandgap could be accounted for if the on-site coulomb repulsion energy of two electrons, γ, was included using a Hubbard Hamiltonian with all bond lengths equal. By fitting the absorption peak in finite polyenes, they found $\gamma = 5.42$ eV, which is much larger than the transfer integral, $|\beta| = 2.4$ eV. Hence, correlation effects ought to be very significant, and it is puzzling as to how one-electron models that neglect this effect can be successful.

The importance of correlation, exchange, and the electron-hole interaction in $(CH)_x$ have been stressed in several recent reports (14, 18, 19). Kertesz (19) has shown that an *ab initio* LCAO Hartree–Fock self-consistent field band structure calculation which includes "exact" nonlocal exchange effects predicts a bandgap of about 7 eV. This drastic overestimation of forbidden gaps is a well-known drawback of many *ab initio* methods and is due to the neglect of electronic correlation. Taking account of correlation, the gap is reduced to below 4 eV. Furthermore, if the electron–hole interaction is included, the lowest energy excitation is an exciton at around 2 eV, in agreement with semiempirical results (14). Nevertheless, the uncertainties in all of the theoretical calculations are such that the true nature of the absorption edge in polyacetylene will only be resolved by experiment.

A second system to which our simple tight-binding model can be directly applied is carbyne, a pure carbon chain with alternating single and triple bonds. Using bond lengths of 1.20 and 1.45 Å (3), a bandgap of 5.2 eV and a filled valence band of width 5.5 eV is predicted. These values are consistent with the transparency of the material and with more detailed calculations on small polyenes which predict a bandwidth of 4.6 eV for a 10-carbon atom chain and *ab initio* calculations on carbyne which yield a 7.5 eV valence bandwidth (20).

For more complicated polymers such as the polydiacetylene structure shown in Fig. 2.4a, the simplest calculations seem to provide best agreement between theory and experiment (21). Wilson has used simple Hückel theory applied to the π-electrons in polydiacetlyene (21). This method is formally similar to the tight-binding method described in Section 2.2.1, differing only in how the transfer integrals are determined (21). In this calculation, the bandgap in polydiacetylene is (21)

$$\Delta = |2\beta_{C-C} - \beta_{C=C} - \beta_{C\equiv C}| \tag{2.13}$$

giving $\Delta = 1.36$ eV. Unfortunately, the onset of optical absorption is ≈ 2 eV. This might be considered a failure of simple one-electron models (12) were it not for the fact that in using Eq. (2.12) we calculate a gap $\Delta = 2.7$ eV for the same

EXTENDED STATES-BAND STRUCTURE

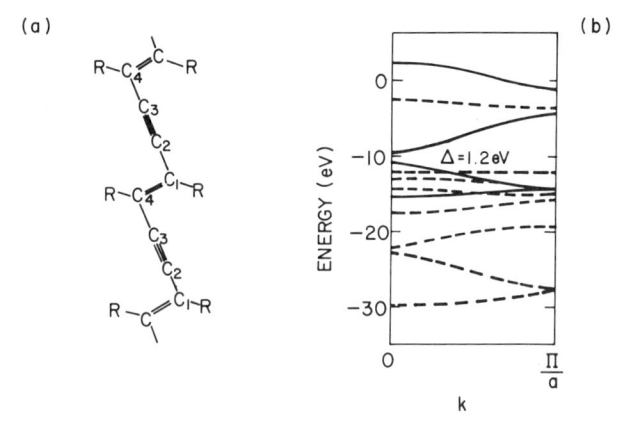

Figure 2.4 Polydiacetylene: (*a*) structure of a single extended chain; (*b*) electronic band structure with hydrogen atoms representing the side groups. (From Parry [22].)

bond lengths. This is in much better agreement with experiment since the absorption edge is known to be dominated by excitonic effects, a point we return to later (22). The conclusion here must be that while simple tight-binding or Hückel methods can indicate the general features of a band structure they should not be considered accurate predictive calculations. As a further example, the more detailed extended Hückel results of Parry (22) shown in Fig. 2.4*b*, while providing an interesting description of the relative shapes and positions of π- and σ-bands, predict a gap Δ of only 1.2 eV.

Ab initio methods should be predictive. However, when applied to two different polydiacetylene backbone structures, the *ab initio* self-consistent-field restricted Hartree–Fock calculations by Kertesz et al. (23) predict band gaps of 8–11 eV, far in excess of the experimental results. As in the case of polyacetylene, the effects of exchange, correlation, and the electron–hole interaction must be taken into account before *ab initio* methods can be expected to yield accurate predictions of the absorption spectrum.

A situation similar to that for conjugated polymers appears to exist in the case of saturated backbone polymers for which polyethylene is the simplest example. Using tight-binding methods, Imamura (24) has calculated the band structure, and it is in good agreement with the extended Hückel results of McCubbin (25–27) shown in Fig. 2.5. The large energy gap between valence and conduction bands can be qualitatively understood in our simple model sketched in Fig. 2.1. The sp^3 hybridization in polyethylene produces directed bonds in which electrons overlap strongly between two particular atoms yielding a large transfer integral $|\beta_1|$. But these same electrons interact only weakly with electrons forming the other bonds along the chain giving a small $|\beta_2|$. Hence from Fig. 2.2*a*, we expect narrow bonding and antibonding bands of width $2|\beta_2|$ separated by a large energy, $2|\beta_1|$.

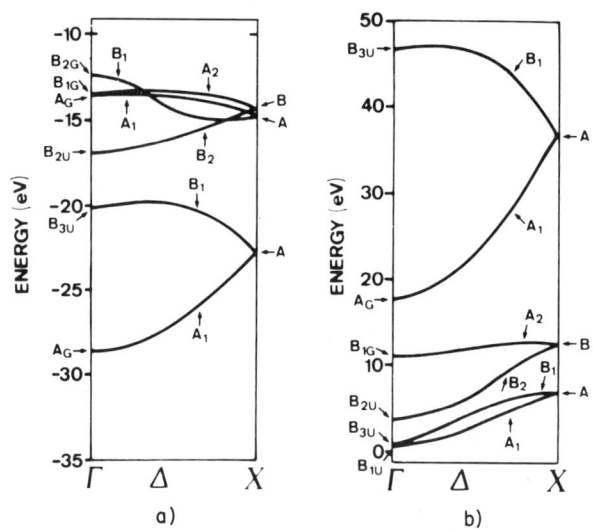

Figure 2.5 Polyethylene band structure: (*a*) filled valence bands; (*b*) empty conduction bands. (From McCubbin [25].)

Bloor (28) has compared the results of several semiempirical and *ab initio* band structure calculations on polyethylene with experimental determinations of the valence bandwidth, the energy gap, and the location of the conduction band edge with respect to the vacuum level. On the whole, the simplest semiempirical calculations appear to do as well as the more detailed *ab initio* results in matching experiment. However, extended Hückel and *ab initio* results considerably overestimate the bandgap (which is expected inasmuch as they did not include correlation and the electron–hole interaction). McCubbin (27) has stressed the limited predictive power of the one-electron band calculations that have been done for polyethylene. Excitations giving rise to optical and magnetic properties are subject to correlation effects which are neglected in one-electron pictures. On the other hand, electrical conduction and the spectrum of hole states are well described by valence band calculations which are in fair agreement for various theoretical calculations (27, 28) so that these quantities should accurately compare with experiment.

Using our model band structure fit to the bandwidth of the highest occupied valence band (24, 25), we calculate $m^* = 0.52m_0$ for the lowest energy holes in polyethylene in good agreement with previous results (27). Assuming $\tau = 10^{-14}$ s, the hole mobility is predicted to be $\mu = 34$ cm^2/V-s. The results of an *ab initio* LCAO SCF method predict $\mu = 25$ cm^2/V-s (29). Unfortunately, experimental measurements range from 10^{-4} to 10^{-11} cm^2/V-s (3), presumably because transport along a single molecule in a macroscopic sample is not possible and the necessity to hop between molecules leads to trapping of

EXTENDED STATES-BAND STRUCTURE

carriers. Thus measurements of electrical conduction and mobility in real polymer samples may not always be useful tests of band structure models.

When applied to polymers other than those considered thus far, band theory often predicts extremely narrow energy bands. For saturated backbone polymers, bands similar to those given in Fig. 2.5 for polyethylene will occur, but new bands associated with pendant groups, for example, will be very narrow. Consider polystyrene. The highest occupied states are the π-electrons of the benzene pendant group. However, since they are about 3.2 Å apart, Eq. (2.12) predicts a bandwidth of only ~ 0.05 eV, consistent with other estimates of this quantity (5). The effects of tacticity could further reduce the estimate of the overlap integral. This bandwidth is so small that the importance of thermal vibrations and disorder can make the concepts and conclusions of band theory meaningless in this case. Thus we must examine the applicability of band theory.

2.2.3 Limits of Band Theory

One criterion for the applicability of band theory comes from studies of organic molecular solids (30). Electrons moving along a polymer chain in an energy band are scattered by the lattice with the electron–lattice scattering time τ. This is a measure of how strongly the electrons interact with lattice vibrations, or phonons. In order to remain within the energy band of width W, the uncertainty principle requires that

$$W\tau > \hbar \qquad (2.14)$$

From the definition of mobility, Eq. (2.11), we conclude that for the band model to make sense,

$$\mu > \frac{\hbar e}{m^* W} \qquad (2.15)$$

Inserting the effective mass we calculated for our tight-binding band model, $m^* = 8\hbar^2/a^2 W$, we conclude that there is a minimum mobility required by band theory:

$$\mu_{min} = \frac{ea^2}{8\hbar} \qquad (2.16)$$

For $a = 3.2$ Å, $\mu_{min} \sim 0.2$ cm^2/V-s. This value is about the minimum mobility attributed to band motion in low-temperature measurements on solid crystalline naphthalene (31). While calculations for polyacetylene, polydiacetylene, and polyethylene all show mobilities considerably higher than μ_{min}, indicating that band models can be applied to these materials, the very narrow bands associated with pendant groups on polymers will make Eq. (2.14) very hard to satisfy. Unrealizable values for τ may be required. Thus when coupling to the lattice is included, the band model may not apply to pendant group polymers.

A second criterion for the applicability of band theory is that the states on each molecular subunit which are involved in building up the band be degenerate and that there be enough subunits joined together so that the concept of a continuous band makes sense. In the latter case, the bonding and antibonding orbitals of a short chain are better described as the discrete levels of a small molecule when the energy spacing between discrete electronic states is large. In the former case, any variation in the energy of a particular molecular orbital due to disorder or any nonperiodic difference in overlap between adjacent molecular subunits in a chain will tend to obstruct band formation by violating the translational invariance implicit in the one electron wave function in Eq. (2.3). In pendant-group polymers, the effects of disorder can be severe, and the electronic states due to the pendant groups are well localized on those groups.

2.3 LOCALIZED STATES

2.3.1 Molecular-Ion Model

Because the electronic states in band theory are extended over the length of the polymer molecule, the addition or removal of an electron has little effect on the energy level structure. Electrons are added to empty conduction bands and holes placed in previously filled valence bands. The simple band calculations do not depend on the number of electrons in the polymer (except to count states), whereas more detailed *ab initio* methods are sensitive to the number of electrons since coulomb and exchange matrix elements are computed (26). Nevertheless, in the case of polyethylene, valence bands calculated by various methods are in good enough agreement so as to indicate that there is little difference between valence states calculated for the neutral system and hole states (26). This expectation appears to be borne out by the photoemission results discussed later.

Such is not the case for pendant-group polymers. The nature of localization in these systems has been described by Duke (5, 32). The electronic structure of the polymer is essentially that of an ensemble of molecular subunits since the bandwidth due to overlap of pendant groups is so small. Charge injected into the polymer exists, as molecular ions of individual pendant groups and the electronic excitations of the solid are just those of the molecular subunit; energy bands are insignificant. The nature of the electronic localization is due to disorder in a fashion similar to that proposed by Anderson (33). There are two kinds of disorder: "diagonal disorder" and "off-diagonal disorder." Diagonal disorder is characterized by a distibution of on-site energies, $\Delta \alpha$. [Recall that on-site matrix elements α appear along the diagonal of matrices like Eq. (2.7) which arise in

LOCALIZED STATES

solving Schrödinger's equation.] Similarly, off-diagonal disorder produces a distribution, $\Delta\beta$, in the transfer integrals. While the specific requirements for Anderson localization depend on dimensionality, localization occurs when $\Delta\alpha \gg \bar{\beta}$, where $\bar{\beta}$ is the average transfer integral. Consistent with the simple calculations of Section 2.2.2, transfer integrals for *intra*molecular interactions are estimated as $\bar{\beta}_{\text{intra}} \lesssim 0.1$ eV, and for *inter*molecular interactions $\bar{\beta}_{\text{inter}} \lesssim 0.01$ eV (3).

Now, in general, the energy of a molecular ion in a polymer is much different than that of the same ion in the gas phase due to the polarization of the surrounding medium. In Duke's theory, there can be large fluctuations in energy due to fluctuations in polarization. Static fluctuations due to surfaces, defects, and local compositional variations as well as dynamic fluctuations due to the coupling of the molecular ion to the excitations of the medium must all be considered. The magnitude of these fluctuations, $\Delta\alpha$, is estimated as ~ 0.5 eV by attributing the anomalous broadening of measured photoemission lines to such fluctuations. Thus $\Delta\alpha \gg \bar{\beta}$ and the charged electronic states of pendant-group polymers consist of electrons (or holes) localized primarily on the pendant group (3).

2.3.2 Dielectric Response

When an ion is placed in a polarizable medium, its energy is always reduced. The change in energy, called the polarization energy P, can be calculated from simple electrostatic theory (2) by considering the ion to be a spherical cavity of radius R embedded in a medium with a zero frequency dielectric constant $\epsilon_1(0)$. Then

$$P = \frac{-[1 - 1/\epsilon_1(0)]\, e^2}{2R} \quad (2.17)$$

Thus if $\epsilon_1(0) = 3$ and $R = 3$ Å, $P = -1.6$ eV. These values are typical for a wide variety of organic molecular crystals (2). However, Duke (3, 32) has shown that in polymers Eq. (2.17) can lead to incorrect results and that the frequency and momentum dependence of the complex dielectric response function $\epsilon(\mathbf{q}, \omega)$ must be considered. Moreover, the strength of the coupling of the ion to the various excitations of the medium is crucial in determining P.

The dielectric response of a polymer consists of excitations Δ_α well separated in frequency. The lowest frequency excitations, $\Delta_1 \sim 10^{-13}$ eV, are associated with motions of the polymer chain and torsional motions of the pendant groups. Infrared active modes have energies Δ_2 of $\sim 10^{-2}$ eV, and valence electronic excitations occur for $\Delta_3 \sim 10$ eV. The core electrons can be excited at $\Delta_4 \sim 300$ eV. The real and imaginary parts of the dielectric function can be measured over this entire energy range (5), an example for poly(2-vinylpyridine) is given in

Fig. 2.6. To simplify analysis, Duke (5, 32) fits the experimental data to the simple parameterized form

$$\epsilon(E, \mathbf{q}) = 1 + \sum_{\alpha=1}^{4} \frac{E_{p\alpha}^2}{\Delta_\alpha^2(\mathbf{q}) - E^2} \qquad (2.18)$$

The zeros of the dielectric response associated with individual excitation modes are labeled $E_\alpha(0)$, and a simple form for the momentum dependence is assumed (5). Treating the molecular ions as spherical charge densities, the strength of the coupling of the ion to various excitations is given by

$$g^2(\alpha) = \left[\frac{E_{p\alpha}}{E_\alpha(0)}\right]^2 \frac{e^2}{2RE_\alpha(0)} \qquad (2.19)$$

which can be calculated readily from Eq. (2.18). Then the "zero-point" polarization energy P_0 is given by

$$P_0 = \sum_\alpha g^2(\alpha) E_\alpha(0) \qquad (2.20)$$

However, P_0 is never observed. Excitations such as valence and core electronic excitations which couple weakly to the molecular ion, $g^2 \ll 1$, contribute to the observed polarization energy P. But torsional and infrared excitations have $g^2 \gg 1$ and do not lower the average value of the ion energy but rather contribute a homogeneous broadening of ~0.3 eV to an otherwise sharp distribution of energy states.

Based on experimental determinations of the dielectric function of polystyrene and poly(2-vinylpyridine), Duke calculates that the polarization energy should be -1.3 eV for both materials (32) despite a difference by a factor of 2 in their static dielectric constants. Equation (2.17) predicts $P = -4.5$ eV for poly(2-vinylpyridine) and -1.3 eV for polystyrene (32). The polarization energy of holes injected into these polymers in a photoemission experiment has been measured as -1.5 eV for both polymers, confirming Duke's theoretical expectations (5).

2.3.3 Relaxation Energy

In an effort to understand metal–polymer contact-charging experiments, Duke and Fabish (34) suggested that electronic states in pendant-group polymers could be subject to a very large relaxation of 4–5 eV when they become charged molecular ions. This relaxation energy could be dissected into four terms: E_{pa} (intra), due to intramolecular nuclear relaxation of the individual pendant-group atoms; E_{pe} (intra), due to electronic distortion and charge redistribution within the pendant-group; E_{pa} (inter), due to intermolecular nuclear relaxation; and E_{pe}

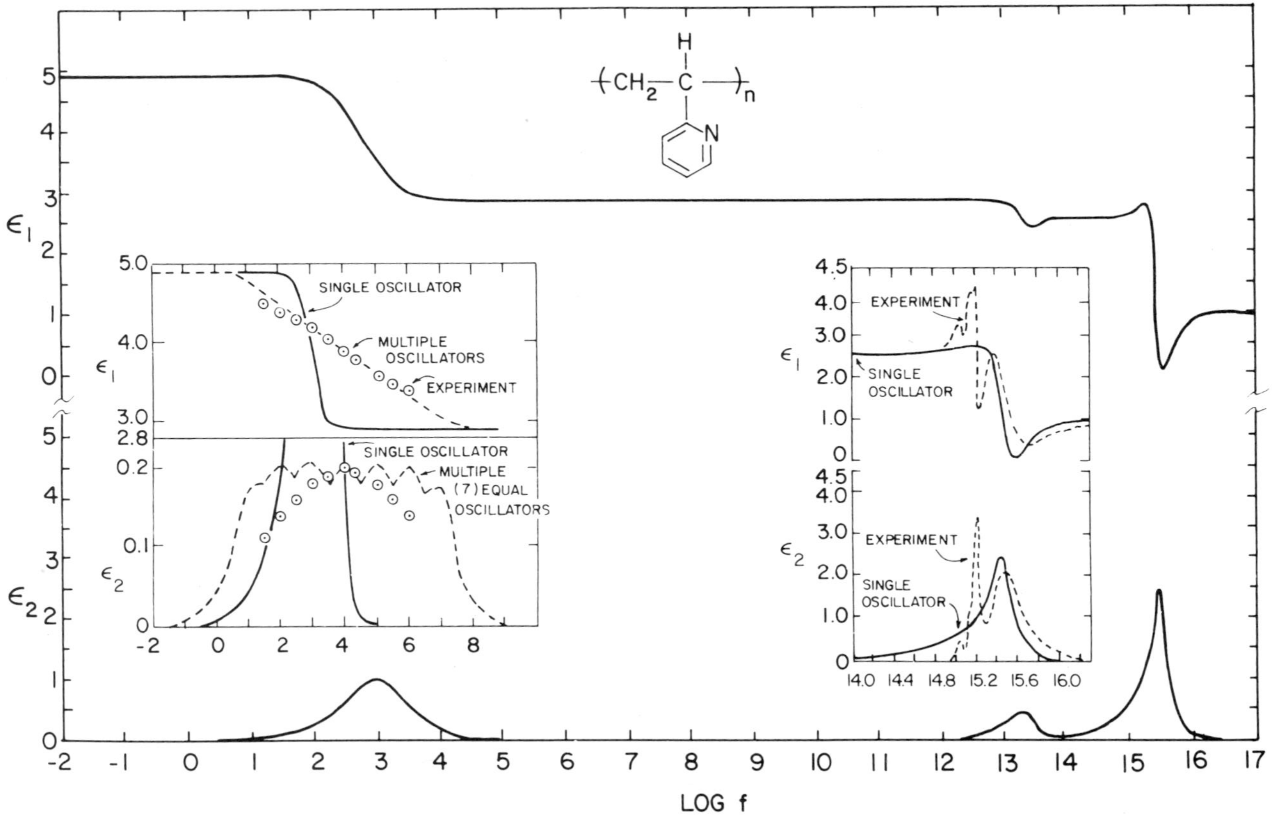

Figure 2.6 Plot of the model dielectric response of poly(2-vinylpyridine) as a function of frequency f, in Hz. Insets show comparisons with the experimental data where available (From Duke et al. [5].)

(inter), due to intermolecular distortion and charge redistribution in the surrounding medium. The last two terms make up the polarization energy discussed in the previous section, that is, $P = E_{pa}(\text{inter}) + E_{pe}(\text{inter})$. This follows because both atomic and electronic branches of the excitation spectrum are included in the dielectric response function (32). As described in the last section, we expect $P \sim -1.5$ eV.

The change in the energy of a charged molecule due to intramolecular nuclear relaxation, $E_{pa}(\text{intra})$, is conceptually similar to polaron formation in crystals (1). The charge distorts the molecule and therefore lowers its energy. The magnitude of $E_{pa}(\text{intra})$ can be computed (32) by calculating the coupling constants g_i of a given molecular orbital to the intramolecular vibrations i: $g_i = (\partial E / \partial Q_i)/\hbar\omega_i \cdot 2^{1/2}$, where E is the energy of the molecular orbital; Q_i is the amplitude of the intramolecular vibration; ω_i is the vibrational frequency; and $\partial E / \partial Q_i$ can be computed using semiempirical molecular orbital calculations for molecules with slightly distorted nuclear coordinates characteristic of the different intramolecular modes. Then $E_{pa}(\text{intra}) = \Sigma_i\, g_i^2\, \hbar\omega_i$. Values of -0.2 ± 0.1 eV are obtained for benzene and similar molecules, and these estimates should be typical for other pendant groups (32, 34).

The change in energy $E_{pe}(\text{intra})$ due to electronic distortion and charge redistribution is the most difficult quantity to estimate. There is no measureable intramolecular dielectric function. For hole states, this quantity can be estimated as the change in energy of a particular molecular orbital on an isolated molecule between its doubly occupied ground state (since there are two spin states) and its singly occupied excited state in which the excited electron has been removed to infinity. The ground-state energy is just the binding energy determined by suitable Hartree–Fock self-consistent field methods (35), and the excited state can be determined using similar open-shell calculations (35). The results show that for styrene, naphthalene, azulene, anthracene, and several other molecules, the binding energy of the hole state is ~ 2 eV above that of the original doubly occupied orbital, indicating for cations that $E_{pe}(\text{intra})$ is ~ -2 eV. Thus the total relaxation energy, $E_n = E_{pa}(intra) + E_{pe}(\text{intra}) + E_{pa}(\text{inter}) + E_{pe}(\text{inter})$, could be nearly -4 eV.

However, when applied to negative molecular-ion states, intramolecular electron repulsion causes a previously unoccupied molecular energy level to rise in energy rather than relax to lower energies (36). In the calculations of Hoyland and Goodman (36), the lowest unoccupied molecular orbital of benzene is 0.5 eV below the vacuum level; but when a negative ion is formed, the level is raised to 1.4 eV above the vacuum level, that is, the electron affinity of benzene is calculated to be -1.4 eV. The model of Duke and Fabish (34) suggests that benzene has a positive electron affinity and that this quantity could be measured by photodetachment experiments. Nevertheless, a series of theoretical calculations and experiments on dilute gases indicate that the electron affinity of benzene is negative (2, 37). Electrons can be bound to benzene molecules in high pressures of nitrogen gas, but in these experiments the relative importance of intra- and intermolecular effects is difficult to determine (37). Thus the relaxation energy of molecular-ion states can indeed be large as suggested by Duke and Fabish (34), but the actual value will depend on the nature of the molecular ion and the dielectric response of the medium. Positive molecular ions will exhibit greater net relaxation than negative ions because of the effect of intramolecular electron repulsion.

2.4 SPECTROSCOPIC TECHNIQUES

2.4.1 Photoemission

A direct measure of the distribution of occupied electronic states in polymers can be obtained by measuring the distribution in energy of electrons emitted from the polymer when it is exposed to monochromatic light. The electrons are emitted into the vacuum with kinetic energy E_i, from which the ionization energy I_i of the occupied states i can be determined, since

$$E_i = h\nu - I_i \qquad (2.21)$$

where $h\nu$ is the incident photon energy. The general aspects of the spectroscopy and results relating to organic molecular crystals have been reviewed by Grobman and Koch (38). A prime advantage to the technique (and also a source of complexity in its interpretation) is its extreme surface sensitivity. Depending on the kinetic energy, the mean free path of the photoemitted electrons can vary between 2 and ~20 Å, and in all cases only the first few molecules near the surface are probed. Thus the question of the relative importance of surface versus bulk electronic structure is always raised in these studies.

The information contained in ultraviolet photoemission spectra of polymers is illustrated in Fig. 2.7, where spectra of pyridine and ethylpyridine in the gas phase are compared with ethylpyridine and poly(vinylpyridine) in the solid state. The spectra of the small gas-phase molecules consist of sharp lines corresponding to the occupied molecular orbitals broadened by the instrument resolution, which is typically 0.1–0.2 eV (5, 38). There are additional broadening and sometimes discrete side bands due to the excitation of intramolecular vibrations in the photoemission process. If there were no intramolecular electronic and nuclear relaxation, the measured ionization energy would correspond to the binding energy of the unexcited ground-state orbital energy. However, such relaxation effects are large (typically ~2 eV as described in the previous section), and the spectrum really measures the energies of relaxed cation states.

On condensation into the solid phase, the ionization energies of all orbitals are reduced by 1–2 eV and the sharp lines are broadened by typically 0.5–1.0 eV (5, 38). Similar effects are seen for condensed small molecules and polymers. The reduction in ionization potential is due to intermolecular relaxation pro-

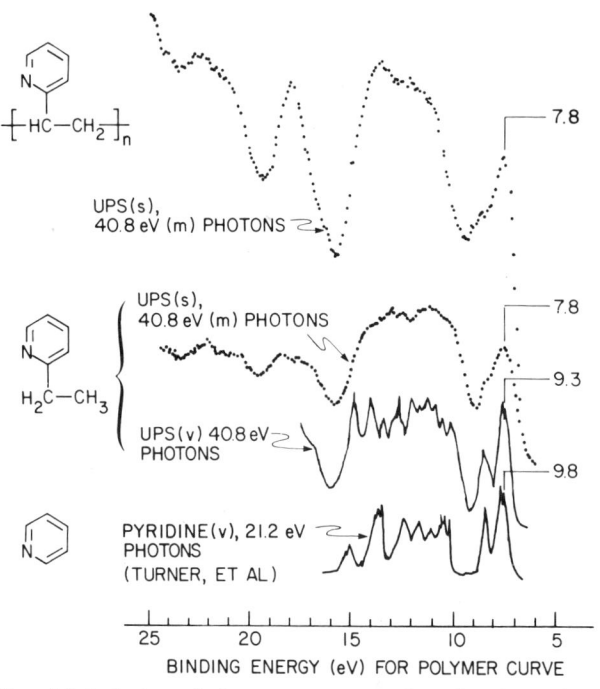

Figure 2.7 Ultraviolet photoemission spectroscopy data for the series of materials pyridine (v), vapor phase; 2-ethylpyridine (v), vapor phase; 2-ethylpyridine (s), a condensed solid film; and poly(2-vinylpyridine) solid. Spectra are aligned relative to the lowest ionization peak the absolute value of which is indicated. (From Duke et al. [5].)

cesses, that is, the polarization energy (discussed in Section 2.3.2), and is a measure of this effect. The additional broadening is the subject of considerable speculation. Mechanisms which could account for all or part of the solid-state broadening are: differential sample charging (38–40), hole lifetime effects due to intermolecular charge-transfer processes (39), thermal vibrations in the solid (39, 40), and incomplete polarization relaxation during the photoemission process (41). A further possibility is that the polarization relaxation is complete and that the broadening is due to differences in the polarization energy of various cations due to static effects, that is, defects, surfaces, and variations in the local structure of the polymer (40). The angular dependence of photoemission from anthracene condensed films indicates that the polarization energy at the top monolayer is 0.3 eV less than that of second and subsequent layers (42). If the entire broadening were due to "static polarization fluctuations" (40), then the measured width of photoemission lines in polymers would be a direct measure of the degree of "diagonal disorder," that is, the actual inhomogeneous distribution of cation states, as is assumed in the molecular-ion model of pendant-group polymers (Section 2.3.1).

SPECTROSCOPIC TECHNIQUES

A detailed comparison between photoemission studies of valence electronic structure and theoretical calculations on specific materials will be carried out in Section 2.5. Here, we will take note of the important information contained in core electron studies carried out by X-ray photoemission spectroscopy. Typical spectra are shown in Fig. 2.8. Core-level photoemission spectra are narrow lines characteristic of particular atoms and hence can be used as a tool for quantitative surface chemical analysis (43, 44). Surface oxidation and hydrocarbon contamination can be detected (43). Shifts in the binding energy of core levels indicate the different charge states of atoms in molecules since the binding energy E_B varies with the net atomic charge q:

$$E_B = E_B^0 + kq + M \qquad (2.22)$$

where k is an empirically determined constant and M is the coloumb energy of all the surrounding charges (43). The effects on the atomic charge distribution due to chemical substitution can be studied. In Fig. 2.8, the main splitting in the carbon 1s core-level binding energy is due to the different charge on the carbonyl carbon compared to the backbone carbons, while further splitting of both carbon and oxygen 1s lines is caused by substitution of a methyl group for a hydrogen atom as indicated. Removed in energy from the main lines shown in Fig. 2.8 are weaker lines called "shake-up" satellites which are due to simultaneous creation of a core hole and an excited valence state. It is possible to obtain further information on the structure and bonding in polymers from observations of the relative intensities and separation from the direct photoionization peak of these "shake-up" satellites (43). For specific applications of core-level spectroscopy to polymers, the reader is referred to recent reviews (43, 44).

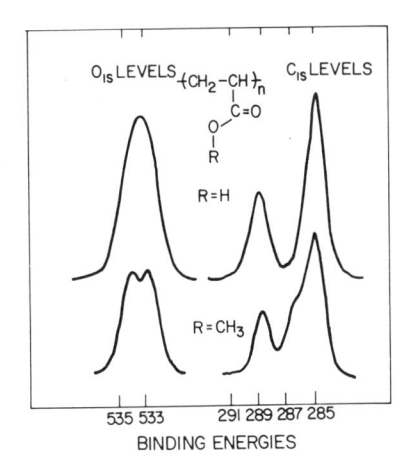

Figure 2.8 X-Ray photoemission core-level spectra for two poly(alkylacrylates). (From Clark and Thomas [43].)

2.4.2 Optical Spectroscopy

The most common technique used to measure the excitation spectrum of solids is optical absorption spectroscopy. If the sample thickness x is known, the absorption coefficient α can be determined by measuring the transmitted intensity I of incident light: $I = I_0 \exp(-\alpha x)$. However, to properly characterize the excitation spectrum, the complex dielectric function ϵ discussed in Section 2.3.2 must be accurately determined. The energies of characteristic excitations correspond to peaks in ϵ_2 (the imaginary part of ϵ), not in α; α is $4\pi k/\lambda$, where k is the imaginary part of the index of refraction and λ is the wavelength of the incident light. If the real part of the refractive index is n, then $\epsilon_2 = 2nk$ and $\epsilon_1 = n^2 - k^2$. Thus if more than one independent optical property can be measured, ϵ_1 and ϵ_2 can be determined. One such often measured property is the magnitude of the normal incidence reflectivity R:

$$R = \frac{(n-1)^2 + k^2}{(n+1)^2 + k^2} \tag{2.23}$$

The reflectivity itself is a complex quantity characterized by an amplitude and a phase which are related by the Kramers–Kronig integral. Once one part is measured, the other can be calculated, and from the complex reflectivity all other optical properties can be determined. Such an analysis has recently been done for pure and doped polyacetylene (45). For highly conjugated polymers the fundamental absorption occurs at sufficiently low energy so that standard laboratory spectrometers can be used. However, for systems with saturated bonds, the absorption occurs primarily in the very far ultraviolet, and novel light sources such as synchrotron radiation are now being used (5).

For studies of polymers, measurements of reflectivity appear to have significant advantages over optical absorption when a quantitative analysis is desired. A thick sample with a specular surface is all that is required for good reflectivity measurements. Optical absorption measurements require uniformly thick samples whose thickness (often ~100 Å) must be accurately known. As described in a careful study of polyethylene by George et al. (46), difficulties can arise with pin holes, scattered light, and sample fluorescence.

2.4.3 Inelastic Electron Scattering

The technique of high-resolution inelastic electron scattering spectroscopy has recently been applied to a number of model polymeric systems (47–49). It provides information about the fundamental electronic structure of solids which is complementary to both photoemission and optical spectroscopies. The experiment as diagramed in Fig. 2.9. consists of shooting a monoenergetic high-energy electron beam (kinetic energy E_0 ~80 keV) through a thin film (<100 nm thick) and measuring the number of electrons per unit time which lose energy E (which

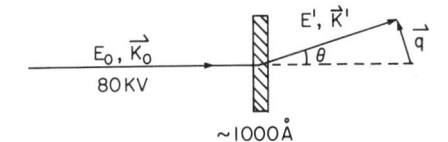

Figure 2.9 Inelastic electron scattering experiment.

is the difference between the incoming and outgoing energy) as a function of the scattering angle θ. In the inelastic collision with the sample, the fast electron loses energy and transfers momentum **q** to the excitation it has created. The energy loss probability, or the differential scattering cross section per unit energy loss E per unit solid angle Ω for fast electrons of velocity $v = \beta c$, is (48)

$$\frac{d^2\sigma}{dE\,d\Omega} = \left[\frac{e}{\pi\beta\hbar cq}\right]^2 \text{Im}\left[\frac{-1}{\epsilon(\mathbf{q},E)}\right] \qquad (2.24)$$

where ϵ is the complete momentum-dependent complex dielectric response function. From the measured energy loss probability, the "loss function" $\text{Im}[-1/\epsilon(\mathbf{q},E)]$ can be computed. The function $\text{Re}[1/\epsilon(\mathbf{q},E)]$ then follows from the Kramers–Kronig relation (49):

$$\text{Re}\left[\frac{1}{\epsilon(\mathbf{q},E)}\right] = 1 - \frac{2}{\pi} P \int_0^\infty \text{Im}\left[\frac{-1}{\epsilon(\mathbf{q},E)}\right] \frac{E'\,dE'}{E'^2 - E^2} \qquad (2.25)$$

where P indicates the Cauchy principle value. Thus the real and imaginary parts of $\epsilon(\mathbf{q},E)$ can be determined. In the limit of small **q** (on the order of the momentum associated with a photon of energy E), E is equivalent to the optical dielectric function. Hence optical properties can be calculated and compared with direct optical measurements.

One of the advantages of electron energy-loss spectroscopy is the ability to measure the electronic excitation spectrum over a wide energy range. Using specially designed spectrometers, energy losses can be recorded from 0.2 to 1000 eV with a resolution of ~0.1 eV. Typical momentum resolution is ~0.06 Å^{-1} corresponding to an angular resolution of $\pm 2 \times 10^{-4}$ rad for an 80 keV beam (49). Reviews of experimental techniques and their applications have been made by Raether (50), Daniels et al. (51), and Schnatterly (52).

In addition to measuring the excitations of valence molecular orbitals in polymers, electron energy-loss spectroscopy has been used to study the excitations of the well-separated atomic core levels (49, 53). Since these states have a very narrow energy spread, the excitation spectrum (near its onset) provides a measure of the distribution of empty valence molecular states. Thus the core excitation spectrum is complementary to photoemission measurements in which the distribution of filled valence molecular states is measured. In addition, since according to Eq. (2.22) the binding energy of an atomic core level depends on the net atomic charge, there will be shifts in the core excitation spectra analogous to those measured in X-ray photoemission spectroscopy (53).

A further advantage to inelastic electron scattering lies in the information contained in the momentum dependence of the energy-loss spectra. For **q** large compared to that of a photon, optically forbidden excitations are observed. The dipole selection rules no longer apply at large **q**, and monopole and quadrupole excitations can be measured (54, 55). In addition, the curvature of energy bands can be directly measured in some cases. As shown in Fig. 2.10, Zeppenfeld (56) was able to map the dispersion of the conduction band in graphite by measuring the onset of absorption as a function of momentum. Figure 2.10a shows the energy bands near the point P in the Brillouin zone. Arrows indicate excitations at an arbitrary finite q_x. Figure 2.10b shows the same excitations drawn with the conduction band shifted by q_x. Figure 2.10c gives the interband gap and hence the location of the conduction band edge as a function of momentum. Hence, nonvertical interband transitions can be measured in electron scattering spectroscopy, whereas only vertical transitions are measured optically. Thus the momentum dependence of energy-loss features can be used as a probe of the

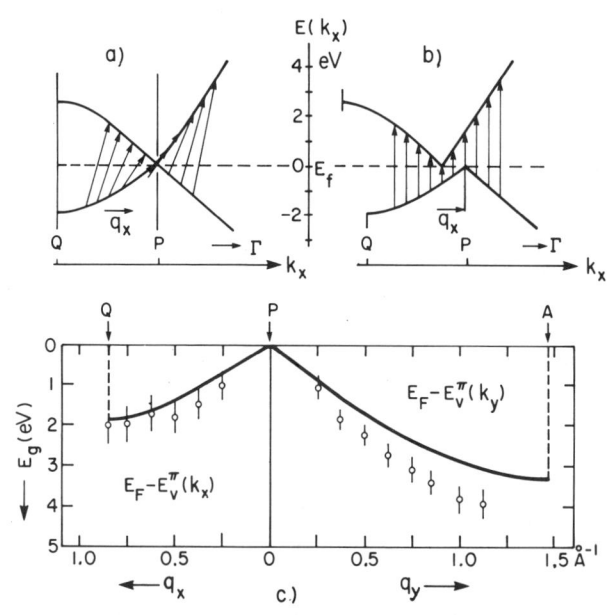

Figure 2.10 Nonvertical interband excitations in graphite. For a given q_x diagrams (a) and (b) are equivalent, but the origin of the interband gap is more clearly seen in (b). Panel (c) shows the dispersion of the onset of interband transitions for momenta along two crystal directions compared with theory (solid curves). (From Zeppenfeld [56].)

RESULTS

nature of the electronic states. Excitations between extended states characterized by relatively wide energy bands will show dispersion with momentum, whereas localized excitations (for example, intramolecular excitons) will not. The most common delocalized excitation in solids is a collective oscillation of the valence electrons called a "plasmon." It occurs when ϵ_1 goes through zero and is seen as a large peak in the energy loss probability, Eq. (2.24). Such delocalized excitations are common in metals and semiconductors and have recently been observed in polyacetylene (57).

In comparison with optical techniques, electron energy-loss spectroscopy is not sensitive to sample thickness nonuniformity, pin holes, fluorescence, or scattered light. However, there will always be multiple inelastic scattering which must be deconvoluted from the measured spectrum (49–54). A more serious drawback is radiation damage to the sample by the probing beam. By using low beam intensities ($\sim 10^{-6}$ A/cm^2) and efficient counting techniques, spectra can be obtained with doses as low as about 2 coulombs/m^2 (48, 49). This is low enough that the initial spectra contain primarily intrinsic electronic excitations, while subsequent spectra taken after considerable irradiation show new features characteristic of the products of radiation damage. By monitoring the spectra as a function of dosage, energy-loss spectroscopy can both cause and record the effects of electron irradiation of polymers (48, 49, 58).

2.5 RESULTS

In this section, the results of spectroscopic measurements will be compared with theoretical predictions of electronic states and their excitations for several representative polymers.

2.5.1 Polyethylene

Photoemission studies of the valence bands of polyethylene have been made by several groups (3, 59, 60). The results of an *ab initio* band calculation and the predicted and measured X-ray photoemission spectra are shown in Fig. 2.11 (3, 60). The calculated spectrum was obtained from the band structure density of states by taking account of differing matrix elements for photoemission from atomic *s* and *p* derived orbitals and the experimental resolution. The theoretical results are in good agreement with the extended Hückel calculation shown in Fig. 2.5. And the experimentally measured spectrum of cation states is in reasonable agreement with the theoretical ground-state density of states as anticipated by McCubbin (26, 27) (see Section 2.2.2). The lowest energy structure in Fig. 2.11 is due to the topmost valence bands, which are ~ 8 eV wide and are made up primarily of carbon $2p$ and hydrogen $1s$ atomic components. The strongest peaks at higher binding energy are due to the high density of states in

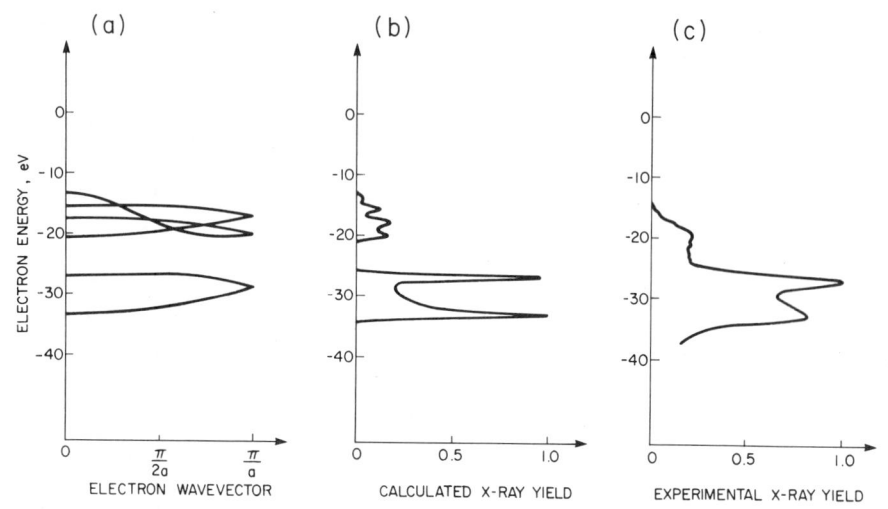

Figure 2.11 Polyethylene photoemission: theoretical valence bands (*a*) are used to predict the X-ray photoemission yield (*b*). The experimental results (*c*) have been shifted to facilitate comparison with theory. (From Fabish [3].)

the deepest valence bands which are mainly of carbon 2s character. Thus the relevance of band calculations for the ground-state electronic structure has been established.

On the other hand, the inadequacy of one-electron band calculations for the prediction of the excitation spectrum is shown both by optical and electron energy-loss measurements. The optical absorption spectrum has been measured by Partridge (61) and later by George (46) up to 12 eV. The strong, broad intrinsic absorption begins just above 7 eV and reaches a plateau between 9 and 11 eV before continuing to rise at higher energy. Unless specifically parameterized to produce this optical gap, band calculations usually predict band gaps greater than 12 eV (3). Using a simple model of molecular excitons in methane and ethane, Partridge (61) has calculated exciton transition energies at 8.5 and 10.3 eV, with a weaker transition at 24.5 eV. This suggests that the leading absorption edge could be excitonic rather than that of a direct interband transition as expected from one-electron band models.

An optical technique which can be used to make the distinction between excitonic and direct interband absorption edges is photoconductivity (62). If the threshold for intrinsic photoconductivity coincides with the absorption edge, then the edge must be a direct interband transition since the electron and the hole, finding themselves in highly mobile extended states, will contribute to the photocurrent (62). Such is not the case in polyethylene, where the photoconductivity threshold is 8.8 eV and the absorption between 7 and 8 eV is thereby attributed to exciton formation. Moreover, the photoemission and photoconduction thresholds coincide, indicating that the bottom of the conduction band lies above the vacuum level (62).

RESULTS

Electron scattering measurements have been used to extend the optical data to 390 eV (48). The measured energy-loss function $Im(-1/\epsilon)$ and the real and imaginary parts of the dielectric function at small q are shown in Fig. 2.12 for the valence excitations. In the region below 12 eV, there is good agreement with the shape and magnitude of the calculated and measured optical absorption spectrum. In particular, an absorption threshold was measured at 7.2 eV in good agreement with optical measurements. Strong peaks in ϵ_2 occur at 9.0 and 12.6 eV; and although it is not apparent from Fig. 2.12, low noise measurements of the $q = 0$ energy-loss spectrum indicate additional weak absorption structure at 13.2, 14.7, 16.6, and 19.6 eV. The 12.6, 13.2, and 16.6 eV structure does coincide with expected interband transitions (25, 48), and the 9.0 eV peak nearly coincides with the expected exciton energy (61). However, since a single theoretical model has not been used to calculate both excitons and interband transitions, a definite assignment of these features is not possible at this time (48).

To experimentally test the excitonic or direct interband nature of the absorption threshold at 7.2 eV, the momentum dependence of this edge was measured. In the case of direct interband transitions, the absorption edge should move with energy, as in the example of graphite cited in Section 2.4.2. In McCubbin's calculation, Fig. 2.5, the lowest energy excitations occur between bands composed primarily of carbon–carbon σ-bonds with a bandwidth of $\sim$3 eV, so that very large ($\sim$3 eV) dispersion of the absorption edge with momentum would be expected for direct interband transitions. Excitons, on the other hand, may be localized excitations and as such could show no momentum dependence. No change with momentum was observed for the absorption threshold at 7.2 eV, indicating that the absorption edge is indeed excitonic as previously surmised by optical absorption and photoconductivity experiments.

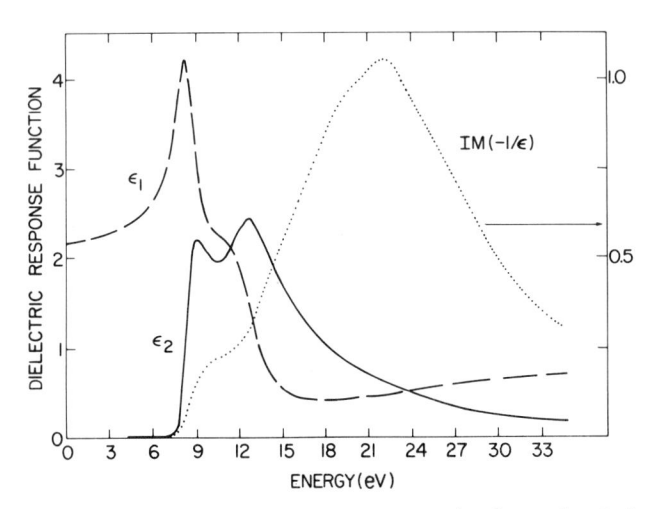

Figure 2.12 Dielectric response function $\epsilon = \epsilon_1 + i\epsilon_2$ for polyethylene and the function $Im(-1/\epsilon)$ determined from electron energy loss data.

2.5.2 Polyacetylene

Polyacetylene and the remarkable changes in its conductivity induced by doping are treated in detail in Chapter 7 of this book on conducting polymers. Here we concentrate on the intrinsic electronic structure of the pristine polymer.

The ultraviolet photoemission spectrum of the transisomer of $(CH)_x$ (14) is shown in Fig. 2.13 along with several recent theoretical predictions. Curve *b* is the result of a semiempirical CNDO calculation on long-chain polyenes extrapolated to apply to the infinite polyene chain (14). Curves *c* and *d* are the results of the valence density of states for an *ab initio* calculation (11) and the extended Hückel calculation (15) of Fig. 2.3, which have been broadened and modified to simulate the matrix element effects of photoemission from *s*- and *p*-derived bands (63). The theoretical curves have been shifted toward lower binding energy by as much as 5 eV so as to line up their strongest peak with the experimental peak at 10 eV. This shift indicates very severe intra- and intermolecular relaxation effects on the cation state measured by photoemission (14) (see Section 2.3.3). All theoretical curves overestimate the relative amplitude of the measured density of states near the low-energy valence band edge. The reasons for this are not known. The extended Hückel calculation provides poor agreement with experiment. While the CNDO method accounts for more spec-

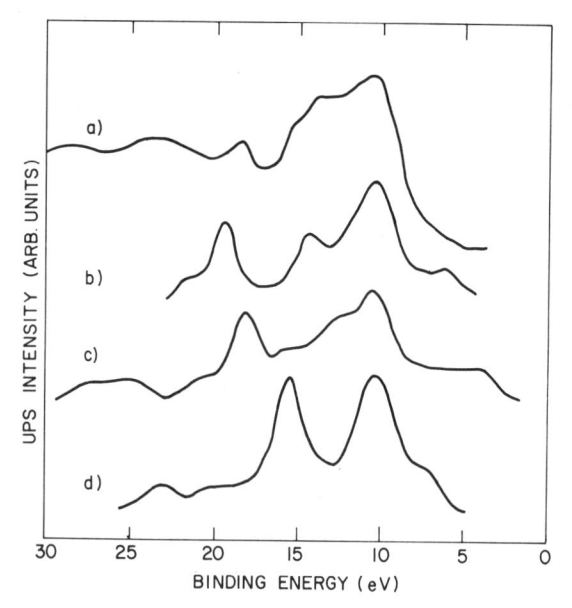

Figure 2.13 Polyacetylene, $(CH)_x$, ultraviolet photoemission spectra: (*a*) experiment (from Duke et al. [14]); (*b*) semiempirical (CNDO/S2) theory (from Duke et al. [14]); (*c*) *ab initio* theory (from Karpfen and Petkov [11] and Vanderbilt [63]); (*d*) extended tight-binding theory (from Grant and Batra [15] and Vanderbilt [63]).

RESULTS

tral features, the overall valence bandwidth is too small. Best agreement as to the positions, relative intensities, and overall valence bandwidth is given by the *ab initio* results (11).

While the hole-state spectrum measured by photoemission is reasonably well described by several theoretical calculations, the excitation spectrum is still the subject of much controversy, as discussed in Section 2.2. The nature of the absorption edge is still not resolved, and calculations are not accurate enough to definitely establish the location of the direct interband absorption edge. As with polyethylene, the relative importance of excitonic or direct interband transitions near the absorption edge must be established. Optical results have been interpreted as indicating an indirect absorption edge (45), as is also suggested by the magnitude and spectral behavior of the photovoltaic response (64).

The momentum dependence of electron energy-loss spectra for $(CH)_x$ are shown in Fig. 2.14 (57). Apart from weak absorption below 0.8 eV attributed to impurities, the onset of intrinsic absorption begins at 1.4 eV, in good agreement with the optical results (16, 45). The spectrum of valence excitations is dominated by the strong peak which occurs at 4.1 eV for $q = 0.1$ Å^{-1}. The results of Kramers–Kronig analyses of both optical and energy-loss data show that ϵ_2 exhibits a strong narrow peak at about 2 eV followed by a weak continuum at higher energies. The ϵ_1 is driven through zero from large positive values by the intensity of the interband transitions and returns through zero near

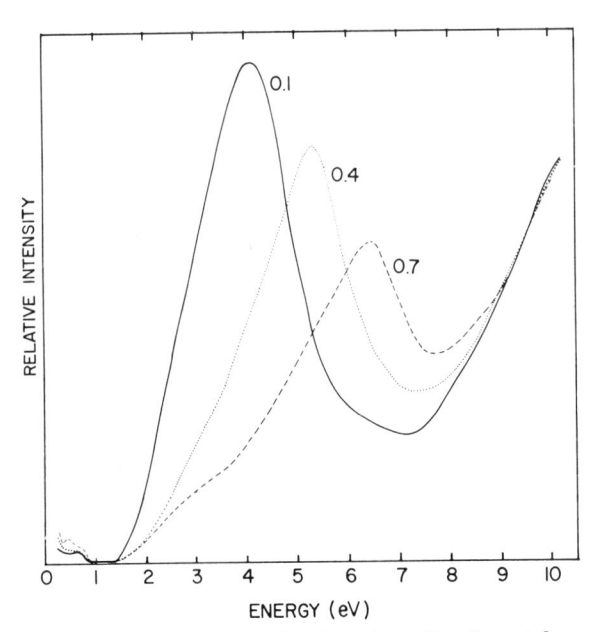

Figure 2.14 Inelastic electron scattering spectra of polyacetylene, $(CH)_x$, as a function of momentum, in Å^{-1} indicated near main peak; spectra are normalized at 10 eV.

4 eV, giving rise to the observed energy-loss peak. Plasmon dispersion is characteristic of the nature of electron states. Excitations of narrow energy bands exhibit plasmons with little dispersion or possibly small negative dispersion (65, 66). Excitations between wide bands produce large positive dispersion. For free-electron metals, semiconductors, and even quasi-one-dimensional metals, the plasmon energy increases quadratically with momentum (66). The spectra of Fig. 2.14 do show a large positive dispersion consistent with the wide energy bands for $(CH)_x$ as given in Figs. 2.2 and 2.3. However, the plasmon energy increases linearly with momentum rather than quadratically (57). This unusual dispersion may be due to the specific nature of the band structure and its excitation spectrum in this material. However, a similar excitation in graphite, which has a related electronic structure, does show quadratic dispersion. The importance of correlation effects which are stronger in $(CH)_x$ than in graphite may significantly alter the plasmon dispersion (12). Further theoretical work will be required to settle this point.

In addition, as seen in Fig. 2.14, the threshold for intrinsic absorption at 1.4 eV does not vary significantly with momentum. As described in the discussion on the fundamental absorption edge of polyethylene, this indicates that the onset of absorption is dominated either by excitons or indirect effects consistent with *ab initio* (19) and semiempirical theories (14) which take account of correlation. Although this interpretation is consistent with existing interpretations of optical absorption (45) and photovoltaic (64) measurements, more detailed investigations must be done to determine the precise location of the direct interband gap and the magnitude of excitonic and correlation effects in $(CH)_x$.

2.5.3 Polydiacetylenes

Depending on the sidegroup R indicated in Fig. 2.4a, the conjugated polymer backbone will assume the polydiacetylene structure shown there or a polybutatriene structure. Many such polymers have been prepared, often as large single crystals (3). Two commonly studied examples are poly-2,4-hexadiyne-1, 6-diolbis(p-toluenesulfonate), abbreviated as PTS (R = —$CH_2SO_3C_6H_4CH_3$), which has the diacetylene structure and poly-5,7-dodecadiyne-1, 12-diolbis-(phenylurethane), TCDU (R = —$(CH_2)_4OCONHC_6H_5$), which has the butatriene structure. Typical X-ray photoemission spectra of PTS are shown in Fig. 2.15, where Knecht et al. (67) have attempted to directly compare a number of bands, indicated by dashed curves in the experimental spectra, with the bands calculated by Parry (22), with hydrogen side groups discussed in Section 2.2.2. Their main conclusions are that agreement with Parry's calculation is reasonable, that the upper valence band has a width comparable to that of an inorganic semiconductor, but that the top of the valence band has an ionization energy of 7 eV versus 11 eV expected. This energy difference is undoubtedly due in part to large relaxation effects discussed in Section 2.3.3 and may also be due to the effects of sample charging and the difficulty of determining the Fermi level

RESULTS

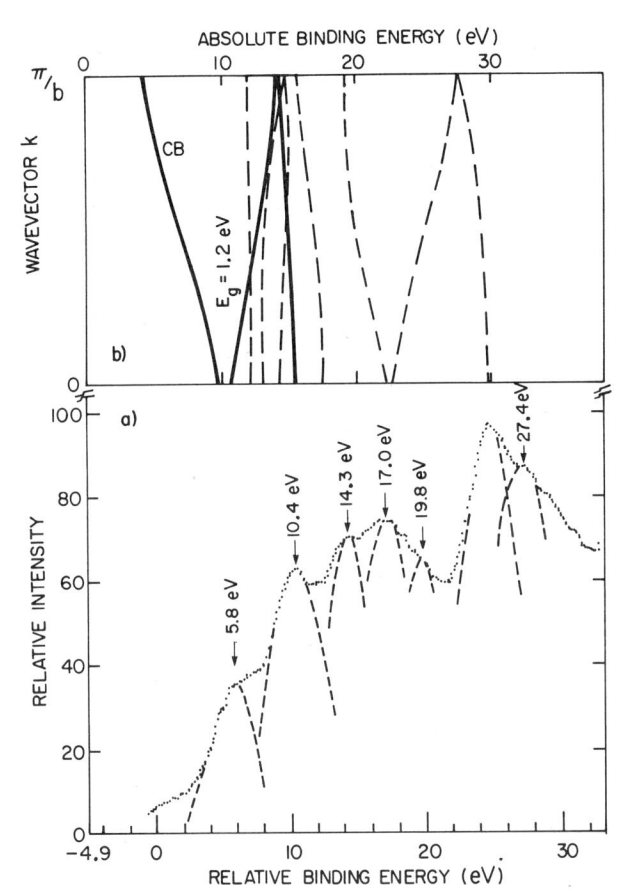

Figure 2.15 X-Ray photoemission spectrum of a polydiacetylene (PTS): (*a*) referenced to the Fermi level of a standard with a work function of 4.9 eV; (*b*) theoretical band structure with hydrogen atom side groups. (From Knecht et al. [67].)

position in insulating samples. The main difficulty with the analysis of Fig. 2.15 is that the subsequent measurements on model unpolymerized molecules have shown that most of the structure in the experimental spectrum is due to the side groups which contain far more electrons than the backbone and hence dominate the spectrum (68, 69). Nevertheless, small changes are observed near the top of the valence band on polymerization which may be identified with formation of the π-bands (69). Until further measurements are performed on polydiacetylenes with much smaller sidegroups and for which the effects of sample charging are better characterized, direct information about the conjugated backbone will be ambiguous at best.

As mentioned in Section 2.2 on band calculations, the nature of the fundamental absorption edge near 2 eV has been the subject of considerable controversy in that both direct interband transitions and excitons have been invoked as sources of the absorption edge (3). Recent optical absorption and photoconductivity results near the absorption edge are shown in Fig. 2.16. Lochner et al. (70) have shown that both spectra for TCDU and PTS are similar in that the absorption maximum corresponds to a minimum in the photocurrent. Detailed measurements of photocurrent versus photon energy indicate that the photoconductivity is due to autoionization of a bound molecular excited state (of unspecified origin) above the first singlet excited state of the molecule. Thus the first excited state and the absorption edge itself is excitonic and the optical (interband) bandgap is a few tenths of an electron volt above the exciton energy (70). The excitonic nature of the fundamental absorption peak in PTS is confirmed by the measured Davydov splittings along different crystal directions (3, 71), which are characteristic of excitonic excitations in organic molecular crystals. Nevertheless, a theoretical understanding of the absorption edge has proven to be illusive. Yarkony (72) has made an explicit comparison of a band theoretical model with an excitonic model based on the band calculation in an effort to elucidate the relative importance of band and excitonic effects. The band model is "approximately self-consistent" and includes electron–electron repulsion. The excitonic model is developed from the band calculation using one-electron Wannier functions in a standard fashion, and configuration inter-

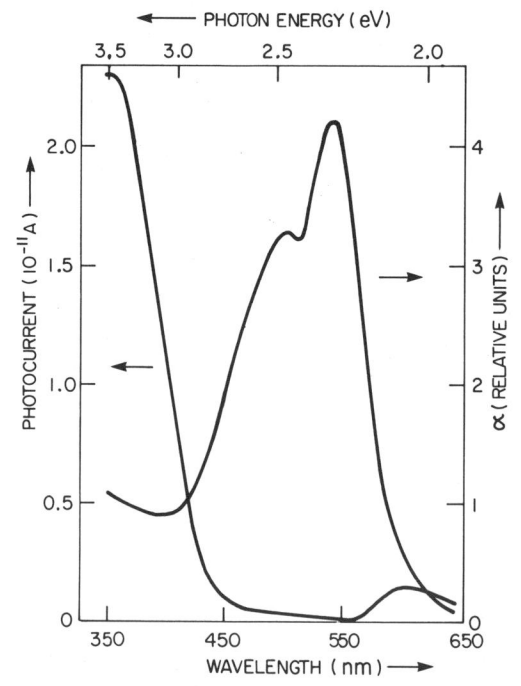

Figure 2.16 Optical absorption α and photoconductivity of a polydiacetylene, TCDU. (From Lochner et al. [70].)

RESULTS

action is included (72). For PTS, the bandmodel predicts a band gap of 7.21 eV, whereas the exciton model with configuration interaction and the Hückel theory give 2.94 and 1.84 eV, respectively, while the measured peak is at 2.0 eV. In calculations for TCDU, which has the butatriene bond configuration, the calculated gaps are uniformly lower: 5.03 eV for band theory, 0.91 eV for the exciton model, and 0.62 eV for Hückel theory. However, as seen in Fig. 2.16, the absorption peak is at 2.3 eV for TCDU. Yarkony concludes that an excitonic interpretation of the absorption edge is favored over band theory when electron–electron repulsion is included, but neither calculation can explain the increase in absorption energy in going from PTS to TCDU. He suggests that the discrepancy may be resolved by including "multiconfiguration stabilization of the ground state and lattice strains or interchain forces." Thus, the nature of the absorption edge as well as the photoconductivity threshold remain to be explained.

Apart from the theoretical difficulties, recent experimental results clearly indicate that, above the photoconductivity threshold, transitions to highly delocalized, highly mobile states occur (73). The results of photoconductivity and total charge transfer in response to fast laser pulses in large single crystals of PTS indicate that photogenerated carriers travel a distance of ~1 mm before being trapped at a chain end. The carrier mobility is estimated as greater than 100 cm^2/V-s, consistent with values for typical wide energy bands, as noted in Section 2.2. Thus the electronic structure of polydiacetylenes appears to be reasonably well characterized by wide energy bands with a relatively small semiconducting bandgap.

2.5.4 Pendant-Group Polymers

The electronic structure, and hence the photoemission spectrum, of pendant-group polymers is dominated by electrons bound in molecular orbitals of the pendant group. Apart from the shift of the spectra to lower binding energy and the broadening discussed in Section 2.4.1, the data in Fig. 2.7 clearly show that characteristic structures in the photoemission spectra are preserved in going from the gas phase to condensed molecules to the polymer. The formation of bands due to backbone bonds is hidden beneath structures associated with the pendant group similar to the case of polydiacetylenes illustrated in Fig. 2.15. Moreover, calculations of the distribution of cation states of small representative molecules using semiempirical techniques are quite successful in explaining the main features of the polymer photoemission spectra (5).

The optical absorption spectra of pendant-group polymers are clearly related to those of the isolated pendant groups, but shifts and broadening of absorption features in the solid state have not yet been completely explained. Weak absorption features such as the 260 nm band in poly(2-vinylpyridine) (5) and the 213 nm band in poly(methyl methacrylate) (49) are red shifted by about 0.1 eV from gas-phase model molecule values and broadened slightly. These effects can be accounted for by intermolecular polarization of the surrounding medium during the electric dipole excitation. This is expected to amount to about 10% of the monopole polarization energy, which is 1.5 eV (5). On the other hand, very strong excitations are much more severely affected. The 6.95 eV E_{1u} $\pi \rightarrow \pi^*$ excitation in benzene is 0.3 eV wide, but in polystyrene the absorption peak is downshifted by 0.6 eV and broadened about 1 eV (74, 75). Such effects may be due to the complex nature of the dielectric response and the measured optical properties of highly polarizable molecules in condensed media. Due to the polarization of surrounding molecules, the local field at one particulatr molecular site will differ from the applied external field. The local field will then depend on the wavelength of the exciting field which can be made quite small in an electron scattering experiment, that is, energy-loss spectra measured at large q (54). Inelastic electron scattering results show that an unexpected negative dispersion of the E_{1u} excitation which has been interpreted as evidence for local field effects (54).

The principal contribution of electron scattering to the study of pendant-group polymers has been to extend the range of optical measurements. Typical energy-loss spectra are shown in Fig. 2.17. Below 9 eV, the spectra agree well with optical measurements and can be accounted for in terms of excitations of electrons from valence to antibonding π-states of the pendant group (53, 74). Above 9 eV, weak structure has been identified with particular excitations of the deep σ-orbitals (53). The large broad peak near 20 eV is a plasmon resonance, since at this energy the magnitude of ϵ is a minimum although ϵ_1 does not actually go

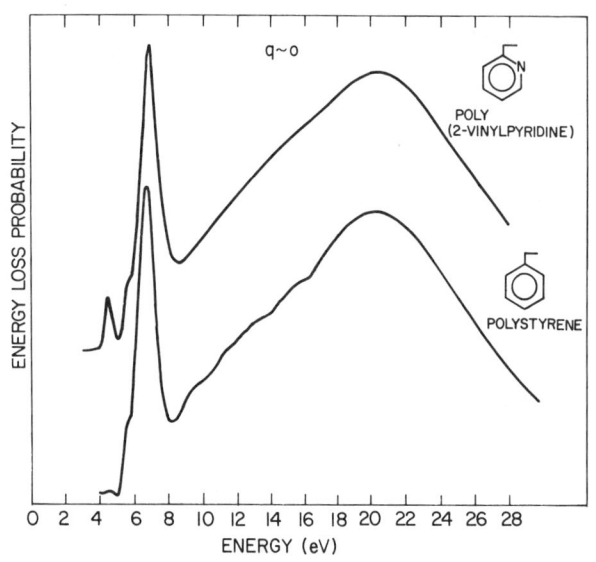

Figure 2.17 Electron energy loss spectra of polystyrene and poly(2-vinylpyridine) for forward scattered electrons $\varphi=0$, $q \sim 0$. (From Ritsko and Bigelow [53].)

RESULTS

through zero. Nearly the entire strength of the energy-loss spectrum is contained in this broad peak.

The real and imaginary parts of the dielectric function derived from energy-loss spectra by a Kramers–Kronig analysis are given in Fig. 2.18, where the minumum in $|\epsilon|$ near 20 eV is clearly seen. In addition to quantifying the position and strength of electronic excitations, this dielectric response function can be used to compute the magnitude of the intermolecular polarization energy as described in Section 2.3.2. The division of oscillator strength between π- and σ-excitations above and below 9 eV is clearly seen in ϵ_2. This division persists in polymers without significant π-excitations (48, 49).

As a check on the accuracy of the Kramers–Kronig calculations, the normal incidence reflectivity and the optical absorption coefficient were computed. Comparison with recent measurements of the reflectivity from 4 to 28 eV using synchrotron radiation is shown in Fig. 2.19. In the case of poly(2-vinylpyridine), the agreement is extremely good (5), thus supporting the dielectric function shown in Fig. 2.18. While the agreement between optical and energy-loss results for other pendant-group polymers is not quite as precise as in Fig. 2.19, the location of strong structure and the magnitude of the reflectivity and absorption coefficient are well accounted for in the energy-loss measurements (48, 49, 53, 54).

The spectrum of valence excitations decreases smoothly above the plasmon energy until, at ~285 eV, transitions from carbon 1s core levels to empty

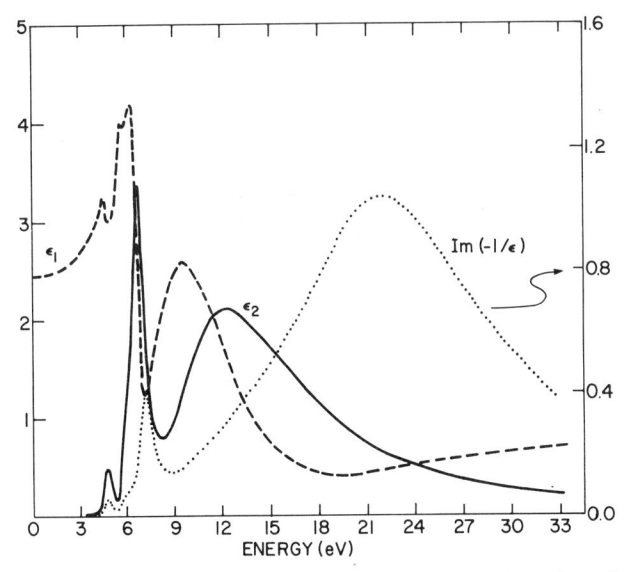

Figure 2.18 Dielectric response function and energy loss function, $Im(-1/\epsilon)$, of poly(2-vinylpyridine) determined from a Kramers–Kronig analysis of electron energy loss data. (From Ritsko and Bigelow [53].)

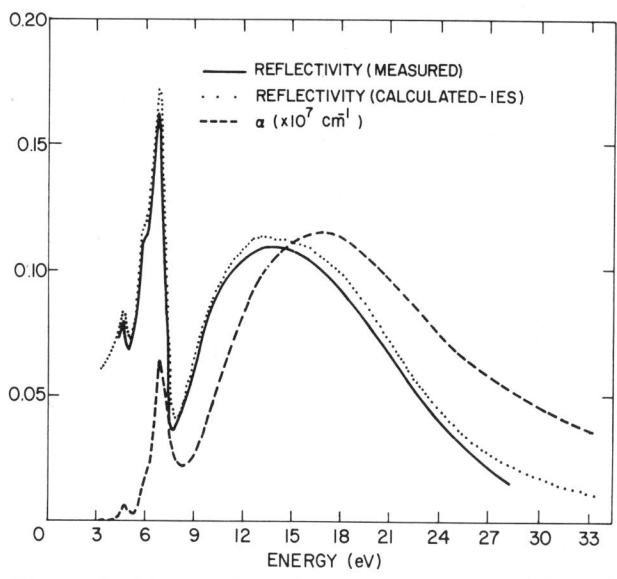

Figure 2.19 Normal incidence reflectivity of poly(2-vinylpyridine). Dotted curve computed from electron energy loss data; solid curve measured using synchrotron radiation. Dashed curve is the computed optical absorption coefficient. (From Ritsko and Bigelow [53].)

molecular and free-electron-like states are observed. The former are seen as sharp peaks in Fig. 2.20. Above 290 eV, a relatively smooth continuum of transitions to free-electron-like states with some modulations due to "X-ray absorption fine structure—EXAFS" are observed (47, 53). While in polymers little use has yet been made of EXAFS, which is caused by the backscattering of the excited electron from neighboring atoms, the sharp peaks shown in Fig. 2.20 for polystyrene have been shown to coincide with the positions of empty molecular orbitals and thus to provide a measure of the distribution of empty states (53). In addition, the 0.7 eV splitting of the leading peak in poly(2-vinylpyridine) was shown to be due to shifts in carbon $1s$ core electron binding energies by inequivalent charge states of carbon atoms around the pyridine ring. These "chemical shifts," similar to those normally measured in X-ray photoemission spectroscopy, and embodied in Eq. (2.22), can thus be measured with inelastic electron scattering spectroscopy as well. The relative importance of both initial and final electronic states in core excitation spectra has been established.

2.6 TRIBOELECTRICITY

Triboelectrification is the process by which two originally uncharged bodies become charged when brought into contact and then separated. This process

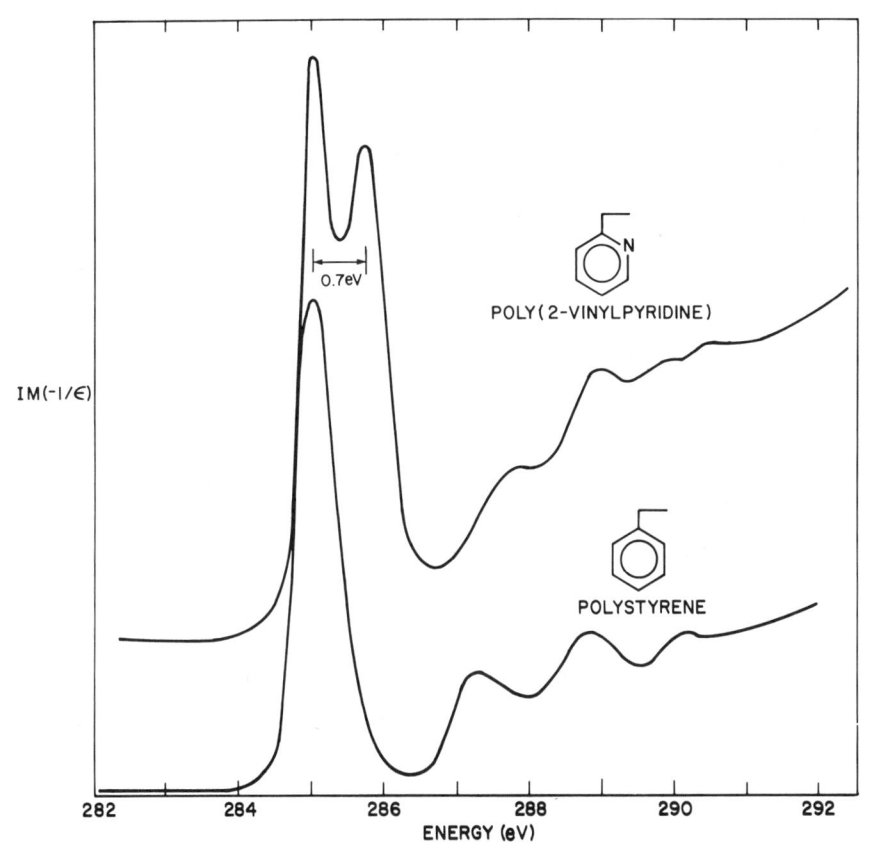

Figure 2.20 Carbon 1s core electron excitation spectra for polystyrene and poly(2-vinylpyridine) measured by electron energy loss spectroscopy. (From Ritsko and Bigelow [53].)

occurs for all highly insulating materials and has considerable practical importance for polymers (76). Here we review the present understanding of the microscopic nature of the electronic states involved in metal–polymer contacts. This system is conceptually well defined and should provide a basis for understanding more general cases of triboelectrification.

2.6.1 Extrinsic Surface States

If insulating polymers are viewed as organic semiconductors with very large energy gaps (see, e.g., the polyethylene band structure of Fig. 2.5), then no charge transfer to a contacting metal will occur. There are no intrinsic states into which carriers at the metal Fermi level, φ_m, can be transferred. Hence the traditional view is that there must be extrinsic states in the forbidden gap which accept or donate a charge to the contacting metal and thereby determine the polymer "work function" φ_p (76–78). The specific microscopic nature of these states is unknown, but unsaturated defects, chain end groups, adsorbed molecules, commercial additives or impurities, or the products of the chemical interaction of the ambient atmosphere with the polymer surface are frequently cited (76–79).

This model is diagramed in Fig. 2.21, where filled and empty extrinsic surface states are indicated as discrete levels and φ_c (φ_{c0}) is the energy difference between the polymer conduction band and Fermi level after (before) contact. Charge is transferred until thermodynamic equilibrium is established, which occurs quickly for surface states. This model is supported by many measurements which show that the sign and amount of charge transferred is dependent on the metal work function (76–80) and that, when a series of metals are contacted to the same polymer, the polymer work function φ_p can be determined; φ_p falls in the range 4–5 eV for a number of polymers with very different intrinsic electronic structures (78). These observations indicate electron rather than ion transfer (78). Moreover, trends observed in polymer–polymer contact measurements agree with metal–polymer contact results and support the general conclusion that triboelectrification involves electron transfer (80). While originally bulk extrinsic states were considered important (78), more recent workers have emphasized the dominance of surface states in view of the very low conductivity of polymers and the shortness of the contact time required (79, 80).

Our understanding of the microscopic nature of triboelectrification is inhibited by the very low charge densities involved. Total charge exchange is typically 10^{-9}–10^{-7} coloumb/cm^2 (76). This density is so low that most studies have been confined to measuring the surface potential changes before and after contact. Few of the more fundamental spectroscopic probes such as those listed in Section 2.4 can be easily applied.

A recent measurement of internal photoinjection from a metal to the polymer conduction band has been carried out by Mizutani et al. (81). The barrier height φ_c was determined for Cu and Al contacts with polyethylene terephthalate. The difference in φ_c between Cu and Al was only 0.07 eV, compared to a work function difference of 1.1 eV for the metals. The existence of surface states was

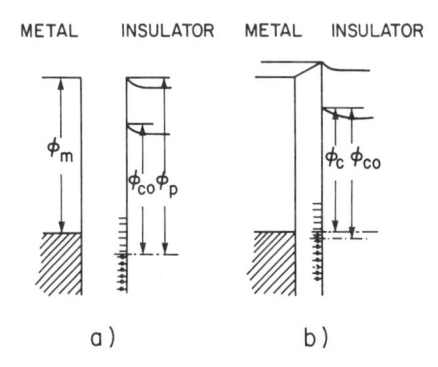

Figure 2.21 Energy level diagram of a contact between a metal and an insulator with surface states: (*a*) before contact; (*b*) after contact.

inferred and their density estimated with simple models. Moreover, exposure to O_2 strongly affected the measured φ_c, suggesting that adsorbed molecules could act as electron traps. Similarly, Hays (82) has shown that products of photo-oxidation localized within 8 Å of the surface can determine the magnitude of the charge transfer in gentle mercury–polystyrene contacts.

Further evidence for filled surface states on polymers was supplied by Murata (80) by means of external photoemission into the vacuum from a free polymer surface. A very weak photoyield was observed for a number of polymers exposed to light with energy as low as 4 eV, which agrees reasonably well with polymer work functions determined in contact charge measurements (78). The main microscopic problem which surface state theories have not addressed is the nature of the surface states and, in particular, how surface states occur near the Fermi levels of metals, that is 4–5, eV below the vacuum level. The electron affinities of O_2 and H_2O are only 0.9 eV (83), and molecules which simulate chain ends or defects such as butadiene or styrene have negative electron affinities (83). The electron affinity of carbonyl groups formed in surface oxidation reactions (82) may also be negative if calculations on small model molecules are accurate (49). Thus gas-phase electron affinities of commonly suggested candidates for extrinsic states in polymers are not sufficient to bind electrons in states near metallic Fermi levels. Additional energy must be supplied as coulomb energy associated with charge transfer and subsequent dipole formation or with a large relaxation energy of molecular ions in polymers, as discussed in Section 2.3. A possible indication of these effects is that the electron attachment rate to O_2 increases dramatically in dense gases (37). If relaxation effects are very large, than even the intrinsic electronic states must be considered as candidates for contact charging.

2.6.2 Intrinsic Bulk States

The whole point to the molecular-ion model described in Section 2.3 is that a semiconducting band picture of the electronic structure of pendant-group polymers is not valid. For such materials, excess charge is localized to individual pendant groups as molecular ions, and the energy of these electronic states is subject to large intra- and intermolecular relaxation effects. Using this theory, Fabish and Duke (2, 5, 32, 34, 30, 84) have interpreted metal–polymer contact charge exchange experiments as involving intrinsic bulk molecular-ion states rather than the extrinsic surface states discussed in the previous section. The experiments on which this explanantion of triboelectricity are based attempted to minimize the influence of possible extrinsic states by using purified resins, solvent casting the polymer films, and heating the films only in vacuum to minimize surface oxidation (85).

Typical results for polystyrene are shown in Fig. 2.22. The charge transfer to the polymer was found to depend on the contacting metal. However, the charge exchange for each metal was independent of previous contacts to other metals.

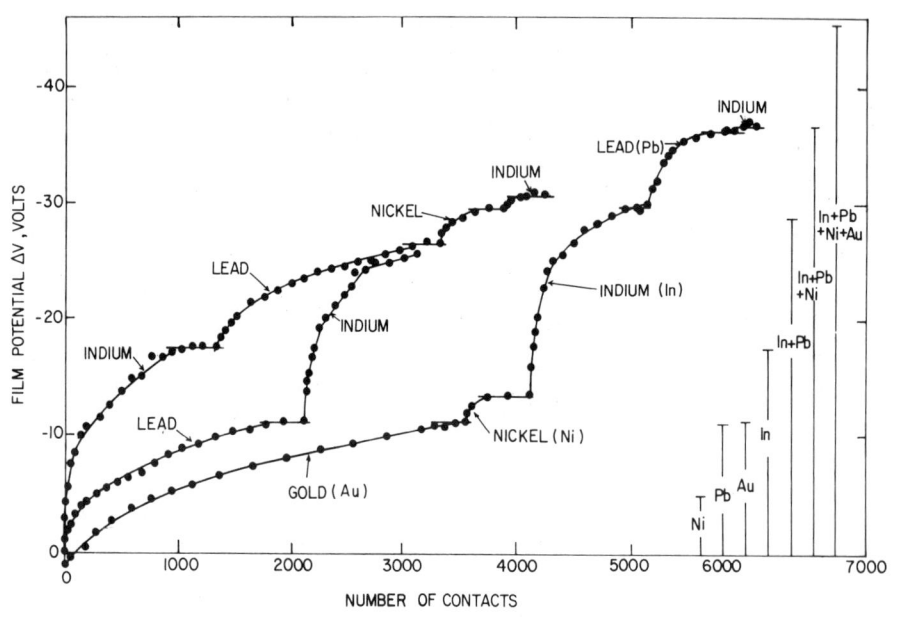

Figure 2.22 Potential of polystyrene films as a function of the number of contacts to various metals. (From Fabish and Duke [84] and Fabish et al. [85].)

Thus it is additive and commutative, since the order of metal contacts does not matter. These results suggested that each metal transferred charge only to particular polymer states which were localized in energy in a narrow "window" (~ 0.4 eV wide) near the metal Fermi level, and that thermodynamic equilibrium is never established during the time of the experiments. Therefore, a metal with a very low work function will inject electrons into polymer acceptor states which are then filled and which never decay or transfer charge to empty states of lower energy due to extreme localization and the inhomogeneous nature of the acceptor state distribution. This distribution is due to static polarization energy differences for different pendant groups (see Sections 2.3 and 2.4.1).

By using metals with different work functions, the distribution of intrinsic donor and acceptor states can be mapped, that is, there is "contact charge exchange spectroscopy." Typical results are shown in Fig. 2.23, where a description in terms of the molecular-ion model is given (3). The fully relaxed intrinsic molecular-ion states are arbitrarily represented as Gaussian functions centered at the nominal molecular anion energy: E(anion), for acceptor states; and the cation energy, E(cation), for donor states which overlap at the minimum of the experimental charge transfer data. The curves are then normalized to produce the maximum charge density which can be experimentally observed ($\sim 130 \times 10^{-9}$ coulomb/cm^2, which amounts to $\sim 4 \times 10^{15}$ states/cm^3) (84). Fitting to the experimental data gives values for the peak position and width of

TRIBOELECTRICITY

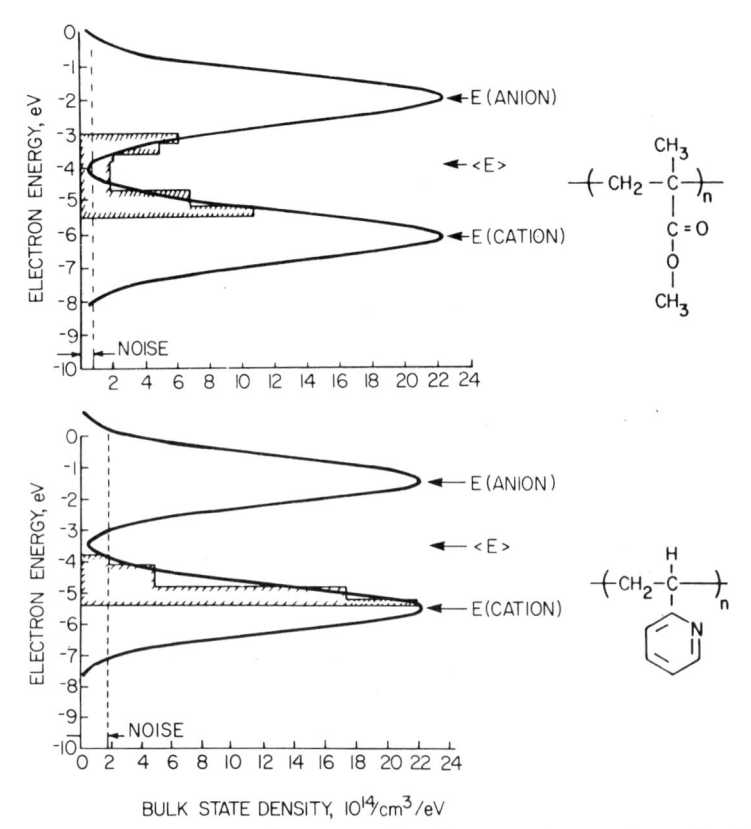

Figure 2.23 Interpretation of contact charging experiments in molecular ion model. Histograms represent measured charge transferred to poly(methyl methacrylate) and poly(2-vinylpyridine) from metals with differing work functions. Smooth curves are distributions of anion and cation states consistent with molecular ion model. (From Duke et al. [5], Fabish and Duke [84] and Fabish et al. [85].)

molecular state distributions. In general, greater relaxation and greater broadening are required to fit the contact charge data than would be expected from photoemission measurements, and the density of states is acknowledged to be far less than the intrinsic site density ($\sim 10^{20}$ states/cm^3), although detailed explanations of these facts are not given (84).

To date, the experiments of Fabish et al. (84, 85) have not been reproduced. On the contrary, in a series of similar experiments, Cottrell et al. (86) found that contact charging with a series of metals was not additive and that the final charge exchange depends only on the most recent contacting metal, thus supporting the idea that thermodynamic equilibrium is established. The conflicting experiments differed in the manner in which samples were prepared and in surface geometry.

They may not be incompatible if the dominant polymer states in Cottrell's samples were extrinsic surface states which rapidly achieved thermodynamic equilibrium, while in Fabish's samples other types of states were observed. Clearly, more careful experiments on well-characterized samples will be required to resolve these discrepancies.

The physical explanation of "contact charge spectroscopy" (84, 85) is based on electron tunneling. An unspecified thin barrier (possibly the metal oxide) is assumed to exist between the metal and polymer on contact. Due to the rapid falloff in tunneling probability with energy below the barrier height, when electrons are injected from a metal into polymer acceptor states, only a narrow region near the metal Fermi level has a large tunneling probability. This is the origin of the injection "window." This tunneling process is elastic in that the metal and polymer states are degenerate in energy. Inelastic tunneling between nondegenerate states which would tend to broaden the window is also possible (87), but these events are considered far less probable. The correlation of the window with the metal Fermi level is less clear when occupied polymer states lie above the contacting metal Fermi level. Furthermore, for all cases tunneling will only occur to polymer states very near the metal surface (usually within 3 nm), and yet the charge penetration depth was estimated at 2–4 μm (84). A detailed description of how triboelectric charge moves through a polymer is lacking although several model calculations have been made (79, 88–90).

Finally, the possibility of intrinsic bulk electronic states other than molecular-ion states might be mentioned. As discussed in Section 2.3.3, the electron affinity of benzene is most likely negative. In that case, even with a 1.5 eV polarization relaxation there will not be molecular-anion states sufficiently close to metal Fermi levels to account for the observed negative charging of polystyrene. The possibility of interstitial electrons or "solvated electrons" should be considered. These electrons would be stabilized by polarizing the surrounding medium similar to molecular ions but without molecular orbital and correlation energy complications. While the mean binding energy of such charges in a dielectric medium might be the polarization energy, $P - 1.5$ eV, the distribution of such states is not known. In organic glasses at low temperatures, a broad distribution of "trapped electrons" has been measured by optical techniques (91). The electron-spin resonance signal of trapped electrons in electron-bombarded polymer electrets has been recently measured (92). After annealing, the signal was shown to be due to excess electrons and not radicals. While of direct importance to charge storage in polymers (Chapter 3), such states may also play a role in triboelectricity.

2.7 SUMMARY AND ACKNOWLEDGMENTS

In this chapter, a selective and critical review of theories of the electronic structure of polymers and how they compare with spectroscopic measurements

SUMMARY AND ACKNOWLEDGMENTS

has been attempted. A simple tight-binding calculation was used to illustrate some concepts of extended electronic band states and to describe the band structure of conjugated polymers. While such simple methods provide a reasonable description of one-electron bands, they should not be considered as accurate predictive theories. The results of more detailed band calculations were compared with photoemission, optical absorption, and inelastic electron scattering data. In general, band theories can provide a good description of the photoemission spectra where the experimentally measured distribution of cation states tends to agree with the density of ground-state valence states. However, no present theories accurately predict the excited-state spectrum. *Ab initio* theories which neglect exchange, correlation, and the electron–hole interaction overestimate the optical absorption gap and simple semiempirical theories do not account for the excitonic nature of the fundamental absorption edge and the delayed onset of excitations to extended states. Moreover, the effects of correlation, which should be large in quasi-one-dimensional systems, are not yet known. Inelastic electron scattering experiments have provided confirmation of the optical absorption and photoconductivity measurements in polyethylene and polyacetylene and have been used to demonstrate the extended or local nature of electronic states and their excitations in these materials.

The limits of band models and localization of states in polymers were described in terms of the molecular-ion model. For the very weakly overlapping electronic states associated with pendant groups, electronic states are localized as molecular pendant-group ions and their energies modified by strong intra- and intermolecular relaxation and polarization effects so that band models are meaningless for these states. The intermolecular polarization energy can be deduced from the experimentally measured dielectric response function and it can be measured directly using photoemission by comparing gas-phase and condensed small-molecule spectra. For insulating polymers, the dielectric response associated with electronic excitations is more easily measured by electron energy-loss spectroscopy than by optical means. Apart from an energy shift due to relaxation and a broadening of structure (which may be caused by an inhomogeneous distribution of relaxation energies due to disorder), the photoemission spectrum of polymers with weakly interacting side groups resembles that of the side groups alone. While the distribution of occupied valence molecular states is probed in photoemission, electron energy-loss spectra from deep core levels provide a measure of the distribution of unoccupied valence molecular states, as well as information about net atomic charges similar to X-ray photoemission from core levels.

Various models of the electronic states involved in triboelectricity were discussed. At present, only the molecular-ion model of bulk intrinsic states provides a microscopic description of the charge transfer process. While aspects of both experimental and theoretical results have been questioned, this model remains as one of the most detailed descriptions of triboelectricity to date. Further advances in our understanding of this phenomenon will occur when new experiments which probe the microscopic nature of the triboelectric charge are carried out.

In the course of working on the electronic structure of polymers, this author has been fortunate to know several remarkable people who have contributed many of the ideas discussed in this chapter. E. J. Mele taught me much about energy band calculations, their applicability, and limitations. C. B. Duke has contributed greatly to the understanding of localization of electronic states in polymers and the consequences thereof. I am grateful to both of them for critically reading this manuscript. I benefited greatly from many discussions with T. J. Fabish and T. Smith, and am particularly indebted to S. E. Schnatterly who introduced me to inelastic electron scattering spectroscopy. I am grateful to J. Mort and H. Scher for their support and encouragement of my work in the area of the electronic structure of polymers.

REFERENCES

1. C. Kittel, *Introduction to Solid Physics*, 3rd Ed. Wiley, New York, 1966.
2. F. Gutman and L. Lyons, *Organic Semiconductors*, Wiley, New York, 1967.
3. T. J. Fabish, *CRC Crit. Rev. Solid State Mater. Sci.*, (Dec. 1979).
4. J-M Andre and J. Ladik, Eds., *Electronic Structure of Polymers and Molecular Crystals*, Plenum, New York, 1975.
5. C. B. Duke, W. R. Salaneck, T. J. Fabish, J. J. Ritsko, H. R. Thomas, and A. Paton, *Phys. Rev. B*, **18**, 5717 (1978).
6. J. M. Ziman, *Principles of the Theory of Solids*, Cambridge University Press, Cambridge, England, 1969.
7. F. Bloch, *Z. Phys.*, **52**, 555 (1928).
8. J. C. Slater and G. F. Koster, *Phys. Rev.*, **94**, 1498 (1954).
9. G. S. Painter and D. E. Ellis, *Phys. Rev. B*, **1**, 4747 (1970).
10. T. Kakitani, *Prog. Theor. Phys.*, **51**, 656 (1974).
11. A. Karpfen and J. Petkov, *Solid State Commun.*, **29**, 251 (1979).
12. A. A. Ovchinnikov, I. I. Ukrainskii, and G. V. Kventsel, *Sov. Phys. Usp.*, **15**, 575 (1973).
13. R. Peierls, *Quantum Theory of Solids*, Oxford University Press, New York, 1955, p. 108.
14. C. B. Duke, A. Paton, W. R. Salaneck, H. R. Thomas, E. W. Plummer, A. J. Heeger, and A. G. MacDiarmid, *Chem. Phys. Lett.*, **59**, 146 (1978).
15. P. M. Grant and I. P. Batra, *Solid State Commun.*, **29**, 225 (1979).
16. C. R. Fincher, D. L. Peebles, A. J. Heeger, M. A. Druy, Y. Matsumura, A. G. MacDiarmid, H. Shirakawa, and S. Ikeda, *Solid State Commun.*, **27**, 489 (1979).
17. W. L. McCubbin and I. D. C. Gurney, *J. Chem. Phys.*, **43**, 983 (1965).
18. N. A. Cade, *Chem. Phys. Lett.*, **53**, 45 (1978).
19. M. Kertesz, *Chem. Phys.*, **44**, 349 (1979), and reference therin.

REFERENCES

20 A. Karpfen, *J. Phys. C*, **12,** 3227 (1979).
21 E. G. Wilson, *J. Phys. C*, **8,** 727 (1975).
22 D. E. Parry, *Chem. Phys. Lett.*, **43,** 597 (1976).
23 M. Kertesz, J. Koller, and A. Azman, *Chem. Phys. Lett.*, **56,** 18 (1978).
24 A. Imamura, *J. Chem. Phys.*, **52,** 3168 (1970).
25 W. L. McCubbin, in *Electronic Structure of Polymers and Molecular Crystals,* J-M Andre and J. Ladik, Eds., Plenum, New York, 1975, p. 171.
26 W. L. McCubbin, *Chem. Phys. Lett.*, **8,** 507 (1971).
27 W. L. McCubbin and R. Manne, *Chem. Phys. Lett.*, **2,** 230 (1968).
28 D. Bloor, *Chem. Phys. Lett.*, **40,** 323 (1976).
29 J.-M Andre and G. Leroy, *Chem. Phys. Lett.*, **5,** 71 (1970).
30 A. A. Bright, P. M. Chaikin, and A. R. McGhie, *Phys. Rev. B,* **10,** 3560 (1974).
31 L. B. Schein, C. B. Duke, and A. R. McGhie, *Phys. Rev. Lett.*, **40,** 197 (1978).
32 C. B. Duke, *Mol. Cryst. Liq. Cryst.*, **50,** 63 (1979).
33 P. W. Anderson, *Phys. Rev.*, **109,** 1492 (1958); *Comm. Solid State Phys.*, **2,** 193 (1970).
34 C. B. Duke and T. J. Fabish, *Phys. Rev. Lett.*, **37,** 1075 (1976).
35 J. R. Hoyland and L. Goodman, *J. Chem. Phys.*, **36,** 12 (1962).
36 J. R. Hoyland and L. Goodman, *J. Chem. Phys.*, **36,** 21 (1962).
37 L. G. Christophorou, *Chem. Rev.*, **76,** 409 (1976).
38 W. D. Grobman and E. E. Koch, in *Topics in Applied Physics,* L. Ley and M. Cardona, Eds., Springer, Berlin, 1979, Vol. 27, Chap. 5.
39 P. R. Norton, R. L. Tapping, H. P. Broida, J. W. Gadzuk, and B. J. Waclawski, *Chem. Phys. Lett.*, **53,** 465 (1978).
40 C. B. Duke, T. J. Fabish, and A. Paton, *Chem. Phys. Lett.*, **49,** 133 (1977).
41 A. I. Belkind, A. M. Brodskii, and V. V. Grechov, *Phys. Stat. Sol.*, **85,** 465 (1978).
42 W. R. Salaneck, *Phys. Rev. Lett.*, **40,** 60 (1978).
43 D. T. Clark and H. R. Thomas, *J. Polym. Sci.*, **14,** 1671 (1967).
44 D. T. Clark, *Phys. Scripta.*, **16,** 307 (1977).
45 C. R. Fincher, M. Ozaki, M. Tanaka, D. Peebles, L. Lauchlan, A. J. Heeger, and A. G. MacDiarmid, *Phys. Rev. B,* **20,** 1589 (1979).
46 R. A. George, D. H. Martin, and E. G. Wilson, *J. Phys. C,* **5,** 871 (1972).
47 J. J. Ritsko, *Jap. J. Appl. Phys.*, **17,** S17-2, 231 (1978).
48 J. J. Ritsko, *J. Chem. Phys.*, **70,** 5343 (1979).
49 J. J. Ritsko, L. J. Brillson, R. W. Bigelow, and T. J. Fabish, *J. Chem. Phys.*, **69,** 3931 (1978).
50 H. Raether, *Springer Tracts Mod. Phys.*, **38,** 85 *(1965).*
51 J. Daniels, C. V. Festenberg, H. Raether, and K. Zeppenfeld, *Springer Tracts Mod. Phys.*, **54,** 77 (1970).
52 S. E. Schnatterly, *Solid State Phys.*, **34,** 275 (1979).
53 J. J. Ritsko and R. W. Bigelow, *J. Chem. Phys.*, **69,** 4162 (1978).

54 J. J. Ritsko, *J. Chem. Phys.*, **70**, 4656 (1979).
55 J. J. Ritsko, N. O. Lipari, P. C. Gibbons, S. E. Schnatterly, J. R. Fields, and R. Devaty, *Phys. Rev. Lett.*, **36**, 210 (1976).
56 K. Zeppenfeld, *Z. Phys.*, **243**, 229 (1971).
57 J. J. Ritsko, E. J. Mele, A. J. Heeger, A. G. MacDiarmid, and M. Ozaki, *Phys. Rev. Lett.*, **44**, 1351 (1980).
58 J. J. Ritsko, Annual Report on Conference on Electrical Insulation and Dielectric Phenomena, 1978, Natl. Acad. Sci., Washington, D. C., p. 109.
59 M. H. Wood, M. Barber, I. H. Hillier, and J. M. Thomas, *J. Chem. Phys.*, **56**, 1788 (1972).
60 J. Delhalle, J. M. Andre, S. Delhalle, J. J. Pireaux, R. Caudano, and J. J. Verbist, *J. Chem. Phys.*, **60**, 595 (1974).
61 R. H. Partridge, *J. Chem. Phys.*, **45**, 1685 (1966); *J. Chem. Phys.*, **49**, 3656 (1968).
62 K. J. Less and E. G. Wilson, *J. Phys. C*, **6**, 3110 (1973).
63 These curves have been computed by D. Vanderbilt.
64 M. Ozaki, D. L. Peebles, B. R. Weinberger, C. K. Chiang, S. C. Gau, A. J. Heeger, and A. G. Macdiarmid, *Appl. Phys. Lett.*, **35**, 83 (1979).
65 J. J. Ritsko, D. J. Sandman, A. J. Epstein, P. C. Gibbons, S. E. Schnatterly, and J. Fields, *Phys. Rev. Lett.*, **34**, 1330 (1975).
66 P. F. Williams and A. N. Bloch, *Phys. Rev. B*, **10**, 1097 (1974).
67 J. Knecht, B. Reimer, and H. Bässler, *Chem. Phys. Lett.*, **49**, 327 (1977).
68 G. C. Stevens, D. Bloor, and P. M. Williams, *Chem. Phys.*, **28**, 399 (1978).
69 J. Knecht and H. Bässler, *Chem. Phys.*, **33**, 179 (1978).
70 K. Lochner, H. Bässler, B. Tieke, and G. Wegner, *Phys. Stat. Sol. B*, **88**, 653 (1978).
71 H. Muller and C. J. Eckhardt, *J. Chem. Phys.*, **67**, 5386 (1977).
72 D. R. Yarkony, *Chem. Phys.*, **33**, 171 (1978).
73 K. J. Donovan and E. G. Wilson, *J. Phys. C*, **12**, 4857 (1979).
74 S. Onari, *J. Phys. Soc. Jap.*, **26**, 500 (1969).
75 E. E. Koch and A. Otto, *Chem. Phys. Lett.*, **12**, 476 (1972).
76 D. A. Hays, in *Digest of Literature on Dielectrics,* Natl. Acad. Sci., Washington, D.C., 1979, Vol. 12, Chap. 9, and references therein.
77 T. Freund, *Adv. Colloid Interfac. Sci.*, **11**, 43 (1979).
78 D. K. Davies, *J. Phys. D (Br. J. Appl. Phys. Ser. 2)*, **2**, 1533 (1969).
79 J. Fuhrmann, *J. Electrostat.*, **4**, 109 (1977/1978).
80 Y. Murata, *Jpn. J. Appl. Phys.*, **18**, 1 (1979), and references therein; Y. Murata and S. Kittaka, *Jpn. J. Appl. Phys.*, **18**, 421 (1979).
81 T. Mizutani, Y., Takai, T. Osawa, and M. Ieda, *J. Phys. D*, **9**, 2253 (1976).
82 D. A. Hays, *Electrostatics 1979 (Inst. Phys. Conf. Ser. 48),* 265 (1979).
83 V. I. Vedeneyev, L. V. Gurvich, V. N. Kondrat'yev, V. A. Medvedev, and Y. L. Frankevich, *Bond Energies Ionization Potentials and Electron Affinities,* Arnold, London, 1966.

REFERENCES

84 T. J. Fabish and C. B. Duke, *J. Appl. Phys.*, **48,** 4256 (1977).
85 T. J. Fabish, H. M. Saltsburg, and M. L. Hair, *J. Appl. Phys.*, **47,** 930 and 940 (1976).
86 G. A. Cottrell, J. Lowell, and A. C. Rose-Innes, *J. Appl. Phys.*, **50,** 374 (1979).
87 J. Lambe and S. L. McCarthy, *Phys. Rev. Lett.*, **37,** 923 (1976).
88 M. Hennecke, R. Hoffman, and J. Fuhrmann, *J. Electrostat.* **6,** 15 (1979).
89 C. G. Garton, *J. Appl. Phys.*, **7,** 1814 (1974).
90 A. Chowdry and C. R. Westgate, *J. Phys. D,* **7,** 713 (1974).
91 J. E. Willard, *J. Phys. Chem.*, **79,** 2966 (1975).
92 M. Legrand, G. Dreyfus, and J. Lewiner, *Phys. Rev. B,* **20,** 1725 (1979).

Chapter 3

CHARGE STORAGE

G. M. Sessler
Technische Hochschule Darmstadt
Darmstadt, West Germany

3.1	Introduction	59
3.2	Charging Techniques	61
	3.2.1 Thermal Methods, 61	
	3.2.2 Liquid-Contact Method, 62	
	3.2.3 Corona and Other Discharge Methods, 62	
	3.2.4 Electron Beam Method, 63	
3.3	Charge-Measuring Techniques	64
	3.3.1 Relations Between Charges, Fields, and Currents, 64	
	3.3.2 Measurement of Charge Density, 66	
	3.3.3 Measurement of Mean Charge Depth and Charge Distribution, 69	
	3.3.4 Thermally Stimulated Current (TSC) Techniques, 70	
3.4	Experimental Results and Interpretation of Real-Charge Storage	77
	3.4.1 Trapping Levels and Trap Densities, 77	
	3.4.2 Spatial Distribution of Trapped Charges, 80	
	3.4.3 Charge Decay, 82	
	3.4.4 Comparison of Experimental and Calculated Charge Decays, 86	
	3.4.5 Radiation Effects, 87	
3.5	Experimental Results and Interpretation of Dipole Polarization	90
	3.5.1 Dependence on Poling Parameters, 90	
	3.5.2 The Dipole Polarization, 93	
	3.5.3 Isothermal Decay of Polarization, 95	
3.6	Applications	96
	3.6.1 Electroacoustic Transducers, 96	
	3.6.2 Electrophotography and Electrostatic Recording, 98	
	3.6.3 Piezoelectric and Pyroelectric Devices, 98	
	3.6.4 Electret Gas Filters, 98	
	3.6.5 Electromechanical and Radiation Devices, 99	
	3.6.6 Biological and Medical Applications, 99	
	Acknowledgments	100
	References	100

3.1 INTRODUCTION

A number of polymers are capable of storing electrical charges for long periods of time. The stored charges may be real charges or polarization charges or a combination of both (see also Chapter 4, Section 4.2). The real charges often consist of layers of positive or negative carriers trapped at or near the surfaces of the polymer (surface charges) or of charge clouds stored in the bulk of the material (volume charges). In some cases, surface and volume charges are simultaneously present. Real charges may also be displaced within molecular or domain structures throughout the solid, resembling a dipole polarization. The polarization charges consist of a frozen-in alignment of dipoles which is either uniform through the material or a function of depth.

The permanency of storage of real and polarization charges in polymers depends on conductivity, carrier mobility in the presence of trapping, and dipolar relaxation frequency. Materials with low conductivity and low carrier mobility are capable of long-term real-charge storage. Examples are the Teflon materials poly(tetrafluoroethylene) (PTFE) and its copolymer fluoroethylene-propylene (FEP). On the other hand, polymers with a small dipolar relaxation frequency can store a polarization for long periods of time. Most prominent in this category is poly(vinylidenefluoride) (PVDF or PVF_2) which is discussed in detail in Chapter 4. There are also a number of materials that store both real and polarization charges. Representatives of this category are polyethylene (PE) and poly(ethylene terephthalate) (PET). The more frequently used polymers in this chapter are listed in Table 1.1 in the introduction to this book with their repeat unit and their abbreviated name.

In charge-storage experiments and applications, the polymers are generally used as thin, flexible films 5–50 μm thick, mostly coated on one or both surfaces with a 500–1000 Å metal layer. Typically, polymers storing real charges can be polarized to charge densities of a few 10^{-8} C/cm^2, which is close to the level of external breakdown. On the other hand, materials capable of polarization charge storage can approach surface charge densities of the order of 10^{-6} C/cm^2. The charged materials are often referred to as "polymer electrets."

While charge storage in dielectrics in general has already been mentioned by Gray in 1732 (1), by Faraday in 1839 (2), and by Heaviside in 1892 (3), systematic research into this topic was not undertaken until 1919, when Eguchi (4) reported on work concerning electret properties of certain waxes. In the 1940s, charge-storage phenomena in polymers began to attract scientific interest (5). Charge retention in thin polymer films was extensively studied since the early 1960s after the first application of film electrets in electret microphones was reported (6).

The present chapter commences with an outline of the charging and measuring techniques of polymer electrets. This is followed by a discussion of experimental results of real-charge storage and dipole polarization. Finally, some of the more important applications of polymer electrets are described. The topic of piezo- and pyroelectricity of polymer electrets is discussed in detail in Chapter 4 of this book and is therefore not dealt with in the present chapter. Due to the limited space available, this review of charge-storage phenomena in polymers has to be restricted to the essential features. For more detailed discussions the reader is referred to recent books on polymer electrets (7–9).

3.2 CHARGING TECHNIQUES

The charging of polymers can be performed by several methods. Controlled charging is achieved by application of electric fields at elevated temperatures, by field exposure through liquid contacts, by corona discharge, or by electron injection. The corresponding methods are briefly described in the following and are reviewed in detail in Refs. 8–12. As is also discussed, these methods are differently suited to achieve real-charge and dipolar polarization.

Other charging techniques depending on phenomena such as contact electrification (13) or ionization by penetrating radiation (14) are often the cause for unwanted or hazardous charging effects, while photoelectret formation (15, 16) applies only to a certain class of materials. These techniques are of lesser importance for the controlled charging of polymers and are not described here.

3.2.1 Thermal Methods

The thermal charging method consists of the application of an electric field to a dielectric at an elevated temperature and subsequent cooling while the field is still applied. Such methods can be used on samples having electrodes deposited on the dielectric or on samples being separated from the electrodes by air gaps (see Fig. 3.1a).

Depending on the geometry, electrode configuration, field strength, temperature cycle, and material parameters, a number of different charging phenomena take place. In the presence of air gaps, breakdown processes can occur in the air, resulting in charge deposition onto the polymer surfaces. When deposited electrodes are used, charges may be injected through the electrodes and/or dipoles may be aligned or charges separated within the dielectric. Upon cooling, all charge motions are frozen in.

To achieve the deposition of charge carriers onto an unelectroded surface of a sample, the presence of an air gap is necessary. For the customary 6–25 μm polymer films, gaps of the order of tens of microns to millimeters are typically used. When applying fields of 0.1–1 MV/cm, breakdown in the air gap causes charge deposition on the polymer surface. A maximum charging temperature somewhat above the glass transition temperature but well below the melting

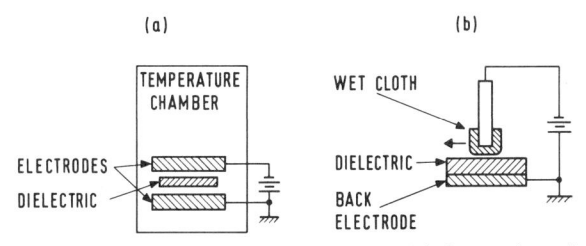

Figure 3.1 Methods for the charging of polymers: (a) thermal method; (b) liquid-contact method.

point (about 150–200°C for Teflon and 100–150°C for Mylar) is maintained during field application (17, 18). Use of the elevated temperature ensures that the deposited charge is deeply trapped and thus thermally stable.

If volume charging, for example dipole alignment, is desired, samples with electrodes deposited on both sides of the film (sandwich structure) are used. For example, polarization of PVDF is achieved at elevated temperatures (80–120°C) by applying fields of up to 2 MV/cm (19, 20). In other cases, charge injection from the electrode material can occur, as observed for PS, PMMA, PVDF, PE, and other polymers (21) or charge may be separated in the bulk, as found in preirradiated dielectrics (see Section 3.6).

3.2.2 Liquid-Contact Method

Liquid-contact charging depends on charge transfer from a soft, wet electrode to a contacting dielectric surface (see Fig. 3.1b). The charge transfer is caused by a potential applied between the electrode and a metal layer on the opposite surface of the dielectric. Many liquids, such as water and ethyl alcohol, can be used for the wetting. During the contact, the surface potential is approaching (but not reaching) the applied potential (22, 23). The charge transfer occurs from a charge double layer which forms at the liquid–polymer interface. This method is particularly suitable for real-charge deposition onto polymer surfaces.

Large areas of the surface may be charged by moving the electrode across the surface. To ensure charge retention on the dielectric, the electrode has to be withdrawn, or the liquid has to evaporate, before the voltage is removed. This method has also been used with nonwetting liquid–insulator contacts to record high-resolution charge patterns onto polymers (24).

3.2.3 Corona and Other Discharge Methods

Corona charging consists of charge deposition onto a dielectric surface by a corona discharge (see Fig. 3.2a). The discharge is generated by application of a voltage of a few kilovolts between a point or knife electrode and a back electrode located behind the sample to be charged. An additional wire mesh located between the point electrode and the dielectric and biased to a potential of at least a few hundred volts is used to control the lateral current distribution and the total current to the sample (25). The current distribution is initially bell shaped with a superimposed "modulation" by the "shadows" of the wires. However, the eventual distribution of the deposited charge is generally uniform if the charging is carried to the point where the entire sample surface has assumed a saturation potential which equals the grid potential (26). For nonconductive materials, the current vanishes at this point. If the surface potential and the compensation current flowing to the rear electrode are continuously monitored, the equivalent surface charge density and the conduction current through the sample can be determined during the charging process (27). The use

CHARGING TECHNIQUES

of heat during corona charging results in electrets of better charge stability (28). Although corona processes are primarily employed to deposit real charges onto the surface of dielectrics, they have also been used to obtain dipolar orientation by means of the field of the deposited charges (29).

Related charging methods depend on the isothermal application of an electric field to a dielectric–air–gap sandwich between parallel disk electrodes. If spark breakdowns occur in the gap, one obtains charge deposition on the dielectric. To avoid destructive arcing, a second dielectric insert may be placed additionally between the electrodes, acting as a series resistor of high resistivity (30, 31). With this method, very large charge densities can be deposited onto polymers (30).

3.2.4 Electron-Beam Method

Here the injection of electrons of proper energy into the dielectric is utilized for charging. If the range of the electrons is smaller than the thickness of the dielectric, the charging is directly achieved by trapping of the injected carriers. Electron-beam methods have been used in the past to charge thick dielectric plates (32) and have been adapted more recently to the charging of polymer films (33).

Electron-beam charging with electrons of range smaller than the dielectric thickness is governed by the range-energy relations. For example, the range of electrons of 10 and 20 keV in Teflon is about 1.5 and 5 μm, respectively (34). Thus application of this method to thin polymer films requires electron beams of correspondingly low energy. Charging of such films is therefore conveniently achieved with small electron accelerators or electron microscopes, as schematically shown in Fig. 3.2b.

The charging process for a sample metalized and grounded on one surface and irradiated by an electron beam through the other (nongrounded) surface can be visualized as follows: When striking the surface, the electrons release some secondary electrons (emission yield at 10–20 keV $\approx$ 0.2), leaving a positively charged surface layer. The electrons then penetrate into the volume and generate secondary carrier pairs which are quickly trapped. The secondaries cause a

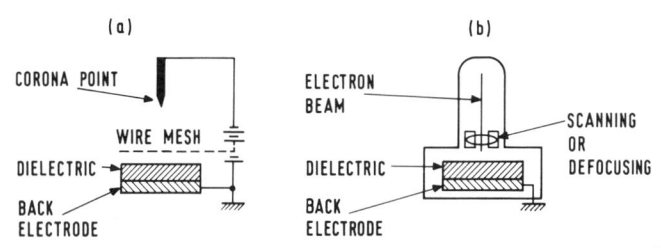

Figure 3.2 Methods for the charging of polymers: (*a*) corona method; (*b*) electron beam method.

radiation-induced conductivity (RIC) several orders of magnitude greater than the intrinsic conductivity. Due to collisions, the primary electrons are eventually slowed down enough to be also trapped, forming initially a distribution of negative charge around the average range, which is about two-thirds of the practical range. The self-field of the charges, which is essentially directed toward the electrode on the rear of the sample, causes currents within the region of RIC which correspond to a further inward motion of electrons. In FEP, the measurable currents cease after time periods of the order of minutes and a stable charge arrangement results (35) (see also Section 3.4.2). Electron-beam charging is most suitable for the generation of trapped volume charges.

3.3 CHARGE-MEASURING TECHNIQUES

Measurements of the properties of charged polymers may be performed with a variety of methods based on the evaluation of external potentials or currents under isothermal or nonisothermal conditions and on optical, X-ray, electron-beam, and other techniques (8). Of greatest importance have been electrical measurements such as charge density studies (often as a function of time), charge distribution experiments, and thermally stimulated current (TSC) measurements. Since these methods depend on an evaluation of charges, potentials, or currents, some important relations concerning these quantities will be derived first.

3.3.1 Relations Between Charges, Fields, and Currents

On dielectrics with one electrode, the charge density is generally measured by means of induction methods utilizing the action of the external field of the sample on an adjacent electrode. A customary arrangement is depicted in Fig. 3.3, which shows a charged dielectric sheet covered with an electrode on one side (lower electrode) and separated by an air gap from another (upper) electrode. The thickness dimensions s_1 and s_2 and the dielectric constants ϵ_1 and ϵ_2 are explained in the figure.

The charge of the dielectric is characterized by a surface charge density σ, located at $x = s_1$ and measured in charge per unit area, and a volume charge density $\rho(x)$, measured in charge per unit volume. Generally, σ and ρ are

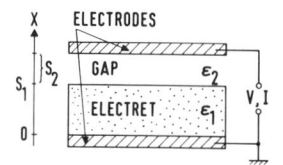

Figure 3.3 Arrangement of charged dielectric with gap and electrodes in parallel-plate geometry.

composed of real-charge and polarization-charge contributions. If the real-charge densities at the surface and in the volume are denoted by σ_r and ρ_r, respectively, and the permanent polarization by P, one has

$$\sigma(s_1) = \sigma_r(s_1) - P(s_1) \tag{3.1}$$

$$\rho(x) = \rho_r(x) - \frac{dP(x)}{dx} \tag{3.2}$$

Assuming initially the existence of merely a surface charge layer of density σ at $x = s_1$ and applying Gauss' law $\epsilon_0(-\epsilon_1 E_1 + \epsilon_2 E_2) = \sigma$, and Kirchhoff's second law $s_1 E_1 + s_2 E_2 = 0$, yields for the fields E_1 in the dielectric and E_2 in the air gap, respectively,

$$E_{1,2} = \mp \frac{s_{2,1}\sigma}{\epsilon_0(\epsilon_1 s_2 + \epsilon_2 s_1)} \tag{3.3}$$

where ϵ_0 is the permittivity of free space. If only volume charges are present, the fields E_1 and E_2 are expressed by more general relations (see, for example, Ref. 8). If surface and volume charges are present, the fields follow by linear superposition of the separate contributions of these charges. For example, the external field E_2 in this case is obtained by replacing σ in Eq. (3.3) by the "effective" or "projected" surface charge density $\hat{\sigma}$ given by

$$\hat{\sigma} = \sigma + \frac{1}{s_1} \int_0^{s_1} x\rho(x)\, dx \tag{3.4}$$

Since the induction charge σ_i on the upper electrode is, under short-circuit conditions, related to the field E_2 by $\sigma_i = -\epsilon_0 \epsilon_2 E_2$, the effective surface charge density $\hat{\sigma}$ of an electret can be determined from σ_i according to Eq. (3.3) (with σ replaced by $\hat{\sigma}$) by

$$\hat{\sigma} = -\frac{\epsilon_1 s_2 + \epsilon_2 s_1}{\epsilon_2 s_1} \sigma_i \tag{3.5}$$

If the upper electrode is removed to infinity ($s_2 = \infty$), Eq. (3.3) with $\sigma = \hat{\sigma}$ yields for the surface potential $V_0 = E_1 s_1$ of the "free" electret

$$V_0 = -\frac{\hat{\sigma} s_1}{\epsilon_0 \epsilon_1} \tag{3.6a}$$

In many measuring setups an external bias V_0 is applied to the upper electrode (Fig. 3.3) such that the external field E_2 of the electret is compensated. Writing

Kirchhoff's law as $s_1 E_1 = V_0$, one obtains together with Gauss' law (see above) for the effective surface charge density

$$\hat{\sigma} = -\frac{\epsilon_0 \epsilon_1 V_0}{s_1} \quad (3.6b)$$

which is independent of the air-gap thickness. The voltage V_0 corresponds to the potential of the free electret surface determined in Eq. (3.6a).

Often dielectrics sandwiched between two electrodes and which store a charge layer of density σ at a certain depth in the material are used. By analogy with Eq. (3.5), one obtains the following relation between σ and the induction charges σ_{i1} and σ_{i2} on the lower and upper electrodes, respectively:

$$\sigma_{i1,i2} = -\frac{s_{12,11}}{s_{11} + s_{12}} \sigma \quad (3.7)$$

where s_{11} and s_{12} are the distances between the charge layer and the lower and upper electrodes, respectively. (Compare also similar derivations in Chapter 4, Section 4.2.)

Charges stored in polymers may give rise to currents due to the application of heat, ionizing radiation, or illumination. If currents flow, the quantities E, P, and σ, so far assumed to be time invariant, become functions of time. Quite generally, the total current density $i(t)$, which is the same everywhere in the circuit, is given by

$$i(t) = \epsilon_0 \epsilon \frac{\partial E(x,t)}{\partial t} + \frac{\partial P(x,t)}{\partial t} + i_c(x,t) \quad (3.8)$$

Here the terms on the right represent, respectively, the densities of the displacement current, the depolarization current, and the conduction current. The latter can be further expressed as

$$i_c(x,t) = [g + \mu_+ \rho_{r+}(x,t) + \mu_- \rho_{r-}(x,t)] E(x,t) \quad (3.9)$$

where g is the intrinsic conductivity; and μ_+ and μ_- are the mobilities of positive and negative carriers, respectively, whose excess densities (over and above the intrinsic equilibrium density) are ρ_{r+} and ρ_{r-}, respectively.

3.3.2 Measurement of Charge Density

According to Eq. (3.5), a measurement of σ_i for volume charged dielectrics always yields the effective surface charge density $\hat{\sigma}$, defined in Eq. (3.4). To measure the total charge density,

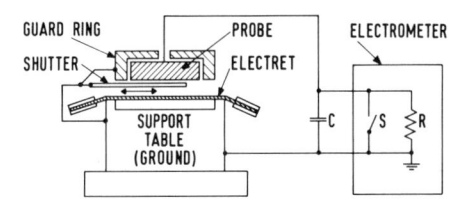

Figure 3.4 Capacitive probe setup for measuring effective surface-charge density. (From Sessler and West [41].)

$$\sigma_T = \sigma + \int_0^{s_1} \rho(x)\, dx \qquad (3.10)$$

a second measurement is necessary, as is shown below.

The classical tool for measuring the effective surface charge density on dielectrics is the dissectible capacitor, which allowed validation of the two-charge theory (36). Over the past three decades, however, it has been replaced by other techniques.

A widely used modern method for determining surface charge densities on dielectrics with one surface electroded is the capacitive probe method. Contrary to the dissectible capacitor method, it uses a well-defined and relatively large air gap between a probe and the charged dielectric (37–41). In one such setup, shown schematically in Fig. 3.4, the probe is shunted by a large capacitor ($C \gg C_{\text{probe}}$) and can be shielded from the field of the dielectric by a shutter. If the probe is exposed to the field, a charge $a\sigma_i = -CV$ flows into it from the capacitor (a is the probe area), charging the latter to the voltage V. Since short-circuit conditions prevail, Eq. (3.5) holds, and one obtains

$$\hat{\sigma} = \left(1 + \frac{\epsilon_1 s_2}{\epsilon_2 s_1}\right) \frac{CV}{a} \qquad (3.11)$$

where all quantities on the right are accurately measurable. Such probes, if made small enough, are also well suited to the measurement of lateral charge distributions (local charge meters).

Also in frequent use for determining charge densities is a compensation method, utilizing a dynamic capacitor (42, 43). If the electret or the opposing electrode is set into mechanical vibrations (for example, by a loudspeaker), an ac voltage is generated by the setup shown in Fig. 3.5. Upon compensating the field in the air gap by application of a dc voltage V_0, the ac signal disappears. The charge density σ can then be determined from V_0 by means of Eq. (3.6b). Thus a measurement of the air gap thickness is not necessary in this case. According to Eqs. (3.6a and 3.6b), the compensating voltage V_0 corresponds to the potential of the free electret surface.

The total charge density σ_T can be determined by the thermal pulse method

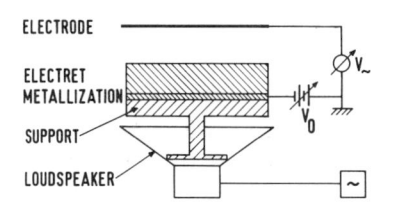

Figure 3.5 Dynamic capacitor with field compensation for measuring effective surface-charge density. (From von Seggern [44].)

(45–47), which is also in extensive use for evaluating the charge centroid (see below). This procedure, schematically shown in Fig. 3.6, is based upon an evaluation of the potential change across the dielectric during diffusion of the thermal energy generated by a short light pulse. The change ΔV of the voltage V is due to thermal expansion of the material and variations of the dielectric constant. If the voltage variations $\Delta V(t_1)$ immediately following excitation and $\Delta V(t_2)$ after the establishment of thermal equilibrium are measured, one has

$$\sigma_T = \hat{\sigma}\frac{\Delta V(t_1)}{\Delta V(t_2)} \qquad (3.12)$$

According to this equation, a measurement of the effective charge density $\hat{\sigma}$ together with thermal-pulse measurements yields the total charge of the electret.

The methods discussed above relate to samples having at least one non-metallized surface. On samples metallized on both surfaces, charge densities may be determined in some cases by measuring polarization and depolarization currents. An example is charge injection by electron-beam irradiation where every electron remaining in the dielectric will cause a positive induction charge to flow into the electrodes. Another example is the depolarization of a uniform dipole alignment P, which will result in a charge transfer equal to $\pm P$ into the two electrodes. For general and unknown charge distributions, however, depolarization techniques generally do not yield an accurate measure of the total stored charge in sandwich-type samples.

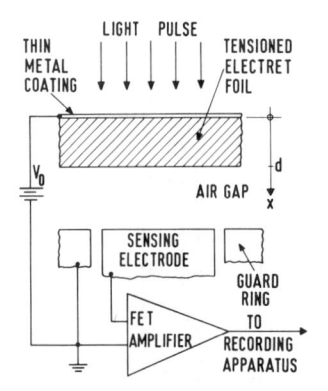

Figure 3.6 Thermal-pulse setup for measuring charge centroid. (From Collins [45].)

CHARGE-MEASURING TECHNIQUES

On unelectroded samples, the net charge may be determined by measuring the induction charge flowing into a Faraday cup upon lowering the electret into it. In cases where this method cannot be applied, a procedure based on measuring the real and imaginary parts of the dynamic capacitance of an arrangement consisting of two electrodes with the sample mounted in between and set into vibrations may be used (48).

3.3.3 Measurement of Mean Charge Depth and Charge Distribution

The classical methods for determining charge distributions in plate electrets are sectioning and planing techniques. These methods are primarily applicable to relatively thick dielectrics but have also been used for thin polymer films (49). For a detailed evaluation of these methods, we refer to the literature (17, 50).

The most convenient method to determine the mean charge depth (or centroid location) on charged polymers is the thermal-pulse technique introduced in the previous section. If the voltage changes $\Delta V(t_1)$ and $\Delta V(t_2)$ [Eq. (3.12)] are measured, the centroid location $\bar{r}$, defined by $\bar{r}/s = \hat{\sigma}/\sigma_T$ (s is the sample thickness) may be obtained by means of Eq. (3.12) from

$$\frac{\bar{r}}{s} = \frac{\Delta V(t_2)}{\Delta V(t_1)} \qquad (3.13)$$

Experiments of this kind are nondestructive since the thermal pulse causes only a minor temperature increase in the sample.

In principle, the thermal-pulse technique can also be used to determine the actual distribution $\rho(x)$ of the charge. However, because of inaccuracies of the numerical deconvolution procedure necessary for such evaluations, a unique solution for $\rho(x)$ is generally not obtained (46, 51). If, in addition, one allows for measuring errors of the order of 0.1%, only 3–10 spatial Fourier coefficients of the charge distribution can be found accurately, the number depending on the location of the charge in the film (52).

A related method is the pressure-pulse technique (53). Here an acoustic pulse of short rise time is propagated through the sample. Again, voltage changes are produced in an external circuit from which the charge distribution can be calculated. Since the pulse propagates with a steep front, a deconvolution is not required in this case and $\rho(x)$ can be determined uniquely and accurately. Compared to the thermal-pulse technique, this method requires more complicated instrumentation. Its resolution is determined by the rise time of the acoustic pulse. Using excitation of the polymer sample by nanosecond pulses obtained with Paseo or quartz-crystal excitation, resolutions of the order of 1 μm have been achieved recently (54).

Electron-beam sampling of electrets also yields the charge distribution (55). If a charged sample electroded on both surfaces as shown in Fig. 3.7 is irradiated with an electron beam of slowly increasing energy, the conductive region in the

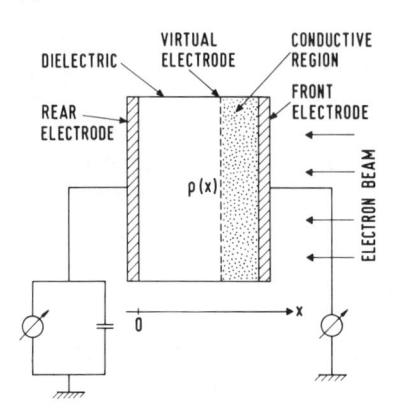

Figure 3.7 Electron beam setup for measuring charge distribution. (From Sessler et al. [55].)

sample extends progressively deeper into the material. Thus all the charges residing originally in the dielectric are gradually removed. The corresponding change of the induction charge q on the rear electrode yields $\rho(x)$ by means of

$$\rho(x) = -\frac{d^2(qx)}{dx^2} \quad (3.14)$$

where x is the thickness of the nonconductive region. This method is capable of a resolution of about 20% of the sampling depth in samples where the Schubweg of the holes in the nonirradiated region is correspondingly small (55, 56).

A method to find the charge distribution $\rho_r(x)$ is based on an evaluation of the time dependencies of the depolarization current and dielectric displacement during isothermal discharge of the sample (57).

A number of other methods are in use to determine charge centroids. One of these, applicable to precharged and electrets with one electrode, is based on a measurement of the effective surface charge and the integrated depolarization current of the electret (8). Another method, designed to measure the charge centroid during electron-beam charging, depends on an evaluation of the induction charges on the two electrodes by means of a split Faraday cup (35).

3.3.4 Thermally Stimulated Current (TSC) Techniques

Thermally stimulated current (TSC) techniques consist of taking measurements of electrode currents from a charged sample while the sample is heated at a constant heating rate. The currents originate from the motion of charges in their own field or from the reorientation of dipoles due to thermal motions. Both processes are stimulated by the temperature increase of the sample. The mobility increase with temperature combined with the eventual charge depletion results in a current–temperature curve characterized by one or more peaks.

The peak location and the structure of the TSC curves is determined by the relaxation frequencies α of the decaying real or dipolar charges. In the simplest case, the decay process is characterized by a single relaxation frequency. In many materials, a distribution of α values is encountered, which is either discrete or continuous and might be due to a distribution of preexponential factors (natural frequencies) or a distribution of activation energies [see Eq. (3.19) below].

In polymers, a number of discrete groups of relaxation frequencies exist which are related to the γ-, β-, α-, and ρ-relaxations. While the γ-relaxation is due to motions within side groups of the molecular chains, the β-relaxation originates from motions of the side groups and the α-relaxation is caused by joint motions of side groups and main chains. These three relaxations can result in the reorientation of dipoles or in the release of real charges. As opposed to this, the ρ-relaxation is exclusively due to the motion of space charges. If dipolar or space charges are present, the above relaxations can result in single or multiple peaks which are referred to as γ-, β-, α-, and ρ-peaks, respectively.

The two most widely used arrangements for TSC measurements on disk-shaped samples are shown in Fig. 3.8. In the short-circuit method, illustrated in Fig. 3.8a, the electrodes of a sandwich-type sample are connected to an electrometer and the depolarization current is recorded as the temperature is linearly increased. For the open-circuit method, samples with only one or no electrodes are required. In the customary arrangement of a one electrode sample, depicted in Fig. 3.8b, a second electrode is placed opposite the unelectroded surface at a distance large compared to the sample thickness. Upon heating, measurements are taken of the minute displacement currents $i(t)$ or of the surface potential $V(t)$ of the dielectric. Since the measuring setup is a capacitor of capacitance C, these quantities are related by $i(t) = C\, dV(t)/dt$. Because $i(t)$ reflects directly any minor change in $V(t)$, measurement of $i(t)$ yields generally more information than measurement of $V(t)$. The short-circuit and open-circuit methods provide different information about the charge stored in the sample, as will be shown.

TSC methods are an important tool for the investigation of charged polymers since they are capable of resolving charge decay processes characterized by a wide range of time constants in a single, relatively short experiment. These

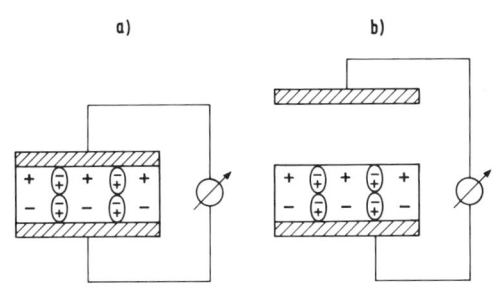

Figure 3.8 Important TSC methods: (a) short-circuit method; (b) open-circuit method.

techniques have also been referred to as thermally stimulated discharge (TSD) methods, ionic thermocurrents (ITC), and other designations. They are similar to thermally stimulated conductivity and thermoluminescence methods and were introduced to the study of charged dielectrics in the 1960s (58, 59). Recent reviews of these methods have been delivered by van Turnhout (17, 60) and Vanderschueren (61). In the following, we give a short review of the basic TSC methods and phenomena, following generally the approach by van Turnhout (60).

Short-Circuit TSC

Thermally stimulated currents can be generated by dipole reorientation or by motion of real charges. In some cases, the currents may be caused by both processes simultaneously. We shall consider, however, the two processes separately and refer to the literature for a discussion of combined motions (see, for instance, Ref. 60).

The depolarization of a dipole polarization $P(t)$, assumed to be spatially uniform so that $E = 0$, yields according to Eq. (3.8) with $i_c = 0$ the external current

$$i(t) = \frac{dP}{dt} \tag{3.15}$$

The decay of $P(t)$, if characterized by a single dipole relaxation frequency $\alpha(T)$, is governed by the Debye equation

$$\frac{dP(t)}{dt} + \alpha(T) P(t) = 0 \tag{3.16}$$

Integration yields the decay of the polarization

$$P(t) = P_0 \exp\left[-\int_0^t \alpha(T)\, dt\right] \tag{3.17}$$

Assuming a linear temperature increase with an inverse heating rate $h = dt/dT$ and using Eq. (3.15), the current is

$$i(T) = -\alpha(T) P_0 \exp\left[-h\int_{T_0}^{T} \alpha(T)\, dT\right] \tag{3.18}$$

For relaxations which involve the rotation of small molecular groups, $\alpha(T)$ obeys an Arrhenius law:

$$\alpha(T) = \alpha_0 \exp\left(\frac{-A}{kT}\right) \quad (3.19)$$

where α_0 is the natural frequency and A is the activation energy. Although Eq. (3.19) does not hold for many other phenomena, such as transition from the glass to the rubber state, the further discussions are based on it.

Substituting Eq. (3.19) into Eq. (3.18) and approximating the integral (60),

$$h \int_0^T \alpha(T) \, dT = \alpha(T) \frac{hkT^2}{A + 1.85kT}$$

one obtains $i(T)$.

Model curves are shown in Fig. 3.9. The bell-shaped current maximum is asymmetric in the sense that the current drop is steeper than the current rise. The figure also indicates that higher activation energies lead to lower and broader peaks located at higher temperatures. Similar shifts are obtained when the heating rate or the relaxation frequency changes. Thus, according to Fig. 3.9, higher heating rates $1/h$ or lower relaxation frequencies result in higher peak temperatures.

Two important relations can be derived from the above equations. First, the activation energy follows from differentiating Eq. (3.18) with respect to $1/T$ for the initial current rise, where $h \int \alpha(T)dT \approx 0$. This yields

$$\frac{d}{d(1/T)} \ln i(T) = -\frac{A}{k} \quad (3.20)$$

The activation energy can also be derived from other parameters of the TSC

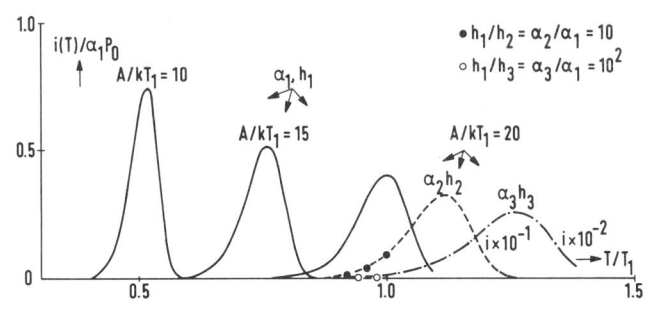

Figure 3.9 Short-circuit TSC curves for dipole polarizations with a single Debye relaxation. The current curves are shown for three different activation energies A, three different natural frequencies α, and three different heating rates $1/h$. Currents represented by dashed curves are reduced as indicated. (From van Turnhout [60].)

curve (62). Second, one obtains for the temperature of the maximum by differentiating Eq. (3.18) the relation

$$\frac{d}{dT}\frac{1}{\alpha(T)} = -h \tag{3.21}$$

This yields for the case of the Arrhenius law, Eq. (3.19),

$$\alpha(T_m)\frac{hkT_m^2}{A} = 1 \tag{3.22}$$

A plot of this relation shows an almost linear dependence between T_m and A.

The depolarization of a space charge $\rho(x,t)$ of unipolar sign yields, according to Eqs. (3.8) and (3.9) for $P = 0$, the current

$$i(t) = \epsilon_0\epsilon\frac{\partial E(x,t)}{\partial t} + [g(T) + \mu(T)\rho(x,t)]E(x,t) \tag{3.23}$$

By integration over the sample thickness s and by considering the short-circuit condition $\int_0^s E(x,t)dx = 0$ and the Poisson equation

$$\frac{\partial E(x,t)}{\partial x} = \frac{\rho(x,t)}{\epsilon_0\epsilon} \tag{3.24}$$

Eq. (3.23) can be cast into the form

$$i(t) = \frac{\epsilon_0\epsilon\,\mu(T)}{2s}[E^2(s,t) - E^2(0,t)] \tag{3.25}$$

The external current is thus entirely due to excess-charge currents while displacement currents and conduction currents make no contribution.

The field $E(x,t)$ required to evaluate $i(t)$ is obtained from the Poisson equation, Eq. (3.24), and from the continuity equation

$$\frac{\partial\rho(x,t)}{\partial t} = -\mu(T)\frac{\partial\rho(x,t)\,E(x,t)}{\partial x} - g(T)\frac{\partial E(x,t)}{\partial x} \tag{3.26}$$

Since these equations depend on space and time, analytical solutions are only possible in a few specific cases. One such case is the so-called box distribution, a charge distribution whose initial density $\rho(x, 0)$ is constant between the surface and a certain depth r_0 and zero elsewhere in the dielectric. With time, the distribution remains a box with advancing front and decreasing charge density (63). For this distribution, the continuity equation is easily particularized at the zero-field plane x_0. With $\partial \rho(x,t) E(x,t)/\partial x = \rho^2(x_0,t)/\epsilon_0\epsilon$ and $\partial \rho(x_0,t)/\partial t = d\rho(x_0,t)/dt$, Eq. (3.26) becomes

$$\frac{d\rho(x_0,t)}{dt} = -\frac{1}{\epsilon_0\epsilon}[\mu(T)\,\rho(x_0,t) + g(T)]\,\rho(x_0,t) \qquad (3.27)$$

Analytical integration for $g(T) = 0$ or otherwise numerical integration yields $\rho(x_0,t)$ and thus $E(x,t)$. This permits the calculation of $i(t)$ with Eq. (3.25).

The currents released due to the discharge of three different charge distributions of box shape are shown in the model curves of Fig. 3.10. The largest currents are generated by distributions which extend halfway through the sample. An increase in the conductivity from zero to finite values diminishes the current.

For general initial charge distributions, Eqs. (3.24) and (3.26) must be evaluated numerically. Calculations of this kind were carried out by van Turnhout (17, 60) and others. Also, the method of characteristics has been used to solve the partial differential equations (64). We refer to the literature for more detail on these evaluations (60).

In the above cases, the conductivity of the dielectric was assumed to be independent of the thickness coordinate and independent of time for constant temperature. This model does generally not apply to dielectrics charged by injection of energetic electrons. This process causes a radiation-induced conductivity (RIC) in the volume penetrated by the electrons (see Section 3.6). Since the RIC decreases with time, its effect on the TSC currents is dependent on storage time. Consideration of this phenomenon leads to an upward shift of TSC peak temperatures with storage time (65).

Open-Circuit TSC

The open-circuit method is mostly used for dielectrics with real-charge storage and only occasionally for samples possessing a dipole polarization. We therefore

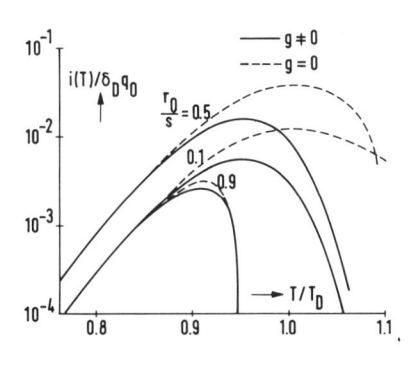

Figure 3.10 Short-circuit TSC curves for space charges in nonconducting and conducting dielectrics. The charge distributions are box-shaped, extending from one surface to a depth indicated by the parameter r_0/s. Curves normalized for equal total charge $q_0 = \rho_0 r_0$. (60).

discuss only the case of real-charge depolarization and assume the unipolar charges to have a finite mobility $\mu(T)$ within a dielectric of conductivity $g(T)$. Trapping is neglected.

Since in open-circuit TSC the current $i(t)$ is zero everywhere in the system, Eq. (3.23) yields the following after integration over the sample thickness s:

$$\epsilon_0 \epsilon \frac{dV(t)}{dt} = -g(t)V(t) - \mu(T) \int_0^s \rho(x,t) E(x,t) \, dx \qquad (3.28)$$

With Poisson's equation, Eq. (3.24), and considering that in an open circuit $E(0,t) = 0$ (free sample surface at $x = 0$), Eq. (3.28) yields

$$\epsilon_0 \epsilon \frac{dV(t)}{dt} = -g(t) V(t) - \tfrac{1}{2} \epsilon_0 \epsilon \, \mu(T) \, E^2(s,t) \qquad (3.29)$$

Solutions of this equation can only be found if $E(s,t)$ is known. Different approaches for evaluating $E(s,t)$ are necessary for times up to and exceeding the transit time t_λ, that is, the time required for a carrier to move across the sample. For $t < t_\lambda$, the charge density $\rho(s,t)$ will be zero. We therefore particularize Eq. (3.23) for $x = s$ and obtain with $i(t) = 0$ after integration over time the following:

$$E(s,t) = E(s,0) \exp\left[-\frac{1}{\epsilon_0 \epsilon} \int_0^t g(T) \, dt\right] \qquad \text{for } t < t_\lambda \qquad (3.30)$$

For $t \geq t_\lambda$, a general expression for the time dependence of $E(s,t)$ cannot be given. For the case of a box distribution, however, the charge has uniform density throughout the sample, and the field is thus given by $E(x,t) = (x/s) E(s,t)$. From the relation $V(t) = \int_0^s E(x,t) \, dx$, one has therefore

$$E(s,t) = \frac{2V(t)}{s} \qquad \text{for } t \geq t_\lambda \qquad (3.31)$$

With Eqs. (3.30) and (3.31), it is possible to calculate $V(t)$ from Eq. (3.29) for times up to and exceeding the transit time, respectively. The evaluations have to be performed numerically except for $g(T) = 0$, where analytical solutions exist.

A few typical open-circuit TSC curves, as calculated from the above equations, are plotted in Fig. 3.11. Shown is the dependence of $V(t)$ and $dV(t)/dt$ (which is proportional to the displacement current) on T for nonconducting dielectrics with different box-shaped charge distributions. The figure indicates that dielectrics with charges close to the free surface ($r_0/s = 0.1$) are more stable in the sense that the charge decay occurs at higher temperatures than with

REAL-CHARGE STORAGE

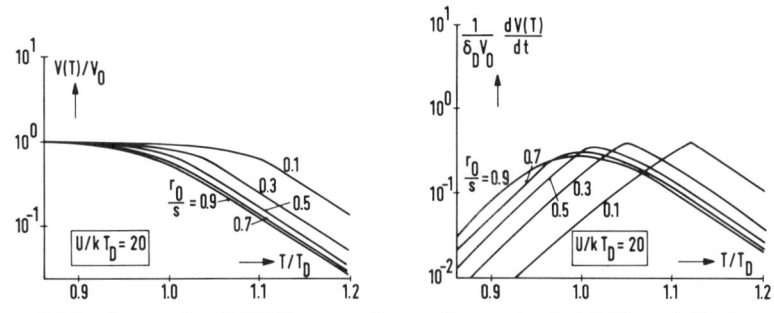

Figure 3.11 Open-circuit TSC curves for surface potential (left) and displacement current (right) in nonconducting dielectrics. The charge distributions are box-shaped, extending from the open surface to the depth indicated by the parameter r_0/s. (From van Turnhout [60].)

dielectrics having charges reaching almost through the sample. For the curves with $r_0/s \leq 0.5$, the cusp in $dV(T)/dt$ indicates the transit time, while for the curves with $r_0/s > 0.5$, the peak occurs after the transit time.

The open-circuit analysis has been extended to nonuniform charge distributions by Zahn (64) and to multiple trapping (66). Recently, the effect of a number of parameters, such as heating rate, surface potential, activation energy, and initial charge distribution on open-circuit TSC has been studied by numerical analysis (67). This particular model is based on trap-controlled transport and assumes a box-shaped initial charge distribution with rounded corners. For more details on these investigations we refer to the original publications.

3.4 EXPERIMENTAL RESULTS AND INTERPRETATION OF REAL-CHARGE STORAGE

3.4.1 Trapping Levels and Trap Densities

Charge storage in dielectrics is due to the presence of trapping states which are capable of holding electrons or holes for long periods of time. Trapping in polymers is often interpreted in terms of a modified energy band model, as depicted in Fig. 3.12a. According to this model, traps are localized states, belonging to certain molecules or molecular groups. Since polymers are amorphous or polycrystalline, the energy levels, which are affected by their environment, are different in different molecular regions of the material. Thus the trap depths are correspondingly distributed (see below). Activation from such local states is by thermal energy.

Apart from these localized traps there are delocalized states, generally referred to as extended states, which are energetically located near the bottom of the conduction band and at the top of the valence band (Fig. 3.12b). They are

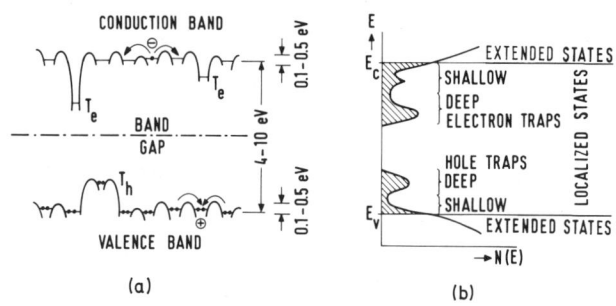

Figure 3.12 (a) Energy band model for a polymer; T_e-electron traps, T_h-hole traps (b) Density of states for a polymer; localized states (traps) are shaded; E_c and E_v-mobility edges. (From Bauser [68].)

separated from the localized states by the so-called mobility edge at which the carrier mobility drops by several orders of magnitude (69). Carriers in such states move by quantum-mechanical hopping. Charge trapping in extended states is generally negligible in electret materials. However, the extended states do play a role in charge transport.

Charge storage in polymers may occur in surface and volume traps. While it is readily possible to distinguish experimentally between these categories (see below), it is difficult to assess the molecular origin of the traps. Surface traps may be due to chemical impurities, specific surface defects caused by oxidation products, broken chains, adsorbed molecules, or differences in short-range order of surface and bulk (13). For volume traps, investigations on substituted polyolefins have indicated the presence of three structural trapping levels but no trapping due to impurity centers (70). While the primary levels are at atomic sites on the molecular chains, the secondary levels are between groups of atoms in neighboring molecules, and the tertiary levels are in the crystalline regions or at crystallite–amorphous interfaces. More detailed theoretical models for the localized states in polymers are discussed in Chapter 2 of this book.

TSC measurements in PET, FEP, and other polymers have shed some light on the nature of the trap centers. In short-circuit TSC of PET, the β-peak was found to be composed of three components (β_1, β_2, and β_3) of which the β_1-relaxation at $-40\,°C$, which is attributed to the motions of carboxyl groups in the main chains, is responsible for a considerable release of trapped charge (71). The active trapping sites are thus apparently primary levels. At higher temperatures, trapped charge is released in the temperature ranges of the α- and ρ-peaks at 80 and 120 °C, respectively (see Ref. 17, Fig. 10.32). Since the α-peak corresponds to joint motions of main chains and side groups, the charges could be released from primary or secondary levels. Finally, the charges released in the ρ-peak might be related to the crystalline structure of the polymer. In electron-beam-charged and thermally charged FEP, short-circuit TSC peaks were found at 80 and 190 °C (α- and ρ-peaks) (17, 65). While the α-peak is probably caused by a radiation-induced conductivity (65) (see Section 3.4.5), indications are that the ρ-peak is again sensitive to the crystalline structure (17). There is, however, still a considerable amount of uncertainty in these and other results on the molecular origin of volume traps in polymers.

The spatial location and thermal activation temperature of surface and volume traps in Teflon FEP have recently been studied by TSC measurements (72). (The spatial location of the actually trapped charge is discussed in Section 3.4.2.) First, surface and volume traps for corona-charged electrets were separated by means of the open-circuit TSC data shown in Fig. 3.13. Analysis of these data yields release temperatures of 155, 170, and 200 °C with the 155 °C peak disappearing for the annealed samples. Comparison with short-circuit TSC data, which does not show the 155 °C peak, enables one to assign the 155 °C peak to surface traps, the 170 °C peak to near-surface states, and the 220 °C peak to volume states. Also, a shallower trap level active under conditions where the deeper traps are filled (trap-filled limit) and discharging at 95 °C was detected in the short-circuit experiments. Finally, information about the spatial distribution of the 155– 200 °C traps was derived from open-circuit TSC experiments with electron-beam-charged samples where the charge penetration is known. The results of this investigation are shown in Table 3.1.

Activation energies of trapping levels in a number of polymers have also been derived from TSC and other measurements. Typical values are 1.9 eV for Teflon (73), 1.8–2.2 eV for Mylar, 1.5 eV for polyethylene, and 1.2 eV for polyimide (60). Some of these trapping levels correspond actually to a distribution of activation energies around these values. Traps of depths between 1 and 2 eV hold charges for long periods of time. One obtains for the trap-release frequency

$$\nu = \nu_0 \exp\left(-\frac{U}{kT}\right) \quad (3.32)$$

values of 10^{-4}–10^{-22} s^{-1} at room temperature assuming an escape frequency $\nu_0 = 10^{13}$ s^{-1}.

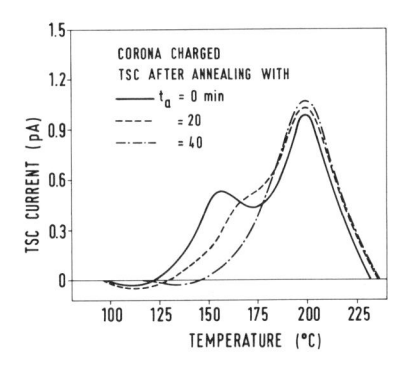

Figure 3.13 Open-circuit TSC curves for one-sided metallized, negatively corona-charged 25 μm Teflon FEP. After charging, the samples were annealed at 145°C for the time periods indicated. (From von Seggern [72].)

Table 3.1 Distribution of Traps for Negative Charges in 25 μm Teflon FEP-A (72)

Peak Temperature °C	Location Relative to Charged Surface μm	Kind of Trap
95	0–25	Energetically shallower trap active under trap-filled-limit conditions
155	0–0.5	Surface trap
170	0.5–1.8	Near-surface trap
200	1.8–25	Bulk trap

Trap densities in polymers have been inferred from maximum stored-charge densities. Some experimental values are given in Table 3.2.

Since most of the data in the tables are limited by breakdown and not by trap saturation, the actual trap densities may be larger than those shown. However, measurements on Teflon FEP show that the value of 0.14×10^{15} cm^{-3} is the actual density of the 190° C electron trap in this material.

3.4.2 Spatial Distribution of Trapped Charges

The methods described in Section 3.3.3 have been used extensively to gain insight into the charge distribution of charged polymer samples. Presently, a host of good data on charge centroid locations are available for polymer electrets, while distribution data are still scarce.

Table 3.2 Greatest Observed Charge Densities and Full-Trap Densities in Some Polymer Materials

Polymer	Thickness	Charging Method: Breakdown B or Electron Beam E	Surface Charge S or Volume Charge V	Projected Charge Density and Sign (10^{-6} C/cm^2)	Full-Trap Density in Volume and Sign (10^{15} cm^{-3})	References
FEP	12.5	B	S + V	0.5 (+,−)		30
FEP	25	E	V		0.14^a(−)	66, 72
					0.18^a(+)	44
PET	3.8	B	S + V	1.2 (+,−)	20 (+,−)	30
PC	2.0	B	?	1.0 (+,−)		30
PS	25	B	V	0.5 (−)	8(−)	74

^{a}Attributable to deep traps with TSC peak temperature of 190° C or higher.

REAL-CHARGE STORAGE

The charge centroid on Teflon films electroded on both surfaces and charged by injection of electrons, was determined by split-Faraday-cup measurements (75). Comparison of the initial charge, end-of-charging, and final charge depths, as shown in Fig. 3.14, indicates a significant charge drift during charging but less drift thereafter. The initial depth is somewhat smaller than the so-called practical range, while the final depth exceeds it substantially. The charge drift is due to the action of the self-field in the presence of a radiation-induced conductivity. Other measurements of the final depth on electron-beam-charged Teflon films, performed with the thermal-pulse method, substantiate and extend the data in Fig. 3.14 (47, 72).

The centroid location was also measured on positively and negatively liquid-charged Teflon films with one electrode by the thermal-pulse method (45). For both polarities, the charges reside originally on the surface of the samples. Injection into the volume occurs at 100 and 180°C for positive and negative carriers, respectively. The experiments further show that most of the positive charges, once released from the surface, move rapidly through the film (indicating a long Schubweg), while the negative charges are trapped in the volume. The large hole Schubweg in the volume is substantiated by hole transit experiments (76); the efficient electron trapping is confirmed by TSC data (72).

Centroid data for the charge distribution were also obtained for corona-, Townsend-, and breakdown-charged polymers with a variety of methods. For Teflon, negative corona charging does not yield measurable charge penetration (27), while negative breakdown charging at high voltages yields charge depths of a few microns (17, 30). On Mylar, even corona-charging and Townsend charging leads to charge depths of 1–2 μm (17). Charge depths of a few microns were also found for breakdown-charged polystyrene (74).

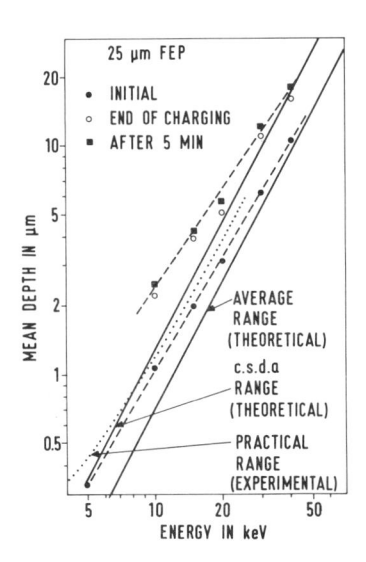

Figure 3.14 Location of charge centroid in two-sided metallized Teflon FEP charged with electron beams of various energies. Dashed lines are best fits to the experimental points. For comparison, various electron ranges are also shown. (From Gross et al. [75].)

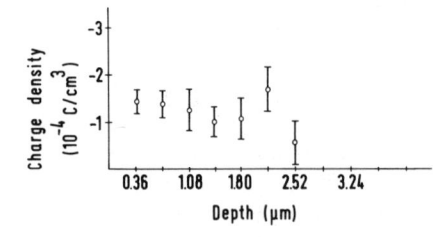

Figure 3.15 Charge distribution in two-sided metallized 10 μm PET, charged by short-circuit irradiation with a 12.1 keV electron beam. (From Tong [56].)

The actual charge distribution in polymer films has been determined with the thermal-pulse, the electron beam and the pressure pulse methods. The data obtained with the thermal-pulse method suffer from a nonuniqueness introduced by the deconvolution process. For a discussion of these data, we refer to the literature (46, 47). Encouraging progress has been achieved with the electron-beam method which is applicable to dielectrics with a short hole Schubweg (77). Recent measurements with this method (78, 79) on electron-beam-charged and thermally charged Mylar samples show charge distributions which peak at the expected penetration depth of the electrons. A distribution in a beam-charged sample is shown in Fig. 3.15. The distribution corresponds to that calculated (80) by taking into account the nonuniform radiation-induced conductivity (see Section 3.4.5). Very recently, accurate distribution data in PET, FEP, and PVDF samples were obtained with the pressure-pulse method (54). A result for electron-beam charged PET is shown in Fig. 3.16.

3.4.3 Charge Decay

According to Eq. (3.6), the effective surface charge density of a dielectric is proportional to its surface potential. Both quantities are used in the following to characterize the charge decay of electrets.

The decay of the real charge on a polymer sample can be due to external or internal causes. If the sample is stored under unshielded conditions such that it

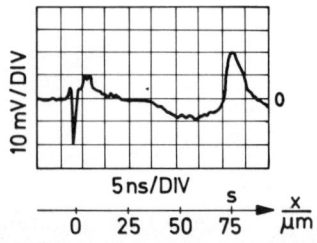

Figure 3.16 Charge distribution in 55 keV electron-beam charged 75 μm PET sample, as measured with pressure-pulse method. Shown is the current response of the sample upon propagation of a pressure pulse in the thickness direction. The current (except for first negative spike which is due to photoionization) corresponds directly to charge density in sample and electrodes. (From Sessler et al. [54].)

REAL-CHARGE STORAGE

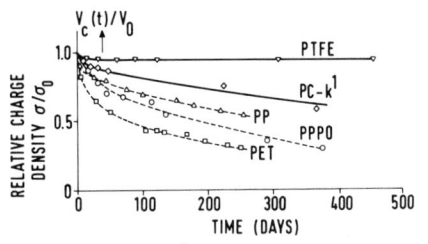

Figure 3.17 Charge decay in dry atmosphere at room temperature on one-sided metallized polymer films: 50 μm Teflon (PTFE), 25 μm polycarbonate (PC-k1), 20 μm polypropylene (PP), 25 μm poly(-2, 6, -diphenyl -1, 4-phenylene oxide) (PPPO) and 25 μm Mylar (PET). All samples thermally charged. (From van Turnhout [17].)

is subject to air circulation, the external charge decay due to ions in the air is accelerated. The effect has been investigated for polypropylene and Teflon electrets with somewhat different results (81). While the polypropylene data show only a minor effect of air flow on charge decay, the Teflon data exhibit a pronounced decay. If air currents are kept away by proper shielding, the external decay is diminished such that internal conduction determines the charge loss.

The charge decay of a number of singly electroded polymer films due to internal causes is shown in Fig. 3.17. The excellent charge stability of negatively charged Teflon is evident from the figure. This material shows also superior charge-storage behavior in high-humidity atmospheres (17) which is attributed to its hydrophobic nature.

Different charging methods lead to different initial charge distributions. This in turn affects the potential decay. The results of a study of TSC curves of corona-charged, liquid-charged, and electron-beam-charged Teflon FEP films are shown in Fig. 3.18 (44). Since the differences expected are small in such experiments, a precision setup allowing simultaneous measurements on several samples was used. The negatively corona-charged or liquid-charged electrets show a decay starting at about 140°C, which is due to the injection from surface to volume states. As opposed to this, the beam-charged samples exhibit initially

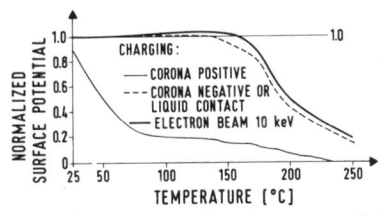

Figure 3.18 Open-circuit TSC curves showing potential decay for negatively corona-charged, liquid contact-charged, and electron beam-charged one-sided metallized 25 μm Teflon FEP samples. (From von Seggern [44].)

a rise in potential due to the injection of holes left at the surface during charging as a result of secondary emission. Since the electrons are already in the volume, the drop around 150°C is absent. Thus, electron-beam-charged Teflon samples have better stability than surface-charged samples. No such difference has been detected for PET electrets (82); this is probably due to the fact that the surface and volume states in this material do not differ significantly (72).

The charge stability of most polymers can be improved by the application of heat before, during, or after charging. For example, better stability has been noted for Teflon electrets annealed at 150°C or quenched from 250°C prior to charging (60, 83). The improvement has been attributed to the generation of deeper traps because of changes in crystalline grain size or crystallinity. The stability was also improved by application of heat up to 250°C to Teflon and other polymers during charging (17, 28, 84, 85). This is attributed to the retrapping of carriers from shallower to deeper levels. The transition could be from surface levels to volume levels or from relatively shallow volume levels to deep volume levels. The same phenomena are responsible for the increase in stability due to annealing after charging (17). The charge loss due to annealing may be made up by repeated charging. Annealed Teflon electrets exhibit room-temperature time constants of the charge decay of about 200 years (73).

The potential decay of electron-beam-charged Teflon at 150° C is shown in Fig. 3.19. At this temperature, two exponential decay sections appear, noticeable as two downward bends in the double-logarithmic plot. The two decay phenomena with very different time constants correspond to the decay regions at about 190 and 240°C, respectively, found in open-circuit TSC experiments on this material. These regions are due to two volume trap levels. A comparison of measured and predicted potential decays at 145°C over a more limited time period (where the deeper volume trap does not yet release) is given in Section 3.4.4.

Electrodes may also affect the charge decay of polymer electrets in certain cases. The customary vacuum-deposited aluminum or gold layers act, in the absence of radiation, as blocking contacts on Teflon and many other polymers and have therefore no influence on charge decay (87, 88). On the other hand, sputtered gold electrodes on Teflon were found to allow hole passage at elevated temperatures, thus causing a lowering of TSC peak temperatures compared to

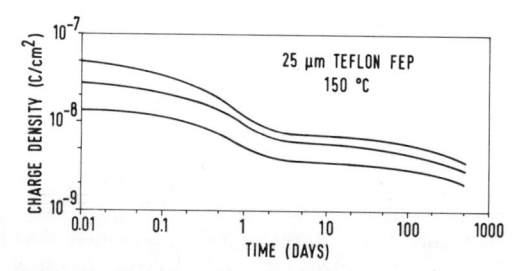

Figure 3.19 Decrease of effective surface-charge density of one-sided metallized 25 μm Teflon FEP at 150°C. Samples negatively charged with 20 keV electron beam. (From Sessler and West [86].)

samples with electrodes deposited by evaporation processes (89). The blocking nature of some contacts is probably caused by charges trapped in the vicinity of the electrode and is thus not an electrode property (90) (see also Section 3.4.5).

The time constants of the charge decay in Teflon at elevated temperatures can be presented as an Arrhenius plot (73). For activated processes, the slopes of such plots yield the activation energies of the trapping levels. For Teflon, an activation energy of 1.9 eV has been derived, in agreement with other methods (see above).

While the charge decay of polymer electrets is generally well behaved in the sense that the potential decay curves of samples of the same material charged initially to different potentials do not cross, exceptions from this have been noted. Most prominent is the case of corona-charged polyethylene, which shows marked potential crossovers (25). The effect has also been found on liquid-contact-charged polyethylene, on corona-charged Teflon PTFE (26, 27), and on electron-beam-charged polyethylene (91). A full understanding of the crossovers has not yet been achieved. It appears that in the more highly charged samples almost full injection into the bulk occurs, where the charges are rather mobile, while in the samples with low charge density the charges remain in deep surface traps. However, the physical reason for this charge-dependent injection process is not clear. Corona light and excited molecules have been held responsible, but the effectiveness of both agents was subsequently questioned (91–94).

Of considerable recent interest are observations of charge transit phenomena in low-conductivity polymer films. The effect has been found in a number of materials, with transit times ranging from the microsecond to the second range (95). Of interest in the context of this chapter are observations of hole transit in Teflon (76), which shows extremely good storage properties for electrons. Typical results of the transit measurements are depicted in Fig. 3.20. (For a

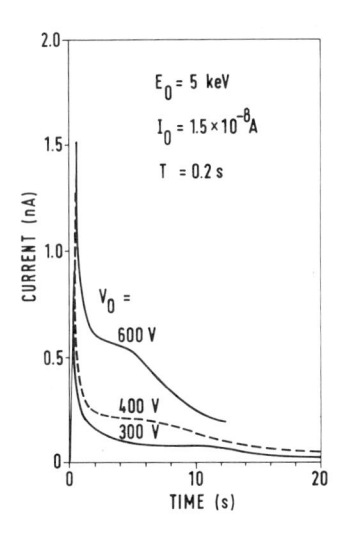

Figure 3.20 Hole currents from two-sided metallized 25 μm Teflon FEP during and after electron beam irradiation. Parameter: voltage applied to sample (positive on side facing the beam). The parameters of the beam are indicated. (From Gross et al. [76].)

detailed description of transit experiments see Chapter 6, Section 6.3.) The hole mobility derived from these data is 2×10^{-9} cm^2/V-s, which corresponds closely to values found from other experiments (96, 97). This has to be compared with an electron mobility of $<10^{-17}$ cm^2/V-s in this material. The curves in Fig. 3.20 and corresponding ones obtained for other polymers display long tails which indicate a disperson of carrier transit times. This phenomenon has been interpreted by two models, namely, the stochastic formalism of dispersive hopping (98) and the theory of multiple trapping (99, 100). Both models are equivalent for the case where the number of trapping events is large. These phenomena, which do not represent charge storage in the usual sense, are discussed in detail in Chapter 6 and in the literature (95, 98–100).

3.4.4 Comparison of Experimental and Calculated Charge Decays

The internal decay of real charges in electrets has been analyzed under a variety of assumptions concerning ohmic conductivity, carrier trapping and mobility, injection from surface states, and so on (8). The simplest model is that of a one-sided metallized dielectric initially carrying a charge on its unelectroded surface. If the carriers have a field- and space-independent mobility μ, the potential $V(t)$ of the nonmetallized surface decays as

$$\frac{V(t)}{V(0)} = 1 - \frac{1}{2}\frac{t}{t_0} \qquad \text{for } t \leq t_0 \qquad (3.33)$$

$$= \frac{1}{2}\frac{t_0}{t} \qquad \text{for } t \geq t_0 \qquad (3.34)$$

where $t_0 = s^2/\mu V(0)$ is the transit time and s is the sample thickness. Fast retrapping of the carriers (Schubweg much smaller than sample thickness) can be considered by substituting for μ a trap-modulated mobility. This model has been successfully used to describe the charge decay in a number of polymers at low fields (101). Its greatest drawback for describing polymer electrets is the neglect of more involved trapping phenomena encountered in these materials.

Recently, a more general model considering surface trapping with a finite release time and trap-controlled volume transport, described by a free time between traps and a capture time in a trap, has been solved numerically (102). The charge is assumed to be initially at the unelectroded surface of the dielectric. The model is particularly tailored to the properties of Teflon electrets. Numerical results of the model are compared in Fig. 3.21 with experimental data for this material obtained at 145°C. Over the range of voltage decays measured, the agreement between experiment and theory is excellent. The theoretical model also yields the evolution of the spatial distribution of free and trapped carriers. According to the results, the free-charge density shows a steep front which moves with constant velocity through the material, reaching the electrode at the

REAL-CHARGE STORAGE   87

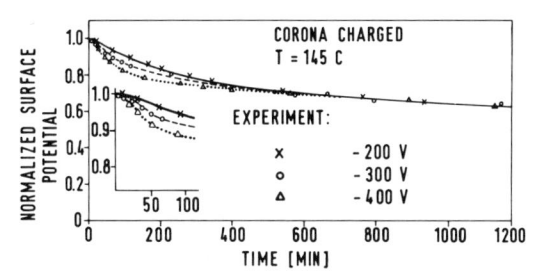

Figure 3.21 Calculated and measured decay of the surface potential of one-sided metallized, corona charged, 25 μm Teflon FEP samples at 145°C. The samples are initially charged to different surface potentials, as indicated. (From von Seggern [102].)

transit time, while the trapped charge has maximum density at the unelectroded side.

For a general description of the charge decay in polymers, a number of other effects have to be considered. Among these are field-dependent mobilities (95), multiple trapping (99), the effect of dielectric loss (103), and the like. Often, however, these phenomena may be neglected.

3.4.5 Radiation Effects

Exposure of polymers to ionizing radiation produces a radiation-induced conductivity (RIC) which is due to the generation of secondary electrons and holes. Since secondaries are lost by recombination, an equilibrium between generation and recombination is eventually reached during irradiation. The secondaries are distributed between various trapping levels and the conduction and valence bands. After termination of the irradiation, a delayed radiation-induced conductivity (DRIC) persists, which falls off with time due to recombination (35, 104).

While the RIC and the DRIC, if present in charged polymers, may be detrimental to the temporal stability of the charges, these phenomena are often unavoidable. Examples are electron-beam-charged electrets and samples exposed to space environments. The RIC and DRIC effects, however, can be used with advantage for the study of the electrical properties of polymers and in radiation dosimetry. Therefore, radiation effects in polymers have been actively studied (see literature in Ref. 14).

The RIC in polymers increases rapidly at the onset of irradiation and reaches a maximum after times of the order of seconds (105). This steady-state conductivity g depends on the dose rate θ of the irradiation by the power law

$$g = K\theta^{\Delta} \tag{3.35}$$

where K is a material-independent constant which for polymers lies between 10^{-19} and 10^{-16} (106) and Δ is an exponent between 0.5 and 1 whose value depends on the energy distribution of the trapping levels and also on dose rate. Upon prolonged irradiation, the conductivity eventually decreases (radiation hardening).

Immediately after termination of the irradiation, the RIC is reduced to $g_0 = kg$, with $k \leq 1$. The time dependence of the DRIC is then given by

$$g_d(t) = g_0(1 + bt)^{-1} \qquad (3.36)$$

where $g_d(0)$ equals g_0 and b is a modified recombination coefficient.

Experimental values of the RIC in Teflon are of the order of 10^{-13} ohm^{-1} cm^{-1} for dose rates of 10^4 rad/s, with $\Delta = 0.79$ (35, 105). The DRIC in this material shows, a few seconds after irradiation, the $1/t$ dependence expected from Eq. (3.36). After 10^4 s, values of about 10^{-17} ohm^{-1} cm^{-1} are reached (35). Even years after irradiation, the DRIC is still noticeable in TSC experiments (65). Electron-beam-charged Teflon films, stored for periods of up to several years at room temperature and up to 1 day at 100°C, show an upward shift of the first TSC peak with storage time. This is seen from the TSC data in Fig. 3.22, depicting data for samples stored at 100°C (107). Together with the peak shift a shift in population from the low-temperature to the high-temperature peaks is observed. Evaluation of the peak shift in Fig. 3.22 demonstrates the validity of Eq. (3.36) over the time periods considered. The DRIC can be cured by annealing since heat accelerates the recombination process (108).

Detailed knowledge of the RIC is also available for other electret-forming polymers, such as Mylar (109) and Polyethylene (80). If the RIC is generated by a partially penetrating beam, such as an electron beam, it depends also strongly on depth in the material (80).

The effect of the RIC on charge storage in polymers has been investgated theoretically by a number of authors (34, 35, 80, 110–114). In one of these studies (111), the charge motion in Teflon films is analyzed by computer simulation using a Monte Carlo model of the polymer to obtain the depth-dependent RIC and charge deposition profiles. Fig. 3.23 shows the energy and charge

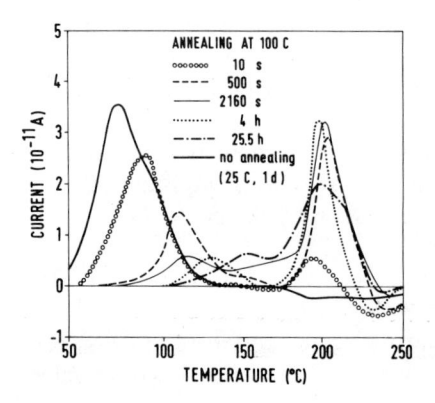

Figure 3.22 Short-circuit TSC curves from 25 μm Teflon FEP pretreated as follows: one-sided metallized samples were charged with 20 keV electrons, stored at 100°C for times indicated, thereafter aluminum coated on irradiated surface, and finally subjected to TSC experiment. (From Sessler and West [107].)

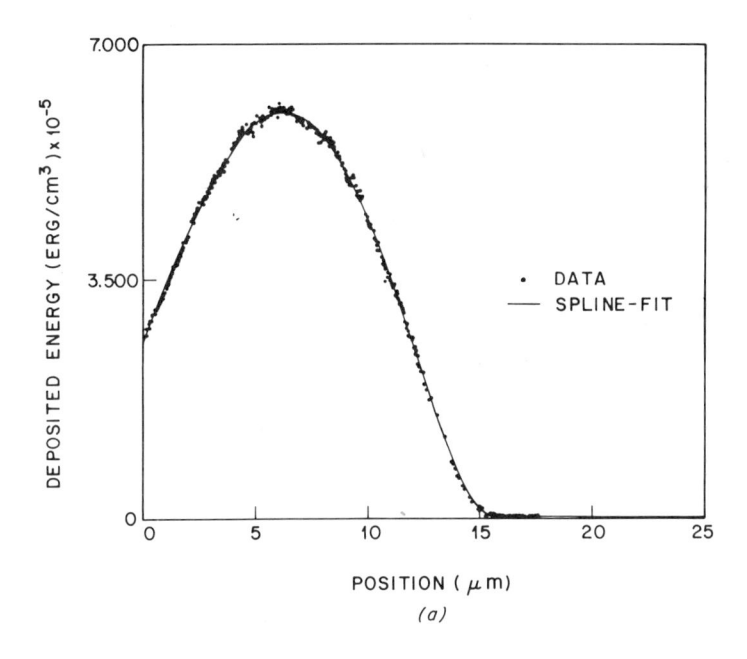

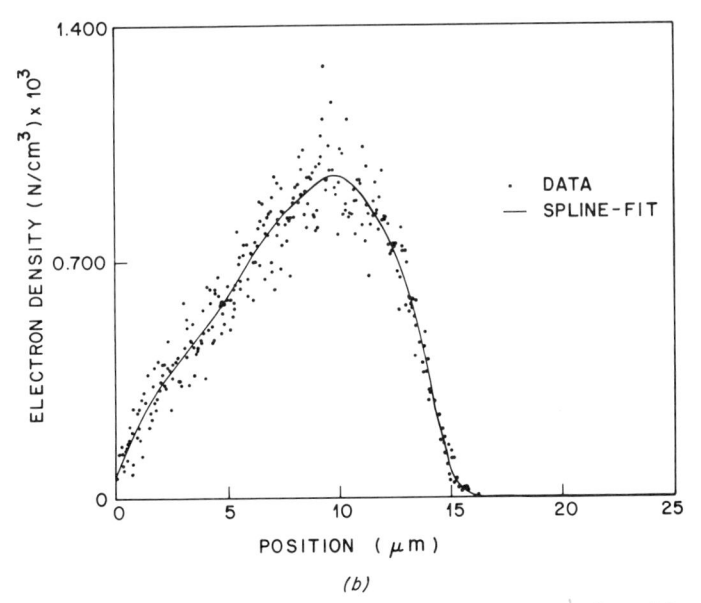

Figure 3.23 Calculated energy-deposition (*a*) and charge-deposition profiles (*b*) due to electron-beam irradiation of a polymer. Data were obtained from a Monte Carlo model of the polymer. (From Berkley [111].)

deposition obtained from such calculations. This and other data were used to calculate charge penetration as function of time, injected charge density, and injected current density. The results agree qualitatively with experimental data.

In another investigation, the time dependence of the voltage across a sample irradiated with floating front or back electrode or the electrode current for a sample irradiated in short circuit was calculated under various assumptions about the temporal and spatial dependence of RIC and injection current (112). The experimental results reported in this study can be explained by theory when radiation hardening is considered.

The RIC affects also the injection of charges through electrodes. In non-irradiated polymers, the injected charges are normally trapped close to the surface. Thus their field disallows further injection and the electrode appears to be blocking. Under irradiation, the RIC prevents the formation of such a charge layer, and continued charge injection is possible (90). Another phenomenon that received attention is the formation of a polarization in irradiated Teflon under an applied field (96, 115).

3.5 EXPERIMENTAL RESULTS AND INTERPRETATION OF DIPOLE POLARIZATION

A number of semicrystalline and a few amorphous polymers are capable of permanent dipole polarization. Representatives of the first group are poly (vinylidene fluoride) (PVDF), which is important because it is strongly piezoelectric (116) and pyroelectric (117) and thus useful in a wide variety of applications, and poly (vinyl fluoride) (PVF). Poly (vinyl chloride) (PVC) belongs to the second group.

In the following, dipolar charge storage in these and other polymers is discussed. The topics of piezoelectricity and pyroelectricity of these materials will be deferred to Chapt. 4. These topics are also extensively discussed in the literature (see, for example, Ref. 118).

3.5.1 Dependence on Poling Parameters

The polarization P is obtained by alignment of the polar groups by an applied field E at elevated temperatures, where the molecular chains are sufficiently mobile (see Section 3.2.1, above). The alignment is governed by the Debye equation which, for a single dipole relaxation frequency $\alpha(T)$ (see Section 3.3.4) can be written (60) as

$$\frac{dP(t)}{dt} + \alpha(T) P(t) = \epsilon_0 (\epsilon_s - \epsilon_\infty) \alpha(T) E \qquad (3.37)$$

where ϵ_s and ϵ_∞ are the static and optical dielectric constants, respectively. Under isothermal polarizing conditions, and if $P(0) = 0$ is assumed, the time dependence of the polarization follows from Eq. (3.37) as

DIPOLE POLARIZATION

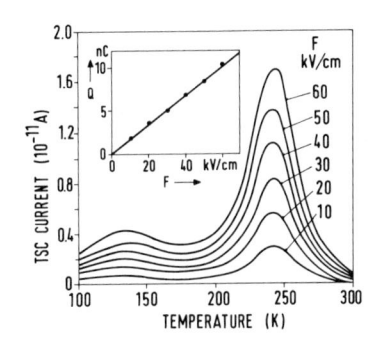

Figure 3.24 Short-circuit TSC curves of low-density polyethylene samples polarized at the fields indicated. The observed peaks are the γ-peak (left) and the β-peak (right). Inset: total released charge. (From Fischer [119].)

$$P(t) = \epsilon_0 (\epsilon_s - \epsilon_\infty) E [1 - e^{-\alpha(T)t}] \quad (3.38)$$

which yields a saturation value of

$$P_s(\infty) = \epsilon_0 (\epsilon_s - \epsilon_\infty) E \quad (3.39)$$

The buildup of the polarization, controlled by Eq. (3.38), is thus dependent upon E, t, and T.

The field dependence has been investigated in many studies with the result that the proportionality between P and E is generally observed. An example is shown in Fig. 3.24, which depicts TSC curves of the γ- and β-peaks (see Section 3.3.4) of polyethylene poled to saturation with different applied fields (119). In the inset of the figure, the total released charge is plotted showing strict proportionality with the polarizing field E. Similar results have been obtained for other polymers, such as polycarbonate (120) and poly(vinyl chloride) (121), while the response of poly(vinylidene fluoride) is more complicated (see Chapt. 4).

One can determine separately the contributions of the various relaxations to the saturation polarization. Experimental values of $P_s/\epsilon_0 E$ for a few polymers, as determined by TSC measurements, are given in Table 3.3. The results are only evaluated for the α- and β-relaxations. A comparison with the differences

Table 3.3 Released Polarization $P_s/\epsilon_0 E$ and Difference $\Delta\epsilon$ for Some Polymers[a]

Polymer[b]	β-Relaxation		α-Relaxation	
	$P_s/\epsilon_0 E$	$\Delta\epsilon$	$P_s/\epsilon_0 E$	$\Delta\epsilon$
PMMA	1.9	1.9	1.7	0.8
PET	0.33	0.37	0.5	0.36
PVC	0.5	0.45		7
FEP	0.011		0.004	
PVDF			~100	~10

[a] See Refs. 17 and 122.
[b] See Table 1.1.

Δε corresponding to these relaxations indicates good agreement for the β-relaxations, while for the α-relaxations $P_s/\epsilon_0 E$ is higher than Δε. This may be partially due to a superposition of space charge currents. According to the table, the permanent polarization of PVDF by far exceeds that of the other polymers. The proportionality between P_s and E ceases when the dipole alignment is approaching saturation. For corona-charged poly(vinylidene fluoride), the proportionality between P_s and E is not observed (29).

The time dependence of the growth of the pyroelectric constant, which is proportional to the polarization, is shown in Fig. 3.25 for poly(vinylidene fluoride). An evaluation of Eq. (3.38) and comparison with the figure indicates that the observed rise of the polarization at 20, 40, and 60°C is more gradual than expected. This suggests the involvement of two or more relaxation times (123), as has also been found in other studies of this material (124, 125). (For details see Chapter 4.)

The dependence on polarization temperature is also seen from Fig. 3.25. Indications are that for times less than the time necessary to achieve maximum poling, the polarization is considerably greater for higher temperatures. However, since all curves appear to reach the same final value, the maximum poling achievable is not dependent on temperature (123). For poly(methyl methacrylate) (PMMA), the dependence of polarization on poling temperature T_p is seen in Fig. 3.26 (60). Since the peak structure depends critically on T_p, a distribution of dipolar relaxation times has to be assumed: the more slowly reacting dipoles are only filled at relatively high poling temperatures. Similar results have been obtained for other polymers by a number of authors (e.g., Refs. 119, 126, and 127). Systematic variation of poling time and temperature allows one to study the gradual orientation of dipoles characterized by different relaxation times. Investigations of the distribution of relaxation times are discussed in the next section.

The effect of the electrode material on the formation of the polarization is generally negligible (128). A possible exception to this are materials where the dipole polarization is significantly affected by space charge injection, such as

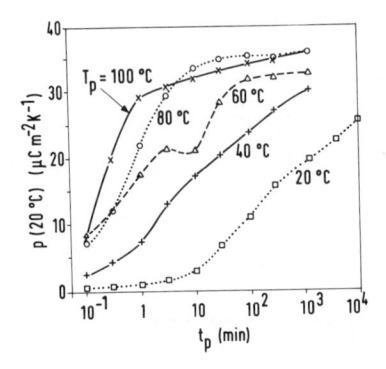

Figure 3.25 Dependence of pyroelectric coefficient of PVDF on poling time. Parameter: poling temperature; poling field 1 MV/cm. (From Blevin [123].)

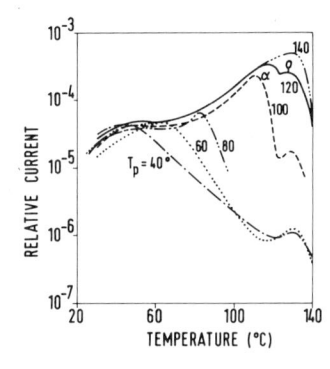

Figure 3.26 Dependence of short-circuit TSC spectra of poly(methyl methacrylate) (PMMA) on poling temperature. (From van Turnhout [60].)

PVDF. In this and other materials, there is also a pronounced effect of stretching on the polarization (see Chapter 4).

3.5.2 The Dipole Polarization

Dependence on Material Parameters

As in the case of real-charge storage, the storage of polarization charges depends on a number of material parameters, such as crystallinity, cross linking, additives, irradiation history, stereoregularity, water absorption, and so on. In the following, the influence of only a few of these is described.

The effect of crystallinity on the dipole polarization is rather severe. Two examples will demonstrate this. In polycarbonate, the β-peak is highest for an amorphous sample and decreases considerably with increasing crystallinity, as shown in Fig. 3.27. In poly(vinylidene fluoride), on the other hand, the α-peak appears to be linked to the crystalline phase (129). This material appears in at least four different crystalline forms which exhibit very different polar behavior (130).

Another structural parameter that affects the dipolar alignment is the steric arrangement of the polar groups (128). In poly(methyl methacrylate), for example, the syndiotactic type (vinyl groups alternately on both sides of the main

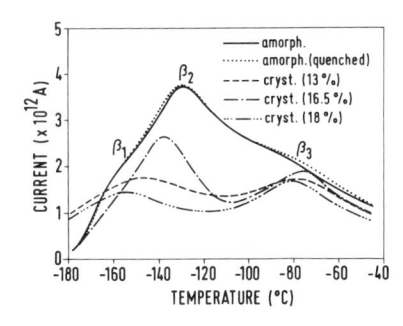

Figure 3.27 Effect of degree of crystallinity on the short-circuit TSC of the β-peak of polycarbonate. (From Vanderschueren [128].)

chain) exhibits a larger β-peak in TSC experiments than the isotactic type (vinyl groups on one side).

The effect of additives and dopants on the polarization of polymers was studied by several authors (131–133). In polyethylene, ionic agents such as NaCl will shift the α-peak to higher temperatures. This is due to an increase in the potential barrier responsible for this relaxation. The lower-temperature β- and γ-peaks are not affected by ionic agents but by antioxidants. The latter increase the activation energy and decrease the preexponential factor (131). Doping with such elements as copper or iodine will also affect the dipolar and space charge relaxations (132, 133).

A number of laboratories have studied the effect of water or humidity on the polarization of polymers (134–137). A general result of these studies is that, in a wide variety of polymers, the α- and β-relaxations are significantly affected by the presence of water in the sense that, with increasing water content, the corresponding TSC peaks shift to lower temperatures (134, 136). This has been attributed to the formation of hydrogen bonds between the water molecules and the macromolecular units, resulting in a plasticizing effect which weakens intermolecular attraction and thus decreases the activation energy. In these cases, the peak amplitude can increase or decrease, depending on the nature of the polymers.

Distribution of Relaxation Frequencies

The peaks observed in TSC experiments with polarized polymer samples are often too broad to be attributable to a single relaxation frequency. Rather, they have to be explained by a distribution in such frequencies. Information about frequency distributions has been derived from studies employing variations of the normal polarizing and TSC techniques described above. Representative methods are fractional (or partial) polarization and multistage TSC (17, 60, 127, 128, 138–140).

Fractional polarization differs from the normal polarization process in that the electric field is not applied continuously during cooling of the sample but in several steps separated by short-circuit periods (119, 138). Thus if a distribution of relaxation frequencies exists, only dipoles with α values within certain ranges are aligned, and subsequent heating will produce a number of partial peaks related to the field application intervals. Evaluation of these peaks yields activation energies E that are a function of $T_C - T_{M_0}$, where T_C is the short-circuit temperature and T_{M_0} is the temperature of the TSC peak of the fully poled material.

Activation energies for the β-peak of a number of polymers subjected to fractional polarization are shown in Fig. 3.28. The data indicate a steady rise of E with $T_C - T_{M_0}$. This suggests that the distribution of relaxation frequencies is caused by a distribution of activation energies. This is also expected from a molecular point of view since the β-relaxation involves motions controlled by the locally differing environment of the side groups. By means of fractional polarization, a distribution of activation energies was also found for the γ- and

DIPOLE POLARIZATION

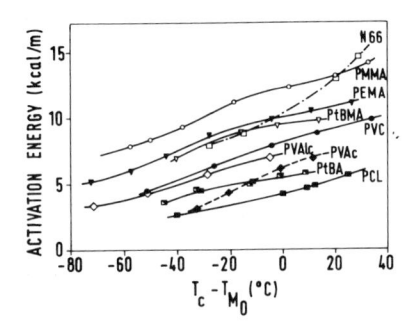

Figure 3.28 Activation energy E as function of (normalized) poling temperature $T_C - T_{M_0}$ for the β-peak of several polymers: T_c = short circuit temperature during poling, T_{M_0} = temperature of TSC peak of fully poled material. (From Vanderschueren [138].)

β-peaks of polyethylene (127). These data were actually used to calculate the density of the distribution.

Similar results were obtained from multistage TSC experiments. Here the heating of a sample during TSC is interrupted and the sample is reheated several times, with the temperature of the interruption increasing. Application of this process to the β-peak of PMMA, for example, produces current rises which steepen from experiment to experiment (140). This indicates an increase in activation energy with depolarization temperature. Distributions in activation energies were also found for relaxations in other polymers (see for example, Ref. 60, Fig. 10.13). It therefore appears that dipolar motions in polymers are characterized by a continuous distribution of activation energies.

3.5.3 Isothermal Decay of Polarization

As pointed out in Section 3.3.4, the decay of a dipole polarization characterized by a single relaxation frequency is controlled by Eq. (3.17). In the isothermal case, this yields with $\alpha(T) = \alpha$

$$P(t) = P_0 \exp(-\alpha t) \tag{3.40}$$

More general formulas hold for distributed relaxations (60).

The isothermal decay of the dipole polarization of polymers has been studied directly and by means of the piezo- and pyroelectric activity of the materials, which can be assumed to be approximately proportional to the polarization (see Chapter 4).

A typical example is shown in Fig. 3.29, which depicts the room-temperature decay of the polarization in PMMA (141). The results in the figure were obtained by integration of TSC spectra which show a particularly severe depletion of the relaxations reacting at lower temperatures. A nonexponential decay is evident from Fig. 3.29 which has to be attributed to a distribution of relaxation frequencies. Similarly, in PET the lower-temperature dipole relaxations were also found to decay faster than the higher-temperature relaxations (see Ref. 17, Fig. 10.44).

Extensive data have also been collected on the decay of the polarization of

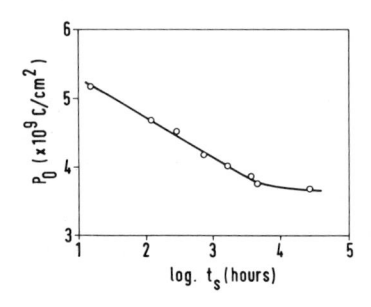

Figure 3.29 Isothermal decay of the dipole polarization in PMMA at room temperature. Polymer was polarized at 135°C. (From Vanderschueren and Linkens [141].)

PVDF (142–144). Results of the decrease of the pyroelectric coefficient of α- and β-phase material at 102°C as function of time are depicted in Fig. 3.30 (143). The data show a pronounced initial decay which is more severe for the β-phase material, although this material exhibits initially about twice the activity of the α-polymer. Eventually relatively stable values are reached. Similar results were obtained in a study of the decay of the piezoelectric d_{31} constant at various temperatures (142). In all cases the decays are nonexponential, suggesting a distribution of relaxation frequencies.

3.6 APPLICATIONS

Charged polymers have been utilized in a wide variety of applications. These range from industrial areas to the biological and medical fields and are in various states of research, development, and production.

3.6.1 Electroacoustic Transducers

The most widely used electret devices are electret microphones, which were first described by Nishikawa and Nukijama in 1928 (145). These transducers, as well as the electret microphones used before and during World War II (146), proved unsatisfactory since they contained wax electrets which have insufficient electrical stability under normal environmental conditions. With the introduction of

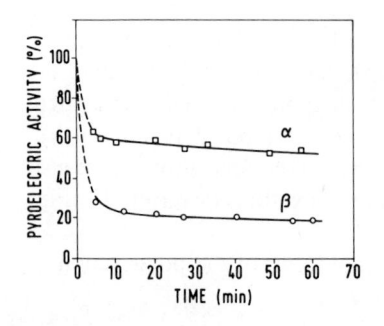

Figure 3.30 Isothermal decay of the pyroelectric coefficient of α- and β-phase PVDF at 102°C. Polymer was polarized also at 102°C. (From Sussner [143].)

APPLICATIONS

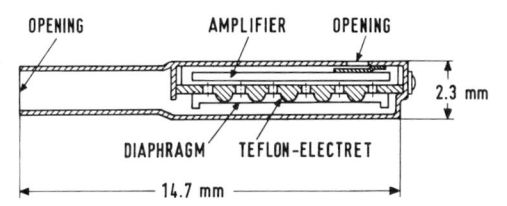

Figure 3.31 Cross-sectional view of an electret microphone with cardioid characteristic for hearing-aid use. (From Carlson and Killion [148].)

the thin-film polymer electret microphone (6, 147), such transducers gained widespread commercial acceptance.

A schematic cross section of such a microphone, consisting of a metallized back electrode onto which a thin Teflon electret is cemented and a metallized Mylar diaphragm, is shown in Fig. 3.31. An incident sound wave causes vibrations of the diaphragm which, in turn, generate an ac signal between the metallizations of backplate and diaphragm. The microphone design shown in the figure (148) has a directional sensitivity pattern. Systems of this kind are used in hearing-aid applications. Technical details of such and other electret transducers are given in the literature (see, for example, Ref. 149).

Electret microphones have the advantage of being insensitive to mechanical vibrations, shock, and electromagnetic pickup. They have all the favorable properties of condenser microphones but are much simpler in design and thus less expensive. Such microphones are being used commercially in cassette recorders, hearing aids, hi-fi setups, sound level meters, noise dosimeters, and movie cameras; and in telephony for operators headsets, speaker phones, and, very recently, handsets (148–156). Worldwide production is about 100 million annually.

An interesting research application of electret microphones is in the detection of air pollution. Due to the low self-noise of such transducers, the detection threshold for pollutants has been lowered by more than an order of magnitude (157). Similarly, electret microphones have been used in photoacoustic spectroscopy (158), for the measurement of rocket exhausts (159), and elsewhere.

Other applications of charged polymers are in earphones, loudspeakers, and ultrasonic and underwater transducers (160–163). As an example, an electret hydrophone is shown in Fig. 3.32. This transducer consists of a structure

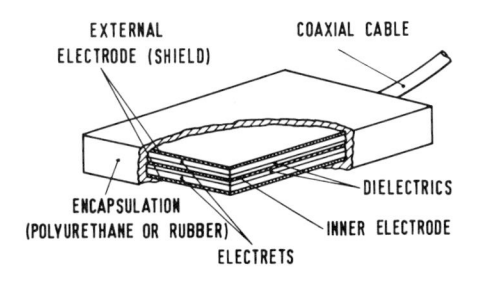

Figure 3.32 Schematic view of an electret hydrophone. (From Hennion and Lewiner [163].)

composed of dielectric and electret layers of different compressibility. Sound waves will therefore cause different deformations of the layers and thus generate an electric output signal between the electrodes.

3.6.2 Electrophotography and Electrostatic Recording

An application of charge-storage phenomena of great practical importance is electrophotography. The basic process used in many electrophotographic methods, namely, the production of a charge pattern on an appropriate photoconducting insulator film and its development with powders, was studied in the early 1930s (164). The breakthrough in this field came a few years later when investigations of photoconductive image formation led to the development of xerographic reproduction methods (165, 166). The initial photoconducting materials used in the electrophotographic process were non-polymeric. More recently the charge-transfer complex of the polymer, poly(N-vinyl carbazole), PVCA and trinitrofluorenone, TNF and other polymer-based systems have been used commercially as electrophotographic photoreceptors. Such polymer materials and the electronic properties which control the formation of the latent electrostatic image in the xerographic process are discussed in Chapter 6 of this book.

Related to electrophotography, but not depending on photographic methods, are electrostatic recording processes. These are used to record electrical signals, digital information, facsimile, or alphanumeric characters on materials such as polymers or paper (167). The recording is performed by electron beam or electrical discharge from a needle electrode. The patterns may be read by a capacitive probe arrangement or with a low-current, low-energy scanning electron beam by monitoring secondary emission from the sample (168).

3.6.3 Piezoelectric and Pyroelectric Devices

A number of promising electret applications are based on the piezoelectric and pyroelectric effects in polarized polymers materials. These are discussed in several recent reviews (149, 169–173) and are summarized in Chapter 4.

3.6.4 Electret Gas Filters

An important commercial application of polymer electrets is in gas filters (81). These filters are made of polypropylene films which are corona charged and fibrillated. Because of repulsive forces, the fibers spread into a broad web. The capture of particles by such electret fibers depends on Coulomb and induction forces which act on charged and neutral particles, respectively. Since the operation of electret filters is based on long-range electrical forces, they can have an open structure with fiber spacings much in excess of the particle size, thus having extremely small flow resistance.

APPLICATIONS

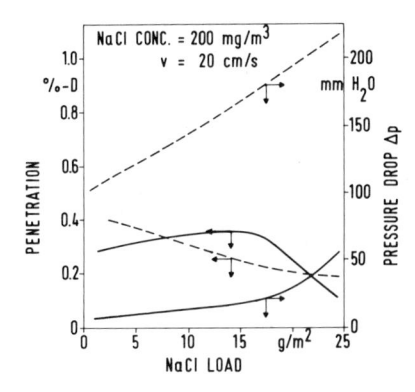

Figure 3.33 Comparison of penetration-load curve and pressure drop of an electret filter (solid lines) and a commercial glass fiber filter (dashed lines). (From van Turnhout et al. [81].)

Filtration results of such a filter are compared in Fig. 3.33 with data for a commercial glass filter. While the penetration through both filters is almost the same, the pressure drop across the electret filter is 4 to 20 times less, thus reducing energy requirements for filtration. The use of finer fibers and of curled fibers has recently led to improved filtering efficiency and dust-holding capacity (174).

3.6.5 Electromechanical and Radiation Devices

Other polymer electret devices include relay switches, optical display panels, phonograph pickups, touch or key sensors, Korotkov sound pickups, electret motors, and radiation dosimeters. While the relay-type switches (175) utilize the external field of electrets to open or close contacts, the optical display devices (176) depend on the opening or closing of illuminated channels by means of hinged and opaque electrets moved by electrostatic forces. Also the phonograph pickups (177, 178) consist of a stylus which actuates an electret transducer, and the touch or key sensors (179, 180) depend on the manual deflection of an electret relative to an electrode. The electret motors (181, 182) are based on the slot effect and use thin electret disks or foil electrets, and the Korotkov sound pickups (183) are microphone-like transducers which detect arterial pressure fluctuations.

In the radiation dosimeters (184), the decay of electret charges in an ionization chamber or the generation of a radiation-induced conductivity in electret materials is employed to measure radiation exposures.

3.6.6 Biological and Medical Applications

Of great future potential are applications of polymer electrets in the biological and medical fields. It is well known that blood vessel walls and human bones have electret properties (185). It is therefore not surprising that it is possible to improve the blood compatibility of polymers by negative charge deposition

(186). This has recently led to the use of charged Teflon implants in human heart and arterial surgery (187). Similarly, Teflon electrets placed in contact with bones of animals *in vivo* cause accelerated growth of callus, necessary for fracture healing (188). Other experiments showed that negatively charged Teflon is also useful in endodontic therapy in that it enhances the formation of osteo-dentin when applied as core-filling material (189). In still another medical test, electret bandages placed on skin incisions considerably improved the tensile strength of the wound over a given period of time and thus accelerated the wound healing process (190).

ACKNOWLEDGEMENTS

The author acknowledges many helpful discussions on the topics of this chapter with Prof. B. Gross and Drs. D. A. Berkley, R. Gerhard, K. Labonte, R. Lerch, H. von Seggern, J. E. West, and R. Zahn at the Technische Hochschule Darmstadt or at Bell Laboratories.

REFERENCES

1. S. Gray, *Phil. Trans. R. Soc. Lond. Ser. 1*, **37**, 285 (1732).
2. M Faraday, *Experimental Researches in Electricity*, Richard and John Edward Taylor, London, 1839.
3. O. Heaviside, *Electrical Papers*, Chelsea, Bronx, N. Y., 1892, pp. 488–493.
4. M. Eguchi, *Proc. Phys. Math. Soc. Jap.* **1**, 326 (1919); *Phil. Mag.*, **49**, 178 (1925).
5. T. A. Dickenson, *Electrical Manufacturing*, 101 (Aug. 1948).
6. G. M. Sessler and J. E. West, *J. Acoust. Soc. Am.*, **34**, 1787 (1962).
7. A. N. Gubkin, *Electrets* (in Russian), Nauka, Moscow, 1978.
8. G. M. Sessler, Ed., *Electrets*, Springer, Heidelberg, 1980.
9. M. M. Perlman, Ed., *Electrets, Charge Storage and Transport in Dielectrics*, Electrochemical Society, Princeton, N.J., 1973; M. S. de Campos, Ed., *International Symposium on Electrets and Dielectrics*, Academia Brasileira de Ciencas, Rio de Janeiro, 1977.
10. R. E. Collins, Proc. IREE **34**, 381 (1973).
11. H. Kiess, *RCA Rev.*, **36**, 667 (1975).
12. K. J. Euler, *J. Electrostat.*, **2**, 1 (1976).
13. J. Fuhrmann, *J. Electrostat.*, **4**, 109 (1978); A. R. Blythe, *Electrical Properties of Polymers*, Cambridge University Press, 1979.
14. B. Gross, in *Electrets*, G. M. Sessler, Ed., Springer, Heidelberg, 1980, pp. 217–284.
15. J. R. Freeman, H. P. Kallmann, and M. Silver, *Rev. Mod. Phys.*, **33**, 553 (1961).

REFERENCES

16 V. M. Fridkin and I. S. Zheludev, *Photoelectrets and the Electrophotographic Process*, Consultants Bureau, New York, 1961.
17 J. van Turnhout, *Thermally-Stimulated Discharge of Polymer Electrets*, Elsevier, Amsterdam, 1975.
18 A. N. Gubkin, T. S. Yegorova, L. M. Kokorin, and N. Y. Zitser, *Vysokomol. soyed.*, **A12,** 602 (1970).
19 G. W. Day, C. A. Hamilton, R. L. Peterson, R. J. Phelan, and L. O. Mullen, *Appl. Phys. Lett.*, **24,** 456 (1974).
20 R. Hayakawa and Y. Wada, *Adv. Polym. Sci.*, **11,** 1 (1973).
21 S. K. Shrivastava, J. D. Ranade, and A. P. Srivastava, *Jap. J. Appl. Phys.*, **18,** 2303 (1979); J. Vanderschueren, Thesis, Liège, 1974; P. C. Mehendru, K. Jain, and P. Mehendru, *J. Phys. D, Appl. Phys.*, **11,** 1431 (1978); Y. Murata, T. Hodoshima, and S. Kittaka, *Jap. J. Appl. Phys.*, **18,** 2215 (1979).
22 R. E. Collins, *AWA Tech. Rev.*, **15,** 53 (1973).
23 P. W. Chudleigh, *J. Appl. Phys.*, **47,** 4475 (1976); *Appl. Phys. Lett.*, **21,** 547 (1972).
24 S. Engelbrecht, *J. Appl. Phys.*, **45,** 3421 (1974).
25 M. Ieda, G. Sawa, and U. Shinahara, *Electr. Eng. Japan*, **88,** 67 (1968).
26 R. Gerhard, to be published.
27 R. A. Moreno and B. Gross, *J. Appl. Phys.*, **47,** 3397 (1976).
28 S. S. Bamji, K. J. Kao, and M. M. Perlman, *J. Electrostat.*, **6,** 373 (1979).
29 P. D. Southgate, *Appl. Phys. Lett.* **28,** 250 (1976); D. K. Das Gupta and K. Doughty, *J. Phys. D, Appl. Phys.*, **11,** 2415 (1978); *J. Electrostat.* **7,** 267 (1979).
30 G. M. Sessler and J. E. West, *J. Appl. Phys.*, **43,** 922 (1972).
31 K. Ikezaki, I. Fujita, K. Wada, and J. Nakamura, *J. Electrochem. Soc.*, **121,** 591 (1974).
32 B. Gross, *J. Polym. Sci.*, **27,** 135 (1958).
33 G. M. Sessler and J. E. West, *Appl. Phys. Lett.*, **17,** 507 (1970).
34 B. L. Beers, H. Hwang, D. L. Lin, and V. W. Pine, in *Spacecraft Charging Technology–1978*, NASA Conference Publication 2071, Washington, D.C., pp. 209–238.
35 B. Gross, G. M. Sessler, and J. E. West, *J. Appl. Phys.*, **45,** 2841 (1974).
36 B. Gross, *J. Chem. Phys.*, **17,** 866 (1949).
37 H. Krämer and D. Messner, *Kunststoffe*, **54,** 696 (1964).
38 D. K. Davies, in *Static Electrification*, A. C. Stickland, ed., Institute of Physics, London, 1967, pp. 29–36.
39 H. Wintle, *J. Phys. E.; Sci. Instr.* **3,** 334 (1970).
40 J. van Turnhout, in *Advances in Static Electricity*, W. de Geest, Ed., Auxilia, Brussels, 1971, Vol. I, pp. 56–81.
41 G. M. Sessler and J. E. West, *Rev. Sci. Instr.*, **42,** 15 (1971).
42 C. W. Reedyk and M. M. Perlman, *J. Electrochem. Soc.*, **115,** 49 (1968).
43 G. M. Sessler and J. E. West, *J. Electrochem. Soc.*, **115,** 836 (1968).

44 H. von Seggern, Dissertation, Darmstadt, 1979.
45 R. E. Collins, *Appl. Phys. Lett.*, **26,** 675 (1975); J. Appl. Phys., **47,** 4804 (1976); *Rev. Sci. Instr.,* **48,** 83 (1977).
46 R. E. Collins, *J. Appl. Phys.*, **51,** 2973 (1980); *1979 Annual Report, Conference on Electrical Insulation and Dielectric Phenomena,* National Academy of Sciences, Washington, D.C., pp. 307–314.
47 R. E. Collins, *Ferroelectrics,* **33,** 65 (1981).
48 P. I. Kuindersma and R. M. van der Heij, in *1979 Annual Report, Conference on Electrical Insulation and Dielectric Phenomena,* National Academy of Sciences, Washington, D.C., pp. 325–333.
49 M. Latour, *J. Phys. Lett.,* **41,** L–35 (1980).
50 B. Gross, in *Static Electrification,* Institute of Physics, London, 1971, pp. 33–43; D. K. Walker and O. Jefimenko, in *Electrets, Charge Storage and Transport in Dielectrics,* M. M. Perlman, Ed., Electrochemical Society, Princeton, N. J., 1973, pp. 455–461.
51 A. S. De Reggi, C. M. Guttman, F. I. Mopsik, G. T. Davis, and M. G. Broadhurst, *Phys. Rev. Lett.,* **40,** 413 (1978).
52 H. von Seggern, *Appl. Phys. Lett.,* **33,** 134 (1978).
53 P. Laurenceau, G. Dreyfus, and J. Lewiner, *Phys. Rev. Lett.,* **38,** 46 (1977); A. Migliori and J. D. Thompson, *J. Appl. Phys.,* **51,** 479 (1980); C. Alquie, G. Dreyfus, and J. Lewiner, *Phys. Rev. Lett.,* **47,** 1483 (1981).
54 G. M. Sessler, J. E. West, and R. Gerhard, *Phys. Rev. Lett.,* **48** 563 (1982). W. Eisenmenger and M. Haardt, *Solid State Commun.,* **41,** 499 (1982).
55 G. M. Sessler, J. E. West, D. A. Berkley, and G. Morgenstern, *Phys. Rev. Lett.* **38,** 368 (1977).
56 D. W. Tong, in *Conference Record of the 1980 IEEE International Symposium on Electrical Insulation,* (IEEE Service Center, Piscataway, N.J., pp. 179–183.
57 L. Badian and J. Klocek, *J. Electrostat.,* **8,** 69 (1979).
58 C. Bucci and R. Fieschi, *Phys. Rev. Lett.,* **12,** 16 (1964).
59 C. Bucci, R. Fieschi, and G. Guidi, *Phys. Rev.,* **148,** 816 (1966).
60 J. van Turnhout, in *Electrets,* G. M. Sessler, Ed., Springer, Heidelberg, 1980, pp. 81–215.
61 J. Vanderschueren and J. Gasiot, *Thermally Stimulated Relaxation in Solids,* P. Bräunlich, Ed., Springer, 1979, pp. 135–223.
62 S. Maeta and K. Sakaguchi, *Jap. J. Appl. Phys.,* **19,** 519 (1980); **19,** 597 (1980).
63 J. H. Calderwood and B. K. P. Scaife, *Phil. Trans. R. Soc. Lond.,* **269,** 217 (1971); H. J. Wintle, *J. Appl. Phys.,* **42,** 4724 (1971).
64 A. Many and G. Rakavy, *Phys. Rev.,* **126,** 1980 (1962). L. Nunes de Oliveira and G. F. Leal Ferreira, *J. Electrostat.,* **1,** 371 (1975); M. Zahn, *IEEE Trans.* El. Ins. **EI–12,** 176 (1977).
65 G. M. Sessler and J. E. West, *Phys. Rev.,* **B10,** 4488 (1974).
66 G. M. Sessler, in *International Symposium on Electrets and Dielectrics,* M. S. de Campos, Ed., Academia Brasileira de Ciencias, Rio de Janeiro, 1977, pp. 321–335.

REFERENCES

67 H. von Seggern, *J. Appl. Phys.*, to be published; *1980 Annual Report, Conference von Electrical Insulation and Dielectric Phenomena*, National Academy of Sciences. Washington, D.C., pp. 345–352.
68 H. Bauser, *Kunststoffe*, **62**, 192 (1972).
69 N. F. Mott and E. A. Davis, *Electronic Processes in Non-Crystalline Materials*, Clarendon, Oxford, 1971.
70 R. A. Creswell, M. M. Perlman, M. A. Kabayama, in *Dielectric Properties of Solids*, F. E. Karasz, Ed., Plenum, New York, 1972, pp. 295–312.
71 D. Ito and T. Nakakita, *J. Appl. Phys.*, **51**, 3273 (1980).
72 H. von Seggern, *J. Appl. Phys.*, **50**, 2817 (1979); **52**, 4086 (1981).
73 P. W. Chudleigh, R. E. Collins, and G. D. Hancock, *Appl. Phys. Lett.* **23**, 211 (1973).
74 R. Singh and S. C. Datt, *J. Electrostat.*, **6**, 297 (1979).
75 B. Gross, G. M. Sessler, and J. E. West, *J. Appl. Phys.*, **48**, 4303 (1977).
76 B. Gross, G. M. Sessler, H. von Seggern, and J. E. West, *Appl. Phys. Lett.*, **34**, 555 (1979).
77 G. M. Sessler, in *1978 Annual Report, Conference on Electrical Insulation and Dielectric Phenomena*, National Academy of Sciences, Washington, D. C., 1978, pp. 3–10.
78 G. M. Sessler, J. E. West, and H. von Seggern, to appear.
79 D. W. Tong, in *1980 Annual Report, Conference on Electrical Insulation and Dielectric Phenomena*, National Academy of Sciences, Washington, D.C., pp. 359–365.
80 S. Matsuoka, H. Sunaga, R. Tanaka, M. Hagiwara, and K. Araki, *IEEE Trans. Nucl. Sci.*, **NS–23**, 1447 (1976).
81 J. van Turnhout, C. van Bochove, and G. J. van Veldhuizen, *Staub-Reinhalt. Luft*, **36**, 36 (1976); E. W. Anderson, L. L. Blyler, G. E. Johnson, and G. L. Link, in *Electrets, Charge Storage and Transport in Dielectrics*, M. M. Perlman, Ed., Electrochemical Society, Princeton, N.J., 1973, pp. 424–435.
82 M. E. Borisova and S. N. Koikov, *Sov. Phys. J.*, **22**, 58 (1979).
83 K. Ikezaki, M. Hattori, and Y. Arimoto, *Jap. J. Appl. Phys.*, **16**, 863 (1977); P. J. Atkinson and R. J. Fleming, in *1980 Annual Report, Conference on Electrical Insulation and Dielectric Phenomena*, National Academy of Sciences, Washington, D.C., pp. 337–344.
84 K. Ikezaki, K. Wada, and I. Fujita, *J. Electrochem. Soc.*, **122**, 1356 (1975).
85 M. M. Perlman and S. Unger, *Appl. Phys. Lett.*, **24**, 579 (1974).
86 G. M. Sessler and J. E. West, *J. Electrostat.*, **1**, 111 (1975).
87 T. J. Lewis, in *1976 Annual Report, Conference on Electrical Insulation and Dielectric Phenomena*, National Academy of Sciences, Washington, D.C., 1976, pp. 533–561.
88 E. A. Baum, T. J. Lewis, and R. Toomer, *J. Phys. D, Appl. Phys.*, **11**, 703 (1978).
89 G. M. Sessler, J. E. West, R. W. Ryan, and H. Schonhorn, *J. Appl. Polym. Sci.*, **17**, 3199 (1973).

90 B. Gross, R. M. Faria, and G. F. L. Ferreira, *J. Appl. Phys.*, **52,** 571 (1981).
91 G. M. Sessler and H. von Seggern, in *1979 Annual Report, Conference on Electrical Insulation and Dielectric Phenomena*, National Academy of Sciences, Washington, D.C., 1979, pp. 160–167.
92 E. A. Baum, T. J. Lewis, and R. Toomer, *J. Phys. D, Appl. Phys.*, **10,** 487; **10,** 2525 (1977).
93 M. Ieda, T. Mizutani, and H. Mizuno, in *Dielectric Materials, Measurements and Applications*, IEE Conference Publication No. 177, London, 1979, pp. 266–269.
94 K. J. Kao, S. S. Bamji, and M. M. Perlman, *J. Appl. Phys.*, **50,** 8181 (1979).
95 J. Mort and D. M. Pai, *Photoconductivity and Related Phenomena*, Elsevier, Amsterdam, 1976.
96 B. Gross, G. M. Sessler, and J. E. West, *J. Appl. Phys.*, **47,** 968 (1976).
97 H. von Seggern, *J. Appl. Phys.*, **52,** 4081 (1981).
98 H. Scher and E. W. Montroll, *Phys. Rev. B*, **12,** 2455 (1975); G. Pfister and H. Scher, *Phys. Rev. B*, **15,** 2062 (1977); G. Pfister and H. Scher, *Adv. Phys.*, 27, 747 (1978).
99 F. W. Schmidlin, *Phys. Rev. B* **16,** 2362 (1977); J. Noolandi, *Phys. Rev. B* **16,** 4466 (1977).
100 R. J. Fleming, *J. Appl. Phys.*, **50,** 8075 (1979).
101 A. Reiser, M. W. B. Lock, and J. Knight, *Trans. Farad. Soc.*, **65,** 2168 (1968).
102 H. von Seggern, *J. Appl. Phys.*, **50,** 7039 (1979).
103 A. K. Jonscher, *J. Phys. D, Appl. Phys.*, **13,** L 137 (1980).
104 J. F. Fowler, *Proc. R. Soc. Lond. A*, **236,** 464 (1956).
105 B. Gross, J. E. West, H. von Seggern, and D. A. Berkley, *J. Appl. Phys.*, **51,** 4875 (1980).
106 J. A. Wall, E. A. Burke, A. R. Frederickson, in *Proceedings of the Spacecraft Charging Technology Conference*, C. P. Pike and R. R. Lovell, Eds., NASA, Cleveland, Ohio, 1977, pp. 569–581.
107 G. M. Sessler and J. E. West, *J. Appl. Phys.*, **50,** 3328 (1979).
108 B. Gross, G. M. Sessler, and J. E. West, *J. Appl. Phys.*, **46,** 4647 (1975).
109 L. H. Beckley, T. J. Lewis, and D. M. Taylor, *J. Phys. D, Appl. Phys.*, **9,** 1355 (1976).
110 B. Gross and G. F. Leal Ferreira, *J. Appl. Phys.*, **50,** 1506 (1979).
111 D. A. Berkley, *J. Appl. Phys.*, **50,** 3447 (1979).
112 K. Labonte, in *1980 Annual Report, Conference on Electrical Insulation and Dielectric Phenomena*, National Academy of Sciences, Washington, D.C., pp. 321–327; *ibid.*, 1981, pp. 52–57.
113 A. R. Frederickson, in *Spacecraft Charging Technology–1978* NASA Conference Publication No. 2071, pp. 554–569.
114 B. Gross and L. Nunes de Oliveira, *J. Appl. Phys.*, **45,** 4724 (1974); L. Nunes de Oliveira and B. Gross, *J. Appl. Phys.*, **46,** 3132 (1975).
115 H. von Seggern and J. E. West, to appear.
116 H. Kawai, *Jpn. J. Appl. Phys.*, **8,** 975 (1969).
117 J. G. Bergman, J. H. McFee, and G. R. Crane, *Appl. Phys. Lett.*, **18,** 203 (1971).

REFERENCES

118 M. G. Broadhurst and G. T. Davis, in *Electrets*, G. M. Sessler, Ed., Springer, Heidelberg, 1980, pp. 285–319.
119 P. Fischer, *J. Electrostat.*, **4**, 149 (1977/78).
120 J. Vanderschueren, A. Linkens, B. Haas, and E. Dellicour, *J. Macromol. Sci.-Phys. B*, **15**, 449 (1978).
121 G. T. Davis and M. G. Broadhurst, in *International Symposium on Electrets and Dielectrics*, M. S. de Campos, Ed., Academia Brasileira de Ciencas, Rio de Janeiro, 1977, pp. 299–318.
122 R. G. Kepler, in *Electrophotography, 2nd International Conference*, D. R. White, Ed., Society of Photographic Scientists and Engineers, Washington, D.C., 1974, pp. 167–170.
123 W. R. Blevin, *Appl. Phys. Lett.*, **31**, 6 (1977).
124 H. Sussner and K. Dransfeld, *J. Polym. Sci. Polym. Phys.*, **16**, 529 (1978).
125 D. Naegele and D. Y. Yoon, *Appl. Phys. Lett.*, **33**, 132 (1978).
126 S. K. Shrivastava, J. D. Ranade, and A. P. Srivastava, *Thin Solid Films*, **67**, 201 (1980).
127 P. Fischer and P. Röhl, *J. Polym. Sci. Polym. Phys.*, **14**, 531 (1976).
128 J. Vanderschueren, Thesis, Liège, 1974.
129 N. Murayama and H. Hashizume, *J. Polym. Sci. Polym. Phys.*, **14**, 989 (1976).
130 G. R. Davies, in *Physics of Dielectric Solids 1980*, Institute of Physics Conference Series No. 58, Institute of Physics, Bristol and London, pp. 50–63.
131 C. Lacabanne and D. Chatain, *Makromol. Chem.*, **179**, 2765 (1978).
132 V. K. Jain, C. L. Gupta, R. K. Jain, and R. C. Tyagi, *Thin Solid Films*, **48**, 175 (1978).
133 C. L. Gupta and R. C. Tyagi, *Ind. J. Pure Appl. Phys.*, **16**, 428 (1978).
134 J. Vanderschueren and A. Linkens, *Macromolecules*, **11**, 1228 (1978).
135 J. Vanderschueren, *J. Polym. Sci. Polym. Phys.*, **12**, 991 (1974).
136 G. Weber, *Angew. Makromol. Chem.*, **14**, 187 (1978).
137 P. C. Mehendru, K. Jain, and P. Mehendru, *J. Phys. D, Appl. Phys.*, **10**, 729 (1977).
138 J. Vanderschueren, *J. Polym. Sci. Polym. Phys.*, **15**, 873 (1977).
139 P. Fischer and P. Röhl, *J. Polym. Sci. Polym. Phys.*, **114**, 543 (1976).
140 P. H. Ong and J. van Turnhout in *Electrets, Charge Storage and Transport in Dielectrics*, M. M. Perlman, Ed., Electrochemical Society, Princeton, N.J., 1973, pp. 213–229.
141 J. Vanderschueren and A. Linkens, *J. Polym. Sci. Polym Phys.*, **16**, 223 (1978).
142 N. Murayama, K. Nakamura, H. Ohara, and M. Segawa, *Ultrasonics*, **14**, 15 (1976).
143 H. Sussner, in *Proceedings of 1979 Ultrasonics Symposium*, J. de Klerk and B. R. McAvoy, Eds., pp. 491–498.
144 L. L. Blyler, Jr., G. E. Johnson, and N. M. Hylton, *Ferroelectrics*, **28**, 303 (1980).
145 S. Nishikawa and D. Nukijama, *Proc. Imper. Acad. Tokyo*, **4**, 290 (1928).

146 F. Gutmann, *Rev. Mod. Phys.*, **20**, 457 (1948).
147 G. M. Sessler an J. E. West, *J. Acoust. Soc. Am.*, **40**, 1433 (1966).
148 E. V. Carlson and M. C. Killion, *J. Audio Eng. Soc.*, **22**, 92 (1974).
149 G. M. Sessler and J. E. West, in *Electrets*, G. M. Sessler, Ed., Springer, Heidelberg, 1980, pp. 347–381.
150 S. P. Khanna and R. L. Remke, *Bell Syst. Tech. J.*, **59**, 745 (1980).
151 J. G. Baumhauer and A. M. Brzezinski, *Bell Syst. Tech. J.*, **58**, 1557 (1979).
152 E. Frederiksen, in *Proceedings of Tenth International Congress on Acoustics*, Sydney, 1980, Vol. 3, Paper L–4.1.
153 A. Boeryd, L. Branden, J. A. Hedman, and O. Larsson, *Ericsson Rev.*, **3**, 118 (1976).
154 N. Sakamoto, M. Matsumoto, H. Naono, and T. Gotoh, *J. Audio Eng. Soc.*, to appear.
155 R. R. Walker and A. J. Morgan, *Post Off. Electr. Eng. J.*, **72**, 15 (1979).
156 W. R. Bevan, R. B. Schulein, and C. E. Seeler, *J. Audio Eng. Soc.*, **26**, 947 (1978).
157 C. K. N. Patel and R. J. Kerl, *Appl. Phys. Lett.*, **30**, 578 (1977).
158 A. Rosencwaig, *Photoacoustics and Photoacoustic Spectroscopy*, Wiley, New York, 1980.
159 M. Susko, *J. Appl. Meteorol.*, **18**, 48 (1979).
160 H. J. Griese and G. Kock, *Funkschau*, **49**, 1250 (1977).
161 N. Sakamoto, T. Gotoh, N. Atoji, and T. Aoi, *J. Audio Eng. Soc.*, **24**, 368 (1976).
162 A. K. Nigam, *J. Acoust. Soc. Am.*, **55**, 978 (1974).
163 C. Hennion and J. Lewiner, *J. Acoust. Soc. Am.*, **63**, 1229 (1978).
164 P. Selenyi, U.S. Pat. 1, 818, 760 (1931).
165 C. F. Carlson, U.S. Pat. 2,221, 776 (1940).
166 R. M. Schaffert, *Electrophotography*, Wiley, New York, 1975.
167 U. Rothgordt, *Philips Tech. Rdsch.*, **36**, 98 (1976/77).
168 J. Feder, *J. Appl. Phys.*, **47**, 1741 (1976).
169 H. Sussner, in *Proceedings of 1979 Ultrasonics Symposium*, J. de Klerk and B. R. McAvoy, Eds., pp. 491–498; H. Sussner and K. Dransfeld, *Colloid & Polym. Sci.*, **257**, 591 (1979).
170 G. M. Sessler, *J. Acoust. Soc. Amer.*, **70**, 1596 (1981).
171 H. J. Shaw, D. Weinstein, L. T. Zitelli, C. W. Frank, R. C. DeMattei, and K. Fesler, in *Proceedings of 1980 Ultrasonics Symposium, IEEE*, Piscataway, N.J., 1980.
172 R. Lerch and G. M. Sessler, *J. Acoust. Soc. Am.*, **67**, 1379 (1980).
173 F. Micheron, *Rev. Tech. Thomson CSF*, **10**, 445 (1978).
174 J. van Turnhout, J. W. C. Adamse, and W. J. Hoeneveld, *J. Electrostat.*, **8**, 369 (1980).
175 D. Perino, J. Lewiner, and G. Dreyfus, *L'Onde Electrique*, **57**, 688 (1977).
176 J. L. Bruneel and F. Micheron, *Appl. Phys. Lett.*, **30**, 382 (1977).

REFERENCES

177 H. Kawakami, *Audio Eng. Soc.*, Prepr. 693 (B3) (1969).
178 Y. Mochizuki, S. Watanabe, M. Kobayashi, N. Yakushiji, and H. Imazeki; *Toshiba Rev.*, 35 (1972).
179 G. M. Sessler, J. E. West, and R. L. Wallace, Jr., *IEEE Trans.*, **COM–21,** 61 (1973).
180 S. F. Demirdjioghlu and R. M. van Dyk, U.S. Pat. 3,668,698 (1972).
181 O. D. Jefimenko, in *Electrostatics and Its Applications,* A. D. Moore, Ed., Wiley, New York, 1973, pp. 131–147.
182 R. Gerhard and J. Kaufhold, to appear.
183 J. E. West, H. von Seggern, J. R. Nelson, and R. A. Kubli, *J. Acoust. Soc. Am.,* **68,** S68 (1980).
184 H. Bauser and W. Ronge, *Health Phys.*, **34,** 97 (1978).
185 S. Mascarenhas, in *Electrets,* G. M. Sessler, Ed., Springer, Heidelberg, 1980, pp. 321–346.
186 P. V. Murphy and S. Merchant, in *Electrets, Charge Storage and Transport in Dielectrics,* M. M. Perlman, Ed., Electrochemical Society, Princeton, N.J., 1973, pp. 627–649.
187 M. M. Perlman, private communication.
188 E. Fukada, T. Takamatsu, and I. Yasuda, *Jpn. J. Appl. Phys.*, **14,** 2079 (1975).
189 N. M. West, J. E. West, J. H. Revere, and M. C. England, *J. Endodont.*, **5,** 208 (1979).
190 J. J. Konikoff and J. E. West, in *1978 Annual Report, Conference on Electrical Insulation and Dielectric Phenomena,* National Academy of Sciences, Washington, D.C., 1978, pp. 304–310.

Chapter **4**

PIEZOELECTRICITY AND PYROELECTRICITY

Yasaku Wada
University of Tokyo
Bunkyo-ku, Tokyo, Japan

4.1	Introduction and Basic Definitions	111
4.2	General Equations of Piezoelectricity and Pyroelectricity in Polymers	114
4.3	Piezoelectricity of Polypeptides	117
4.4	Formation and Structure of Poly(vinylidene fluoride) Electrets	121
	4.4.1 Formation, 121	
	4.4.2 Structure, 122	
	4.4.3 Dipole Orientation, 123	
	4.4.4 Space Charge Effects, 126	
4.5	Piezoelectricity and Pyroelectricity of Polar Crystals	127
4.6	Piezoelectricity and Pyroelectricity of Electrets of Poly(vinylidene fluoride) and Other Polymers	132
4.7	Piezoelectricity and Pyroelectricity of Polymer–Ceramic Composites	140
4.8	Piezoelectric Relaxation	142
4.9	Measuring Techniques	145
4.10	Applications	147
	Acknowledgements	150
	References	151

LIST OF SYMBOLS

A	area of electrode
C	capacitance
D	electric displacement
E	electric field
F	free energy per unit volume
G	elastic modulus
K	compressibility
P	electric polarization

P_r	residual polarization (P at $E = 0$, $X = 0$, and $T = T_0$)
P_s	spontaneous polarization (P at $E = 0$)
Q	charge on the electrode
S	strain
S_e	entropy per unit volume
S_r	residual strain (S at $E = 0$, $X = 0$, and $T = T_0$)
T	temperature
T_g	glass transition temperature
V	voltage across electrodes
X	stress
Y	Young's modulus
d	piezoelectric strain constant (d constant)
e	piezoelectric stress constant (e constant)
i	$\sqrt{-1}$
k	electromechanical coupling constant
k_B	Boltzmann constant
l	thickness of film
m	Poisson's ratio ($m_{ij} = -S_i/S_j$)
p	pyroelectric constant at zero stress
p'	pyroelectric constant at constant strain
t	time
v	volume of crystal or dispersed particle
x,y,z	Cartesian coordinates
υ	volume thermal expansion coefficient
β	linear thermal expansion coefficient
γ_G	Grüneisen constant
Δ	increment from natural state
δ	relaxation strength
ϵ	permittivity
ϵ_0	permittivity of free space
ζ	second-order susceptibility ($\zeta = \partial x^S/\partial E$)
κ	electrostriction constant ($\kappa = \partial \epsilon^S/\partial S$)
λ	temperature coefficient of permittivity ($\lambda = \partial \epsilon^S/\partial T$)
μ	dipole moment
ρ	charge density
τ	relaxation time
ϕ	volume fraction of crystal or dispersed particle
χ	electric susceptibility [$\chi = (\epsilon - \epsilon_0)/\epsilon_0$]
ω	angular frequency

Superscripts

D, E, P, X, and *S* indicate that these quantities are kept constant (e.g., $\epsilon^S = (\partial D / \partial E)_{S=\text{constant}}$).

Subscripts

a	amorphous phase
b	backbone
c	crystal
s	side chain phase
(1, 2, 3)	(*x, y, z*)
(1, 2, 3, 4, 5, 6)	(*xx, yy, zz, yz, zx, xy*)

Conversion Factors

$d(\text{C/N}) = d(\text{cgs esu})/(3 \times 10^4)$
$e(\text{C/m}^2) = e(\text{cgs esu})/(3 \times 10^5)$
$p(\text{C/m}^2\,\text{K}) = p(\text{cgs esu})/(3 \times 10^5)$

4.1 INTRODUCTION AND BASIC DEFINITIONS

The existence of piezoelectricity in certain synthetic and biological polymers has been known for a long time (1–11). In particular, piezoelectricity in polymers of biological origin such as wood, bone, and tendon has been extensively studied (12, 13). The piezoelectric effects in these polymers, however, are rather small, and the interest in this subject remained mostly within the scientific community. This situation drastically changed when Kawai (14–16) in 1969 and others demonstrated that substantial piezoelectric as well as pyroelectric activity could be generated in synthetic polymer films previously subjected to a strong dc electric field at elevated temperatures. Due to their flexibility and availability in the form of films of large area, these materials opened up possibilities for new applications and device concepts which could not be realized with conventional crystalline piezo- and pyroelectric substances. The best known and commercially most attractive example to date is poly(vinylidene fluoride) (PVDF), which has been utilized as the active element in many applications ranging from infrared detector technology to loudspeakers.

This chapter discusses the fundamental mechanisms of piezo- and pyroelectric properties in polymeric materials in order to provide the reader with the general background necessary to understand these phenomena, develop new materials, and select new applications. Reviews on this subject have been written by Hayakawa and Wada (17, 18) and, more recently, by Broadhurst and Davis (19).

In this introductory section, the basic definitions of piezo- and pyroelectricity are given. The formulas will be expressed in S.I. units. A list of symbols used and the conversion factors between the S.I. and esu units are given in the List of Symbols. Abbreviations of the polymers referred to most often in this chapter are given in Table 1.1 of the Introduction to this book.

Piezoelectricity and pyroelectricity refer to the changes in the electric polarization P with strain S (or stress X) and temperature T. The basic relationships are (20)

$$P - P_r = \chi^S \epsilon_0 E + e(S - S_r) + p'(T - T_0) \tag{4.1}$$

or

$$P - P_r = \chi^X \epsilon_0 E + dX + p(T - T_0) \tag{4.2}$$

where the symbols used are defined in the List of Symbols. The difference between the pyroelectric coefficients p and p' is that the former includes the secondary effect due to thermal expansion.

The piezoelectric effect relates the dielectric and elastic properties of the material. Using the strain S and the applied field E as independent variables, $T = T_0$, observing that $\Delta P = \Delta D - \epsilon_0 E$, and using Eqs. (4.1) and (4.2), one obtains the fundamental definitions

$$\left. \begin{array}{l} \Delta D = e\,\Delta S + \epsilon^S E \\ X = G^E\,\Delta S - eE \end{array} \right\} \quad T = \text{constant} \tag{4.3}$$

Or, taking X and E as independent variables,

$$\left. \begin{array}{l} \Delta D = dX + \epsilon^X E \\ \Delta S = \dfrac{1}{G^E} X + dE \end{array} \right\} \quad T = \text{constant} \tag{4.4}$$

Here

$$d = \frac{e}{G^E} \tag{4.5}$$

and

$$\frac{\epsilon^S}{\epsilon^X} = \frac{G^E}{G^D} = 1 - k^2 \tag{4.6}$$

where the electromechanical coupling constant k is defined by

$$k^2 = \frac{e^2}{\epsilon^S G^D} = \frac{G^E d^2}{\epsilon^X} \tag{4.7}$$

INTRODUCTION AND BASIC DEFINITIONS

The pyroelectric effect relates the dielectric and thermal properties of the material. Using the temperature T and the applied field E as independent variables, one obtains the fundamental definitions

$$\left. \begin{array}{l} \Delta D = p\,\Delta T + \epsilon^T E \\ \Delta S_e = \dfrac{c^E}{T}\Delta T + pE \end{array} \quad X = 0 \right\} \quad (4.8)$$

Again the reader is referred to the List of Symbols for the definition of the symbols used in Eqs. (4.3)–(4.8). The electrocaloric effect is the inverse of the pyroelectric effect. This reciprocal relationship was confirmed for PVDF by Brossat et al. (21).

Since e and d are tensors of third rank, intrinsic piezoelectricity is only observed for a system without symmetry center. However, piezoelectricity due to a strain gradient (e.g., bending) (22–26) or a uniform strain under an applied dc field (15, 27–29) should be observed in all materials irrespective of structural symmetry. These higher-order effects are not treated in this chapter. Pyroelectricity is expressed by a vector p in materials with spontaneous polarization.

Cartesian coordinates of the polymer film are defined in Fig. 4.1. Different definitions are used for uniaxially oriented optically active polymers (Fig. 4.1a) and polymer electrets (Fig. 4.1b) because these coordinates are widely employed in the literature. For the latter, the $+z$ axis is taken as the direction of the poling field.

Since the piezo- and pyroelectric constants of polymer films are, in most cases, measured with electrodes fixed to the film surfaces, the electrode area changes with applied strain and temperature variation. The observed e constant, for example, is therefore

$$\bar{e} = \frac{1}{A}\left(\frac{\partial Q}{\partial S}\right)_{E=0} \quad (4.9)$$

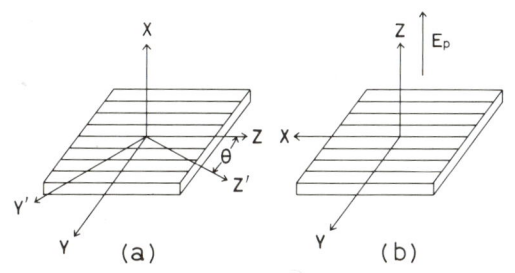

Figure 4.1 Cartesian coordinates for uniaxially oriented polymer films: (*a*) optically active polymer (z-axis: orientation direction, z'-axis: direction of tensile stress) and (*b*) polymer electret (x-axis: orientation direction, z-axis: direction of poling field).

which is related to the true e constant as

$$e = \left(\frac{\partial P}{\partial S}\right)_{E=0} = \left(\frac{\partial (Q/A)}{\partial S}\right)_{E=0} = \bar{e} - \frac{P_s}{A}\frac{\partial A}{\partial S} \tag{4.10}$$

Equation (4.10) holds for d and p by interchange of S with X and T, respectively.

The observed constants are $\bar{e}$, $\bar{d}$, and $\bar{p}$, except those obtained by high-frequency measurement. Theoretical expressions are derived for either e or $\bar{e}$.

Caution should be exercised when evaluating the tensor components $e_{ij} = (\partial P_i/\partial S_j)$ for $S_{k \neq j} = $ const., because this condition is not realized in the typical experiment. In the case of PVDF, P_3 is a function of S_1, S_2, and S_3 (Fig. 4.1b). Hence $\bar{e}_{31}$ observed in the conventional measurement is

$$\begin{aligned}\bar{e}_{31} &= \frac{\partial P_3}{\partial S_1} + \frac{\partial P_3}{\partial S_2}\frac{\partial S_2}{\partial S_1} + \frac{\partial P_3}{\partial S_3}\frac{\partial S_3}{\partial S_1} + \frac{P_s}{A}\frac{\partial A}{\partial S_1} \\ &= e_{31} - m_{21}e_{32} - m_{31}e_{33} + (1 - m_{21})P_s \end{aligned} \tag{4.11}$$

where $\bar{e}_{31}$ is equal to $\bar{d}_{31}$ multiplied by Y_1. In the following sections, we do not differentiate between e and $\bar{e}$ and denote them both e for simplicity.

4.2 GENERAL EQUATIONS OF PIEZOELECTRICITY AND PYROELECTRICITY IN POLYMERS

In the following, the general expressions for e and p for polymer materials will be derived adopting the formalism by Hayakawa and Wada (17, 30, 31). Consider a point charge q in the bulk of a polymer sample (Fig. 4.2). If the electrodes—in intimate contact with the sample—are short-circuited, the induced charges on the electrodes are, respectively,

$$Q_1 = -q\frac{C_1}{C_1 + C_2} \qquad Q_2 = -q\frac{C_2}{C_1 + C_2} \tag{4.12}$$

where C_1 (C_2) is the capacitance between upper (lower) electrode and the plane passing q parallel to the electrodes. Equations (4.12) indicate that, if the ratio C_1/C_2 changes with applied strain, Q_1 and Q_2 change, resulting in a piezoelectric current in the external circuit.

There are three possibilities for Q_1 and Q_2 to change with strain: (a) The charge q is not fixed to the macroscopic plane in the material and moves out of the plane as a function of strain, that is, the displacement of the ion is not the same as the macroscopic displacement in the material. Strain of this kind is called internal strain. (b) The charge q is fixed to the material (embedded

GENERAL EQUATIONS OF PIEZO- AND PYROELECTRICITY

charge), but C_1/C_2 changes with strain owing to the heterogeneity of the material. Since the capacitance is proportional to ϵ/l (l is the separation of the two planes constituting the capacitor), heterogeneity in electrostriction constant and/or Poisson's ratio induces the change in C_1/C_2. (c) The material possesses a spontaneous polarization which changes either due to internal or external strain. Pyroelectricity is related to mechanisms (b) and (c), replacing strain with temperature.

The electric displacement, Eq. (4.3), may be written as

$$D(z) = \epsilon(z) E(z) + \overline{e}_i(z) \overline{S}(z) + P_r(z) \tag{4.13}$$

where $\overline{e}_i(z)$ represents the intrinsic piezoelectric constant due to mechanisms (a) and (c). The z-coordinate is perpendicular to the plane of the film. The quantities in Eq. (4.13) are averages in the xy-plane and, since a displacement of the charge in the xy-plane does not induce charges on the electrodes, one only needs to consider the strain in the z-direction.

The potential V across the electrodes is then

$$V = -\int_{-l/2}^{l/2} E\, dz = -\int_{-l/2}^{l/2} (D - P_r)\frac{dz}{\epsilon} + \int_{-l/2}^{l/2} \overline{e}_i \overline{S}\, \frac{dz}{\epsilon} \tag{4.14}$$

where l is the thickness of the film and $z = 0$ at the point of symmetry between the electrodes (Fig. 4.2).

From the Maxwell equation, we have

$$D(z) = \frac{Q}{A} + \int_{-l/2}^{z} \rho_t(z)\, dz \tag{4.15}$$

where Q is the charge on the lower electrode and ρ_t is the true charge density in the material. The total charge density ρ in the material is the sum of ρ_t and the polarization charge $-\partial P_r/\partial z$. Assuming that $V = 0$ when the electrodes are short-circuited, we have from Eqs. (4.14) and (4.15)

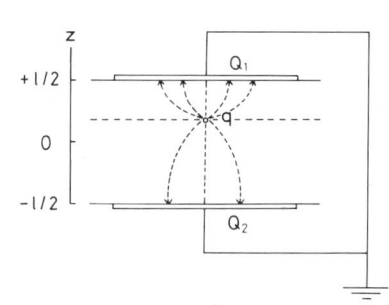

Figure 4.2 Point charge q embedded in polymer matrix of thickness l.

$$-\frac{Q}{A}\int_{-l/2}^{l/2}\frac{dz}{\epsilon} - \int_{-l/2}^{l/2}\frac{dz}{\epsilon}\int_{-l/2}^{z}\rho\,dz + \int_{-l/2}^{l/2}\overline{e}_i\overline{S}\frac{dz}{\epsilon} = 0 \qquad (4.16)$$

The above equation is written for the strained state. We transform Eq. (4.16) into an equation relating quantities with respect to the unstrained coordinate. The transformations are (subscript 0 stands for unstrained state)

$$dz \rightarrow [1 + \overline{S}_0(z_0)]\,dz_0, \qquad \overline{S}(z) \rightarrow \overline{S}_0(z_0), \qquad l \rightarrow l_0$$

$$\epsilon(z) \rightarrow \epsilon_0(z_0) + \overline{\kappa}_0(z_0)\,\overline{S}_0(z_0), \qquad \overline{e}_i(z) \rightarrow \overline{e}_{i0}(z_0) + O(S_0)$$

$$\int_{-l/2}^{z}\rho(z)\,dz \rightarrow \int_{-l_0/2}^{z_0}\rho_0(z_0)\,dz_0 \qquad \text{(conservation of charge)}$$

The transformation yields, neglecting higher-order terms and the subscript 0,

$$-\frac{Q}{A}\left[\int_{-l/2}^{l/2}\frac{dz}{\epsilon} + \int_{-l/2}^{l/2}\left(1 - \frac{\overline{\kappa}}{\epsilon}\right)\overline{S}\frac{dz}{\epsilon}\right] - \int_{-l/2}^{l/2}\frac{dz}{\epsilon}\int_{-l/2}^{z}\rho\,dz$$

$$-\int_{-l/2}^{l/2}\frac{dz}{\epsilon}\left(1 - \frac{\overline{\kappa}}{\epsilon}\right)\overline{S}\int_{-l/2}^{z}\rho\,dz + \int_{-l/2}^{l/2}\overline{e}_i\overline{S}\frac{dz}{\epsilon} = 0 \qquad (4.17)$$

The relationship between local strain $\overline{S}$ and macroscopic strain S applied to the film is assumed to be

$$\overline{S}(z) = -m(z)\,S \qquad (4.18)$$

where m is Poisson's ratio. Similarly, we define

$$e_i(z) = -m(z)\,\overline{e}_i(z), \qquad \kappa = -m(z)\,\overline{\kappa}(z) \qquad (4.19)$$

The piezoelectric constant defined by $e = \partial(Q/A)/\partial S$ [see Eq. (4.9)] at $V = 0$ is finally given by

$$e = \langle e_i \rangle + \left\langle\left\{\left(\frac{\kappa}{\epsilon} + m\right) - \left\langle\frac{\kappa}{\epsilon} + m\right\rangle\right\}\left\{\int_{-l/2}^{z}\rho\,dz\right\}\right\rangle \qquad (4.20)$$

where the angular brackets denote the average,

$$\langle A(z)\rangle = \int_{-l/2}^{l/2} A\frac{dz}{\epsilon}\Big/\int_{-l/2}^{l/2}\frac{dz}{\epsilon} \qquad (4.21)$$

The term of the form $A - \langle A \rangle$ within the braces of Eq. (4.20) is the result of sample heterogeneity, that is, it represents the piezoelectricity due to mechanism (b). The intrinsic piezoelectricity appears in the average $\langle e_i \rangle$.

PIEZOELECTRICITY OF POLYPEPTIDES

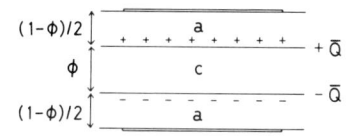

Figure 4.3 Three-layered composite of polymer films with trapped charges on the interfaces.

The pyroelectric constant is analogously given by

$$p = \langle p_i \rangle + \left\langle \left\{ \left(\frac{\lambda}{\epsilon} - \beta \right) - \left\langle \frac{\lambda}{\epsilon} - \beta \right\rangle \right\} \left\{ \int_{-1/2}^{z} \rho \, dz \right\} \right\rangle \quad (4.22)$$

where the linear thermal expansion coefficient β is defined by $\beta(z) = \partial \bar{S}(z)/\partial T$.

Due to heterogeneity and embedded charges in the bulk, most polymers show piezo- and pyroelectricity to some extent, even in the absence of any external poling process (17, 23, 32). Pyroelectricity due to heterogeneity of the material coupled with embedded charges was discussed by Salomon et al. (33).

A simple example of a heterogeneous system with embedded charges is illustrated in Fig. 4.3, where a film c is laminated on both surfaces to films a of equal thickness and charges $\pm \bar{Q}$ per unit area are trapped on the interfaces, respectively (18). Equation (4.20) yields in this case

$$e = \frac{(\kappa_a/\epsilon_a - \kappa_c/\epsilon_c) + (m_a - m_c)}{\epsilon_a \epsilon_c [(1 - \phi)/\epsilon_a + \phi/\epsilon_c]^2} \bar{Q} \, \phi(1 - \phi) \quad (4.23)$$

which becomes a maximum for $\phi = \epsilon_c/(\epsilon_a + \epsilon_c)$,

$$e_{\max} = \frac{1}{4} \left[\left(\frac{\kappa_a}{\epsilon_a} - \frac{\kappa_c}{\epsilon_c} \right) + (m_a - m_c) \right] \bar{Q} \quad (4.24)$$

An artificial piezoelectric and pyroelectric film can be thus composed of non-piezoelectric and nonpyroelectric component films. Broadhurst et al. (19) indicated that Eq. (4.20) gives the operating principle of the electret microphone when applied to a double-layer structure consisting of a monopolar polymer film and an air gap.

4.3 PIEZOELECTRICITY OF POLYPEPTIDES

Uniaxially oriented systems of optically active polymers exhibit piezoelectricity due to internal strain, mechanism (a) in Section 4.2. Piezoelectricity of biologic tissues, wood, and natural fibers belongs to this category. The symmetry of this system is expressed by the space group D_2 and the piezoelectric tensor (13) is, using the coordinate system of Fig. 4.1a,

$$\begin{pmatrix} 0 & 0 & 0 & d_{14} & 0 & 0 \\ 0 & 0 & 0 & 0 & -d_{14} & 0 \\ 0 & 0 & 0 & 0 & 0 & 0 \end{pmatrix} \quad (4.25)$$

The piezoelectric polarization P_1 is induced by shear stress X_4. P_1 and the elongational stress X in the film plane are related by

$$d = \frac{P_1}{X} = \frac{1}{2} d_{14} \sin 2\theta \quad (4.26)$$

where θ is the angle between X and the z-axis. The absolute value of d becomes a maximum at $\theta = 45°$, in agreement with the experiment as shown in Fig. 4.4.

The piezoelectricity of various polymers belonging to this group has been studied, including poly(propylene oxide) (34, 35), cellulose derivatives (34, 36–39), and so on. Oriented films of poly(γ-benzyl L-glutamate) and poly (γ-methyl L-glutamate) are among those most systematically investigated (34, 40–46). The piezoelectric constant was found to be proportional to the degree of orientation of the α-helical backbones (45) and the sign of the constant to be opposite for L and D polymers (40). Piezoelectricity of these polypeptides was thus concluded to arise from α-helical backbones which are aligned parallel along the z-direction.

The side chains of the polypeptides surrounding the backbone are in a disordered state at room temperature and are nonpiezoelectric. Molecular motion of the side chains is responsible for the remarkable temperature and frequency dependence of the piezoelectric constant of these films (see discussion in Section 4.8).

A molecular theory of piezoelectricity of the α-helical backbone was devel-

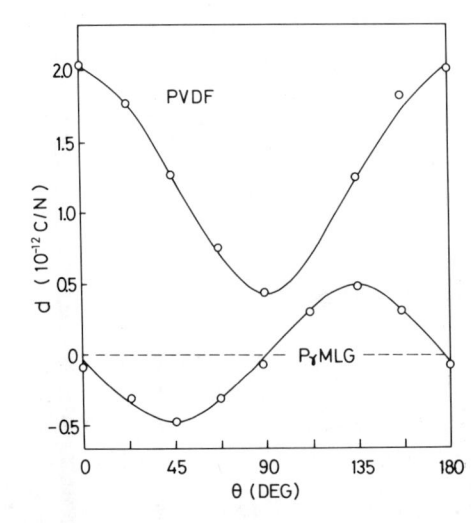

Figure 4.4 Piezoelectric constant of uniaxially oriented PVDF electret (top) and PMLG (bottom) vs. angle between orientation axis and tensile stress. (PMLG, from Fukada et al. [10]; PVDF, from Fukada and Sakurai [136].)

oped by Namiki, Hayakawa, and Wada (46). They assumed that the stress field around the backbone is uniform and a force acts on the α-carbon atom (carbon to which the side chain is attached) when the stress is applied to the film. The magnitude of the force is equal to the stress X times the surface area of the cylindrical backbone allocated to the α-carbon (projection on the plane normal to X). This model enables one to replace the stress field tensor by a force vector.

Two six-component vectors **A** and **B** are defined in terms of four three-component vectors, **E**, $\mathbf{E}_v$, **M**, and $\mathbf{M}_v$, as

$$\mathbf{A} = \begin{pmatrix} \mathbf{E} \\ \mathbf{E}_v \end{pmatrix} \qquad \mathbf{B} = \begin{pmatrix} \mathbf{M} \\ \mathbf{M}_v \end{pmatrix} \qquad (4.27)$$

where **E** is the true electric field, $\mathbf{E}_v$ is the virtual electric field which is equal to the vector stress **X**, **M** is the true dipole moment of a fiber period of the helix, and $\mathbf{M}_v$ is the virtual dipole moment (elastic dipole moment) of a fiber period defined by the product of the area allocated to the α-carbon (projection on plane normal to X) and the position vector of the α-carbon. The replacement of elastic variables by virtual electric ones enables one to extend the conventional fluctuation formalism of electric susceptibility to the elastic compliance and the piezoelectric constant.

The linear relation between **A** and **B** is written as

$$\langle \Delta B_j \rangle = \sum_{k=1}^{6} Y_{jk} A_k \qquad j = 1, 2, \ldots 6 \qquad (4.28)$$

Here $\Delta B_j = B_j - \langle B_j \rangle_0$, where the angular brackets with and without subscript 0 mean statistical averages in the absence and presence of the field A, respectively. Subscripts $(1, 2, \ldots 6)$ stand for (x, y', z', x, y', z'), respectively, where z' is the direction of **X** (Fig. 4.1a).

Since **B** is conjugate to **A**, Y_{jk} in Eq. (4.28) may be related to the fluctuations of **B**,

$$Y_{jk} = \frac{\langle \Delta B_j \, \Delta B_k \rangle_0}{k_B T} \qquad (4.29)$$

where Y_{jk}/v (v is the volume of a fiber period) expresses the electric susceptibility for $j, k = 1, 2, 3$, the elastic compliance for $j, k = 4, 5, 6$, and the piezoelectric d constant for $j = 1, 2, 3$ and $k = 4, 5, 6$, or vice versa. The $\langle \Delta B_j \, \Delta B_k \rangle_0$ value can be calculated from molecular dynamics, using the normal coordinate method familiar in vibrational analysis of the helical chain. The results are: $\epsilon_b/\epsilon_0 = 6.69$, $G_b = 10.6 \times 10^9$ N/m^2, $d_b = 6.3 \times 10^{-12}$ C/N, and $e_b = 6.6 \times 10^{-2}$ C/m^2 for a tensile stress applied at $\theta = 45°$ with respect to the x-axis (Fig. 4.1a).

To finally calculate d and e of the polymer film consisting of backbones and a side chain phase, we use the equations for a composite in which circular rods are aligned parallel in the matrix (46):

$$d = \frac{2G_b}{(G_b + G_s) + (G_b - G_s)\phi} \frac{2\epsilon_s}{(\epsilon_b + \epsilon_s) - (\epsilon_b - \epsilon_s)\phi} \phi d_b \quad (4.30)$$

$$e = \frac{2G_s}{(G_b + G_s) - (G_b - G_s)\phi} \frac{2\epsilon_s}{(\epsilon_b + \epsilon_s) - (\epsilon_b - \epsilon_s)\phi} \phi e_b \quad (4.31)$$

where ϕ is the volume fraction of rods [$\phi = 8.1\%$ for poly(γ-benzyl-L-glutamate) (PBLG)]; and ϵ_s and G_s are determined from observed values of ϵ and G for PBLG at low temperature where motion of side chains is frozen and from calculated values of ϵ_b and G_b so as to satisfy the composite equations for ϵ and G (not shown here). Values of d and e calculated from Eqs. (4.30) and (4.31) are respectively $d = 0.43 \times 10^{-12}$ C/N and $e = 0.22 \times 10^{-2}$ C/m², which agree roughly with observed values of PBLG extrapolated to 100% orientation, $d = 0.83 \times 10^{-12}$ C/N and $e = 0.43 \times 10^{-2}$ C/m².

Displacements of each atom of the backbone for tensile stress at $\theta = 45°$ were calculated and are illustrated in Fig. 4.5. One notes that piezoelectricity of the

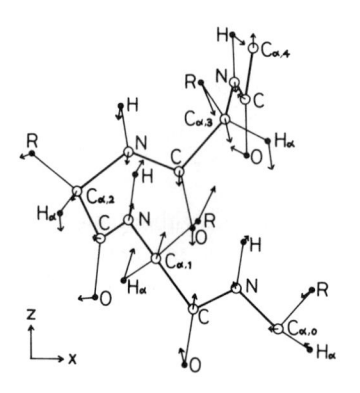

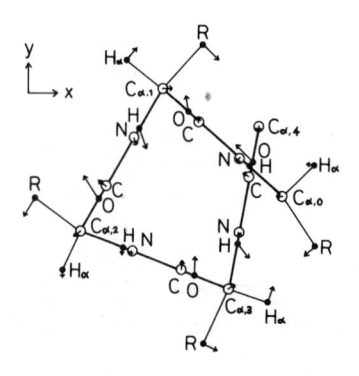

Figure 4.5 Displacements of atoms of α-helical polypeptide backbone in an arbitrary scale when a tensile stress is applied along 45° direction against orientation axis. Top: projection on the xz-plane. Bottom: projection on the xy-plane. Displacements due to piezoelectrically-active (infrared-active) modes alone are shown. (From Namiki et al. [46].)

α-helical backbone arises mainly from the twist of the CN bond of the CO—NH group. From this calculation, the intrachain hydrogen bond was found important for this twist. The fluctuation formalism of the piezoelectric constant described in this section may provide a useful tool which is generally applicable in calculating the piezoelectric properties for polymers with known chain conformation. For a more detailed discussion of this topic, the reader is referred to Ref. 46.

4.4 FORMATION AND STRUCTURE OF POLY(VINYLIDENE FLUORIDE) ELECTRETS

4.4.1 Formation

Electrets are defined as dielectrics which produce an electric field around them. When the electrodes of an electret are short-circuited, charges are induced on them. These effects result from frozen-in dipolar and/or monopolar charges in the dielectric (47).

The conventional way to form an electret is by poling at a high temperature (thermoelectret): the film is heated to the poling temperature T_p, at which a high dc electric field E_p is applied during an extended period of time t_p. The film is then cooled to room temperature before the applied field is switched off. Usually the dc field is applied across the electrodes which are deposited onto the film surfaces. Corona and plasma poling of unelectroded films are useful alternative poling methods in order to increase E_p without risking destructive electrical breakdown of the sample (48). A more elaborate description of various methods used to form polymer electrets is given in Chapter 3, Section 3.2.

The resulting charges in the electret are usually classified into four groups: (1) molecular dipolar orientation, (2) charge separation on a microscopic scale, (3) charge separation on a macroscopic scale, and (4) charge injection from the electrodes. Charges due to (1) to (3) are denoted "heterocharge," because the polarity of the resulting surface charge is opposite to that of the poling electrode. The charge due to (4) is called "homocharge."

The poling-induced polarization of the electret, P_s, can be estimated either from its surface potential (49, 50) or from the integral of the thermally stimulated depolarization current observed on heating the electret to temperatures above T_p to complete the depolarization ($P_s \rightarrow 0$) (51). P_s can be obtained also by integrating the polarization current during the poling process, if conduction through the sample can be neglected (52, 53), or from the surface charge after corona or plasma poling (53).

Piezoelectricity and pyroelectricity have been studied for electrets of a number of synthetic polymers since Kawai's report in 1969 (14). The polymer that shows the largest effects and has found use in many novel devices is poly (vinylidene fluoride) (PVDF). The main body of this section is devoted to results obtained for this polymer. Before discussing the piezo- and pyroelectric properties of PVDF in Section 4.6, its structure and possible mechanisms of the formation of the induced stable polarization will be described.

4.4.2 Structure

PVDF crystallizes from the melt below 150°C into spherulitic structures (54) consisting of form II crystals (α-phase). The structure of form II crystal is monoclinic ($\beta = 90°$) as illustrated in Fig. 4.6a (55, 56). The chains take on a glide-type TG$\overline{\text{T}}\overline{\text{G}}$ conformation and possess a dipole moment with components normal and parallel to the chain axis. The dipole moments of two chains in the unit cell, however, are antiparallel, and the crystal therefore has no spontaneous polarization (antipolar crystal). The degree of crystallinity is approximately 50%.

When the film is drawn at a temperature below 130°C to several times of the original length, the morphology is changes to a c-axis-oriented structure with a new crystal form (form I). The transformation mechanism was studied by small-angle light scattering (57), stress–strain relationship, and x-ray analysis (58). From x-ray analysis, the long period and the lamella thickness along the draw axis (x-axis, Fig. 4.1b) were determined to be 100–200 Å and 60–100 Å (59–61), respectively. The morphological change induced by drawing is similar to that of low-density polyethylene, which has also a degree of crystallinity of approximately 50%.

The new crystal form produced during the draw process, the form I (β-phase), has the orthorhombic structure illustrated in Fig. 4.6b (55, 56, 62) in which the chains take on a planar zigzag (TTTT) conformation. The chain has a large dipole moment ($\mu_0 = 7.06 \times 10^{-30}$ C-m per monomeric unit) normal to the chain axis and is packed so that the crystal possesses a spontaneous polarization P_c (polar crystal). P_c equals 0.23 C/m² for an ideal crystal. In the real form I crystal, P_c may be smaller due to head-to-head and tail-to-tail defects (19). The conversion from form II to form I is evident in infrared spectra (49) (510 cm^{-1} for form I and 530 cm^{-1} for form II). The conversion fraction depends on the draw ratio and the draw temperature.

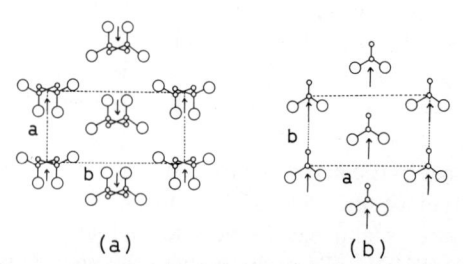

Figure 4.6 Crystal structures of PVDF: (a) form II; (b) form I. Large circle: F atom; small circle: C atom; H atoms are omitted. (From Hasegawa et al. [56].)

FORMATION AND STRUCTURE OF PVDF ELECTRETS

The form I crystal has the highest molecular packing among various crystal forms of PVDF. The separation between atoms belonging to neighbor chains is small compared to the sum of their van der Waals radii, and, to avoid steric hindrance, the chain makes a statistical deflection about the c-axis by $\pm 7°$ (56).

Some other crystal modifications have been reported for PVDF. When form II PVDF is poled in fields E_p ranging between 100 and 300 MV/m, the unpolar form II crystal is transformed into a polar form II (polar α-phase) in which the central chain in Fig. 4.6a rotates about the chain axis by 180°. The chain conformation and lattice constants remain unaltered, and therefore the form II unit cell acquires a net dipole moment (53, 63–72). The polar form II crystal may be transformed into form I by using poling fields exceeding 300 MV/m (68, 69).

A third crystal form, the form III, or γ-phase, is obtained by crystallization under atmospheric pressure at a temperature just below the melting point (54, 73, 74). The form III crystal is also generated by melt crystallization under high pressure (55). Lando, Hopfinger, and their co-workers proposed an orthorhombic unit cell of the form III crystal with chain conformation approximately TTTGTTT$\overline{\text{G}}$ (75–79). Takahashi and Tadokoro (80) prepared a highly oriented form III sample and proposed a monoclinic structure consisting of a TTTGTTT$\overline{\text{G}}$ conformation including some disorder. The form III crystal readily transforms into form I by mechanical deformation.

The equilibrium melting point for the form II phase is in the range 170–190°C, depending on the content of heterojunctions (54, 59, 73, 81) and approximately 220°C for forms I and III (59). Real crystals with finite size melt below these temperatures (82), and the melting point increases in the order of forms II, I, and III. Crystallization, melting, and phase transformation under high pressure and under tensile stress were studied in detail (83–87).

The PVDF electret is prepared by annealing and subsequent poling the c-axis-oriented films containing form I crystals. The induced surface charge is opposite in sign to the poling charge (heterocharge). Typical poling parameters are $T_p \sim 80–120°\text{C}$, $E_p \sim 50–150$ MV/m, and $t_p \sim 30$ min. The resulting polarization is stable for heating cycles below T_p except for an irreversible part which vanishes on the first heating cycle. The mechanism inducing both reversible and irreversible polarization has been a subject of much discussion, and arguments have been brought forward for all mechanisms outlined above, that is, dipole orientation, microscopic and macroscopic charge separations, and charge injection. In the following, the evidence for these different poling mechanisms is discussed.

4.4.3 Dipole Orientation

We begin with the dipole reorientation model. In the as-drawn sample, the polar b-axis (i.e., the dipole moment) is preferentially oriented along the $\pm y$-axis, that is, in the film plane. In the strong poling field, the dipole axis reorients to align predominantly along the poling z-axis, that is, perpendicular to the film plane. Early efforts based on Raman spectroscopy were not successful to establish the speculated reorientation of the PVDF dipoles (88), but more recently the alignment of dipoles could be established on the basis of x-ray diffraction (89–92) and infrared dichroism data (93–96). The field-induced dipolar alignment in form I crystals suggests that the crystal is ferroelectric.

Ferroelectricity of PVDF was first speculated on by Nakamura and Wada (97) and Bergman et al. (98) in 1971. Since then, various experimental findings in agreement with that speculation have been reported, in particular the hysteresis curves of P_s, e, p, and infrared absorbance against bias field (Fig. 4.7) (52, 93, 95, 96, 99–101). The domain structure and Curie point of PVDF have, however, yet to be documented. The detailed mechanisms of dipolar orientation in PVDF crystals due to poling are still controversial. Possible mechanisms are: (a) rotation of individual chains or cooperative rotation of a small number of chains around the chain axis, (b) motion of domain walls, and (c) rotation of individual crystals.

In case (a), the rotation should not be continuous but discrete (switching

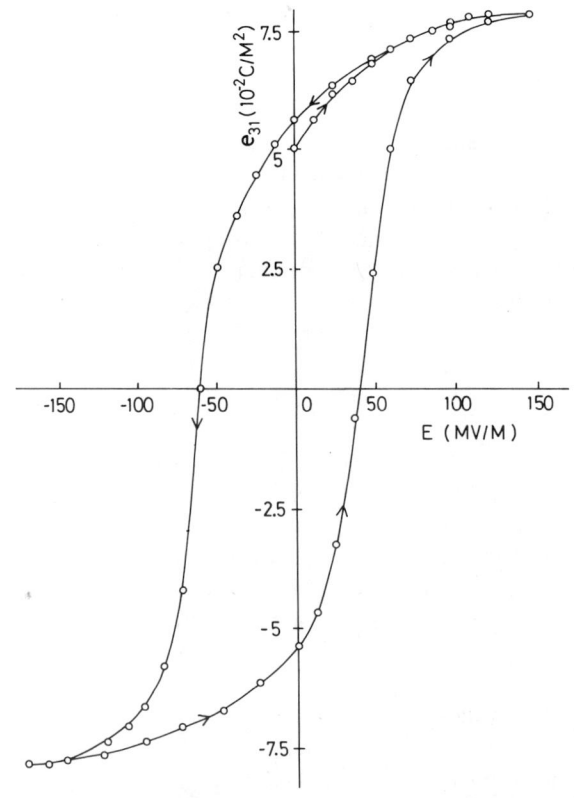

Figure 4.7 Hysteresis curve of piezoelectric constant of PVDF electret vs. electric field. (From Tamura et al. [100].)

model) due to the discrete minima of the interchain potential (92, 102). Since it is difficult to rationalize that in the ideal crystal single chains can rotate while the adjacent chains remain fixed, Takahashi and Odajima (90) proposed the presence of defects in the real PVDF crystal. These defects, a chain rotation by 130–180° accompanied by a $c/2$ translation along the c-axis, are eliminated by proper poling conditions.

Case (b), the motion of domain walls, is the most popular explanation for ferroelectric crystals; but in PVDF, the domain structure has not been established. The rotation of whole entities of form I crystals, case (c), in an amorphous matrix is a widely accepted mechanism of relaxational behavior of semicrystalline polymers, denoted "grain boundary relaxation" (103, 104). It is doubtful, however, that the rotation is stable when the poling field is removed and the sample thermally cycled. According to Furukawa et al. (52), the coercive field in the D–E hysteresis curve increases smoothly with decreasing temperature through the glass transition temperature at $-50°C$, indicating that the dipolar orientation is not affected by the amorphous phase.

The degree of dipole orientation in the crystal induced by poling, $\Gamma = \langle \cos \varphi \rangle$, where φ is the angle between the dipole orientation and the z-axis, was discussed by Kepler and Anderson from their x-ray data (92). They assumed the 60° switching model and the probability for each site to be determined by the Boltzmann factor $\exp(\mu E \cos \varphi / k_B T)$. They obtained $\Gamma = 0.84$ for PVDF poled at $E_p = 60$ MV/m and $T_p = 100°C$. According to Takahashi and Odajima (61), Γ should be smaller if the integrated intensity of the x-ray diffraction profile is used in place of the peak intensity. They showed that the 60° switching model is plausible based on the biaxial orientation distribution obtained from the x-ray analysis. The assumption of thermal equilibrium, however, seems doubtful as suggested from the P–E hysteresis. Furthermore, one must take into account the intrachain energy due to chain twist and the cooperative rotation among neighbor chains. The simplest model is to assume four possible orientations of the dipoles, namely, along the $\pm z$- and $\pm y$-axes with an equal population along the $\pm z$-axes before poling. After poling, some fraction of dipoles along the $\pm y$-axes has rotated to the $+z$-axis. Then, Γ equals half the increment of $\langle \cos 2\phi \rangle$ (ϕ is the angle between the b- and z-axes) due to poling. From the data by Takahashi and Odajima (90, 91), we get $\Gamma = 0.15$ for a sample poled at $E_p = 70$ MV/m and $T_p = 120°C$.

A theoretical model for dipolar alignment in the polarizing field was first proposed by Aslaksen (105). Dvey-Aharon et al. (102) developed a theory of kink propagation as a model for poling. Broadhurst and Davis (19, 106) presented a cooperative dipolar orientation model. The polarization switching time was measured by Furukawa and Johnson (107).

The amorphous phase of PVDF has a glass transition temperature T_g between -40 and $-50°C$ (108–111) and is thus in the rubbery state at the room temperature. The reorientation of dipoles associated with the amorphous phase is observed as a peak in the thermally stimulated current at around the T_g in samples previously cooled to below T_g in an applied electric field (109). In oriented PVDF films, the amorphous phase is partly oriented, too (112). Tasaka et al. (112) showed that the degree of orientation of the amorphous phase reaches 0.3 for a draw ratio of 4. Dipolar orientation in the oriented amorphous phase (often denoted as intermediate or pseudo-crystalline phase) was suggested as part of the origin of P_s (113–115).

4.4.4 Space Charge Effects

Space charges exert two effects on the induced polarization P_s. Firstly, space charges contribute to the polarization, and, secondly, they modify the local electric field during the poling process and therefore give rise to local heterogeneity (116). Coupled with an appropriate heterogeneity, space charge can contribute to the piezo- and pyroelectricity of the sample [see Section 4.2, Eqs. (4.20) and (4.22)]. Murayama (117) concluded that the dominant contribution to P_s arises from microscopic charge separation rather than from dipole reorientation. Indeed, it is well known that the folded chain surface of the crystalline lamella provides efficient trapping sites. In the poling field, the positive and negative surface charges would be redistributed to form a dipole moment which is parallel to the dipolar orientation of the crystal. A space-charge distribution on the lamella surfaces has not yet been directly tested by experiment, but there are several experimental findings suggesting a contribution of space charge to P_s.

Fukada et al. (94, 118) compared the infrared dichroic ratio (a measure of dipolar orientation) with d_{31} for PVDF prepared under various conditions. The result indicates that d_{31} is not a unique function of the degree of dipolar orientation and that some part of d_{31} does not depend on the degree of orientation but on T_p.

Many workers have suggested the importance of charge injection from the metal electrodes in the poling process (119–121). One notes, for instance, that the piezo- and pyroelectric activities of the PVDF film decrease when charge injection from the positive electrode is prevented. This can be achieved by inserting a blocking layer, for instance, Mylar, between the metal electrode and the PVDF film. In these experiments, care was taken to assure that the applied field did indeed exist in the PVDF film during the poling process. Also, the current–voltage (I–V) relationship for the stationary current during the high-temperature poling obeys Schottky's law (log I is proportional to $V^{1/2}$), suggesting that field-induced emission from the electrode is the rate-limiting process for the poling current (119). Furthermore, the activity of the poled film depends on the work function of the electrode metal (121).

A nonuniformity of the piezo- and pyroelectric activities along the thickness direction (z-axis) of the poled film has been reported by many authors (122–125): the activities are higher near the surface which was in contact with the positive poling electrode. For a nonuniform film, the pyroelectric response is different for heat radiaton incident on the front and back surfaces, and the piezoelectric response exhibits an anomalous resonance mode when excited in the thickness expanding mode. The nonuniformity is quantitatively obtained from the dependence of the pyroelectric response on the modulation frequency of incident light (124, 125), which arises from the interaction between the nonuniformity and the thermal diffusion constant in the film. The nonuniformity is usually expressed by a function $\exp[-\xi(l-z)/l]$, where $z = 0$ at the surface which was in contact with the negative poling electrode.

Ibe (122) measured the potential distribution in a PVDF film when the poling voltage is applied. The result shows that the potential versus z curve is concave upward and the poling field is very high near the positive electrode. The nonuniform field may be attributed to negative space charges originally existing in PVDF. This nonuniform field accounts for the nonuniformity of the activity of poled film.

According to Day et al. (125), ξ, which is the measure of nonuniformity, equals 2 at $t_p = 5$ min ($E_p = 42$ MV/m, $T_p = 70°$C), but ξ decreases to almost zero at $t_p = 30$ min. For $E_p = 210$ MV/m, $\xi \simeq 0$ even at $t_p = 5$ min. These results indicate that the poling field becomes uniform with increasing poling time and that the rate increases with increasing E_p. These observations may be interpreted in terms of neutralization of negative space charges by the homocharge injection from the positive electrode. The long poling time required to reach a steady value of the piezo- and pyroelectric constant (126, 127) may be the time required to build up the uniform field. The injected charges may neutralize an existing macroscopic space charge distribution and, furthermore, generate space charge dipoles on a microscopic scale.

Takahashi et al. (119) speculated that the holes injected from the positive electrode break the hydrogen bond between the F and H atoms in the crystal, which could facilitate dipolar alignment.

In summary, the result obtained for polarized PVDF films indicate that both dipolar alignments and space charges contribute to the polarization P_s. The long poling times required may be due to the achievement of space charge equilibrium for a given set of poling parameters (E_p, T_p, electrodes). The hysteresis loop observed under a sinusoidal excitation would be generated by the reorientation of the dipolar contribution to P_s while the space charge dipoles remain frozen on the experimental time scale.

4.5 PIEZOELECTRICITY AND PYROELECTRICITY OF POLAR CRYSTALS

The previous section indicated that the poling-induced polarization P_s in PVDF films derives from alignment of intrinsic molecular dipoles associated with the polar form I crystal phase and, possibly, space charge dipoles resulting from charge redistribution on the lamella surfaces. Furthermore, dipole alignment in the oriented amorphous phase which exists between the crystal lamellae may provide an additional source to the polarization P_s. Before further discussing the general properties of PVDF in the next section, theoretical expressions for the piezo- and pyroelectricity of polar crystals, such as form I of PVDF, will now be derived. In the calculations the subscript c for "crystal" will be deleted for simplicity.

The following thermodynamic theory and molecular model can be applied also to glassy electrets in which dipoles are preferentially oriented (128).

The form I crystal of PVDF belongs to the space group C_{2v} for which the piezoelectric tensor is

$$\begin{pmatrix} 0 & 0 & 0 & 0 & d_{15} & 0 \\ 0 & 0 & 0 & d_{14} & 0 & 0 \\ d_{31} & d_{32} & d_{33} & 0 & 0 & 0 \end{pmatrix} \quad (4.32)$$

The coordinates in the above expression are the (x,y,z)-film coordinates which correspond to the (c,a,b)-axes of the crystal, respectively.

A thermodynamic description of the piezoelectric constant of polar crystals was given by Hayakawà and Wada (17, 129), starting from the free energy F per unit volume as a function of P and S:

$$F(P,S) = F_0 + \tfrac{1}{2}G^P S^2 + f(P) - \gamma S P^2 \quad (4.33)$$

where F_0 is independent of P and S, $f(P)$ is the polarization energy, and γ is the electrostrictive coupling coefficient between P and S. Equation (4.33) does not include a linear cross term SP which would yield piezoelectricity due to internal strain. The derivation is thus restricted to mechanism (c) described in Section 4.2. The $f(P)$ term is expanded into a power series of P starting with P^2:

$$f(P) = P^2 \sum_{j=0}^{\infty} \frac{a_j}{j+2} P^j \quad (4.34)$$

Since $E = (\partial F/\partial P)_S$ and $X = (\partial F/\partial S)_P$, one may derive fundamental equations describing P and X as functions of S and E. These are, neglecting terms of order higher than quadratic,

$$\begin{aligned} P - P_r &= \chi^S \epsilon_0 E + e(S - S_r) + \kappa(S - S_r)E + \tfrac{1}{2}\zeta \epsilon_0 E^2 \\ X &= -eE + G^E(S - S_r) - \tfrac{1}{2}\kappa E^2 \end{aligned} \quad (4.35)$$

Here P_r and S_r are residual quantities defined for $X = 0$, $E = 0$,

$$P_r^2 = \frac{G^P}{2\gamma^2} g(P_r), \qquad S_r = \frac{\gamma}{G^P} P_r^2 \quad (4.36)$$

where

$$g(P) = \sum_{j=0}^{\infty} a_j P^j \qquad (4.37)$$

Furthermore,

$$\chi^S = \frac{1}{\epsilon_0 P_r g'(P_r)}$$

$$G^E = G^P - \frac{e^2}{\chi^S \epsilon_0}$$

$$\zeta = -2 \frac{(\chi^S)^2 \epsilon_0}{P_r} \left[1 + \frac{g''(P_r) P_r}{2g'(P_r)} \right] \qquad (4.38)$$

$$\kappa = 2\gamma[(\chi^S)^2 \epsilon_0 + \zeta P_r] \epsilon_0$$

and

$$e = \frac{\kappa \chi^S P_r}{(\chi^S)^2 \epsilon_0 + \zeta P_r} \qquad (4.39)$$

Equation (4.39) is essentially the same as that derived by Mason (130).

ζ and κ, which depend on $g''(P)$, are very sensitive to the functional form of $g(P)$. If $g(P)$ is such that $|\zeta P_r| \ll (\chi^S)^2 \epsilon_0$, Eq. (4.39) simplifies to

$$e = \frac{\kappa P_r}{\epsilon^S - \epsilon_0} \qquad (4.40)$$

On the other hand, for the simple form $g(P) = a_0 + a_M P^M$ ($M > 0$), one obtains

$$e = -\frac{\kappa P_r}{M(\epsilon^S - \epsilon_0)} \qquad (4.41)$$

For PVDF, one may assume that below T_g, ζ of the film is approximately equal to that of the crystal which then would lead to the estimates $\zeta \sim 10^{-10}$ m/V (113) and $\zeta P_r/(\chi^S)^2 \epsilon_0 \simeq 0.2$–$1.0$, suggesting that Eq. (4.40) might be just valid for this material.

We now proceed to a molecular model. The spontaneous polarization of a polar crystal is given by

$$P_{sc} = \frac{N}{v} \mu \langle \cos \varphi \rangle \qquad (4.42)$$

where N is the number of dipoles in a crystal of volume v and φ is the angle of the dipole moment with respect to the polarization axis. Because of dipole–dipole interaction, the dipole moment μ is the sum of the intrinsic dipole moment μ_0 and the dipole momet induced by the internal field. For an isotropic medium (128),

$$\mu = \left(\frac{\epsilon + 2\epsilon_0}{3\epsilon_0} \right) \mu_0 \qquad (4.43)$$

where ϵ is the permittivity of the crystal (contribution of electronic polarization and atomic vibration). Therefore, piezo- and pyroelectricity of polar crystals is due to the strain and temperature dependence of v, ϵ, μ_0, and $\langle \cos \varphi \rangle$. Using Eqs. (4.42) and (4.43), one obtains for the maximum spontaneous polarization of the ideal form I PVDF crystal $P_c = 0.23$ C/m^2 using $\epsilon/\epsilon_0 = 3.4$ and $\langle \cos \varphi \rangle = 1$.

To calculate $e = (1/v)\, (\partial(vP_{sc})/\partial S)$ for hydrostatic pressure ($S = d \ln v$) and $p = (1/v)\, [\partial(vP_{sc})/\partial T]$ from Eqs. (4.42) and (4.43), Broadhurst et al. (131) assume μ_0 and $\langle \cos \varphi \rangle$ to be constant (rigid dipole model). The result is

$$e = -P_{sc} \frac{\epsilon - \epsilon_0}{3\epsilon_0} \qquad (4.44)$$

$$p = -\alpha P_{sc} \frac{\epsilon - \epsilon_0}{3\epsilon_0} = \alpha e \qquad (4.45)$$

where κ and λ have been obtained from the Clausius–Mossotti equation

$$\kappa = \frac{(\epsilon + 2\epsilon_0)(\epsilon - \epsilon_0)}{3\epsilon_0} \qquad \lambda = \alpha\kappa \qquad (4.46)$$

The same result, Eq. (4.44), is also obtained by inserting Eq. (4.46) into Eq. (4.40).

As an additional effect to the rigid dipole model, Broadhurst et al. (131) considered thermal vibration of the dipoles about their orientation. In this case, μ in Eq. (4.43) is replaced by $\mu J_0(\phi_0)$, where ϕ_0 is the rms angular amplitude of vibration and J_0 is the zero-order Bessel function of the first kind. Since ϕ_0 depends directly on temperature and indirectly on strain, which modifies the potential for vibration, orientational vibration provides an important contribution to the piezo- and pyroelectricity of the crystal. These are (131)

$$e(\text{thermal}) = -\tfrac{1}{2} \phi_0^2\, \gamma_G\, P_{sc} \qquad (4.47)$$

$$p(\text{thermal}) = -\tfrac{1}{2}\alpha\phi_0^2\left(\gamma_G + \frac{1}{2\alpha T}\right) P_{sc} \qquad (4.48)$$

where γ_G is the Grüneisen constant defined by $\gamma_G = -\partial \ln\Omega/\partial S$ (Ω is the characteristic frequency of the dipole vibration).

In the form I crystal of PVDF, however, in which the chains are very densely packed, it is unlikely that the rigid dipole model is a reasonable assumption. The chain conformation and consequently μ_0 and $\langle\cos\varphi\rangle$ in Eqs. (4.42) and (4.43) are expected to be changed by deformation of the unit cell.

Tashiro et al. (132) calculated the elastic modulus and the piezoelectric constant of the form I crystal on the basis of the atomic configuration in the unit cell. As a first approximation, they neglected the effect of the internal field and calculated displacements of each atom under the applied stress, taking into account the force on each atom from neighbor atoms located within four angstroms apart. The forces include those due to chemical bonds, van der Waals interactions, and Coulomb interactions. The results are

$$d\,(10^{-11}\,\text{C/N}) = \begin{pmatrix} 0 & 0 & 0 & 0 & -3.070 & 0 \\ 0 & 0 & 0 & -0.428 & 0 & 0 \\ -0.025 & -0.405 & -2.519 & 0 & 0 & 0 \end{pmatrix}$$
$$(4.48)$$

$$e\,(\text{C/m}^2) = \begin{pmatrix} 0 & 0 & 0 & 0 & -0.14 & 0 \\ 0 & 0 & 0 & -0.01 & 0 & 0 \\ -0.13 & -0.14 & -0.28 & 0 & 0 & 0 \end{pmatrix}$$
$$(4.49)$$

For piezoelectricity of polar crystals, however, the effect of the internal field arising from long-range interaction among dipoles may be important. A variation in the spontaneous polarization by stress changes the internal field, which in turn implies a stress-dependent internal field correction of the dipole moment in Eq. (4.43). These effects are not included in the calculation by Tashiro et al. (132).

It is difficult at present to compare in more detail the various theories with the experimental results because our knowledge of the physical properties of the form I PVDF crystals is insufficient. Takahashi et al. (133, 134) measured the variation of lattice constants under an applied electric field and obtained $d_{33} = -2 \times 10^{-11}$ C/N. They assumed, however, that the electric field in the crystal equals the average field in the film. Assuming the crystal to be spherical and the permittivity of the crystal to be much lower than that of the film [see Eq. (4.51)], d_{33} is corrected to be -1.3×10^{-11} C/N. The stress constant is estimated to be $e_{33} = -13 \times 10^{-2}$ C/m^2, using for the elastic modulus perpendicular to the chain axis the value 10×10^9 N/m^2 (132).

As is discussed in the next section, the intrinsic piezoelectricity and pyroelectricity of crystals described in this section play an important role in the piezoelectricity and pyroelectricity of the electret film. The thermodynamic theory correlates various macroscopic properties of crystal. If the function $g(P)$ were known from a microscopic standpoint, it would be possible to evaluate κ and ζ and thus e from the thermodynamic theory. The theories by Broadhurst et al. and Tashiro et al. are complementary to one another because the former takes into account the internal field but neglects the changes in μ_0 and $\langle\cos\varphi\rangle$, whereas the latter considers only the changes in μ_0 and $\langle\cos\varphi\rangle$.

4.6 PIEZOELECTRICITY AND PYROELECTRICITY OF ELECTRETS OF POLY(VINYLIDENE FLUORIDE) AND OTHER POLYMERS

Numerous publications have been devoted to experimental research on piezoelectricity (14–16, 28, 49, 50, 52, 67, 71, 86, 96, 97, 100, 107, 112, 113, 115, 118, 123, 126, 135–145), pyroelectricity (64, 98, 119–121, 124, 125, 146–152), and both (48, 51, 53, 68, 93, 114, 153–158) of PVDF.

It has been well documented that the piezo- and pyroelectric constants d and p, respectively, are closely related to the poling induced polarization P_s. d, p, and P_s are mutually proportional and increase with increasing E_p, T_p, and t_p (50, 51, 148). The dependence on E_p tends to level off at high E_p. The dependence on poling time indicates a fast initial process followed by a slower approach to saturation with a time constant, depending an T_p (125–127, 148). Figures 4.8–4.10

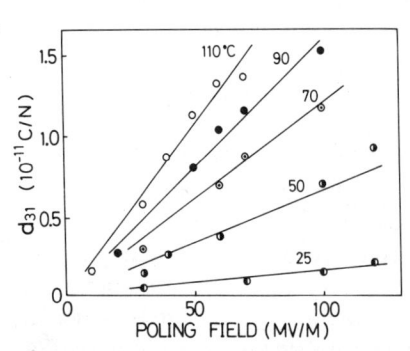

Figure 4.8 d_{31} of PVDF electret at room temperature vs. poling field at various poling temperatures (poling time = 30 min). (From Murayama [50].)

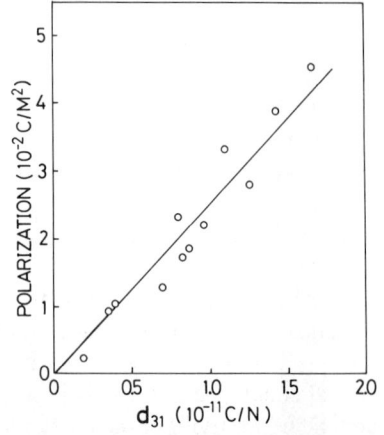

Figure 4.9 Relation between P_s and d_{31} for PVDF electrets. P_s was obtained from integrated depolarization current up to 170°C. (From Murayama and Hashizume [51].)

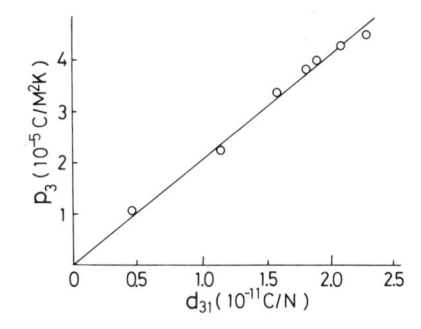

Figure 4.10 Relation between pyroelectric and piezoelectric constants of PVDF electrets. (From Ogasawara et al. [155].)

show some of these relationships. Furthermore, d increases with increasing content of the form I crystalline phase of the film. The induced piezo- and pyroelectric properties are reversible except for a contribution from an irreversible component of P_s which vanishes on the first heating cycle (148). The relationships between d, p, and P_s shown in Figs. 4.9 and 4.10 indicate that d and p derive from a common origin closely related to the properties of P_s.

The anisotropy of d of uniaxially oriented PVDF for tensile stress in the film plane is given by (135)

$$d = d_{31} \cos^2 \theta + d_{32} \sin^2 \theta \qquad (4.50)$$

where θ is the angle between the stress and the x-axis (Fig. 4.1b); d_{31} and d_{32} are positive and d_{31} is larger than d_{32} by one order of magnitude (157) and d_{33}, on the other hand, is negative and possesses an absolute value similar to d_{31} at room temperature (154, 157). The temperature dependence of relevant quantities measured for various experimental conditions is illustrated in Figs. 4.11 and 4.12. Before discussing each tensor component, we shall develop a general theory of piezo- and pyroelectricity in semicrystalline polymers containing polar crystals.

Several theories to model such systems have been developed (105, 131, 132, 159–161). The following approach is based on a model which assumes that N spheres of volume v with spontaneous polarization P_{sc} are dispersed in a matrix (Fig. 4.13). These spheres, referred to as "crystals," represent crystal lamellae with a surrounding oriented amorphous phase. The origin of P_{sc} may be due to dipolar orientation in the crystal, dipolar orientation in the oriented amorphous phase, and dipoles arising from trapped charges on the fold surfaces of the lamellae.

Since the dipole moment of the crystal is vP_{sc} and its external moment (including the dielectric polarization of the medium) is $[3\epsilon_a/(2\epsilon_a + \epsilon_c)]vP_{sc}$, P_s of the film is given (31) by

$$P_s = \frac{N}{Al}\left(\frac{3\epsilon_a}{2\epsilon_a + \epsilon_c} vP_{sc}\right) = \phi \frac{3\epsilon_a}{2\epsilon_a + \epsilon_c} P_{sc} \qquad (4.51)$$

when the volume fraction of the crystals, $\phi = Nv/Al$, is small.

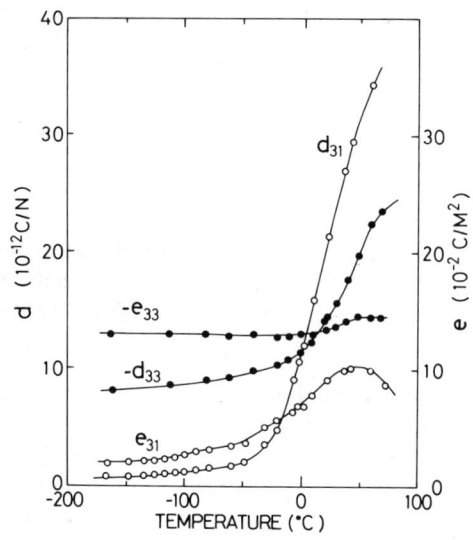

Figure 4.11 Temperature dependence of piezoelectric constants of PVDF electret (E_p = 78 MV/m, T_p = 120°C). e_{33} and d_{33} were measured at ca. 15 MHz, and e_{31} and d_{31} at ca. 45 kHz. (From Ohigashi [138].)

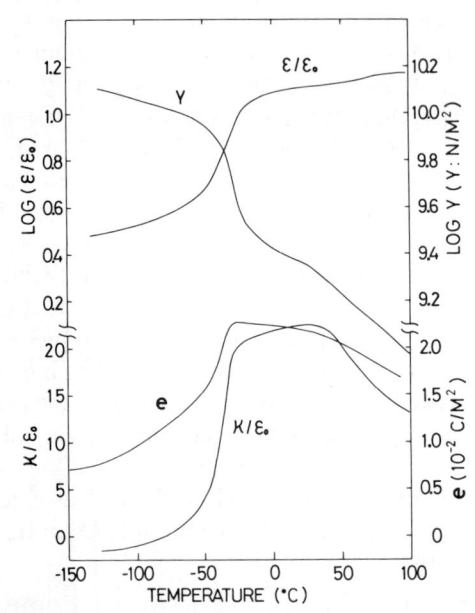

Figure 4.12 Temperature dependence of permittivity (ϵ_3), electrostriction constant (κ_{31}), Young's modulus (Y_1), and piezoelectric constant (e_{31}) for PVDF electret (E_p = 30 MV/m, T_p = 120°C) measured at 10 Hz. (From Furukawa et al. [144].)

PIEZO- AND PYROELECTRICITY OF ELECTRETS

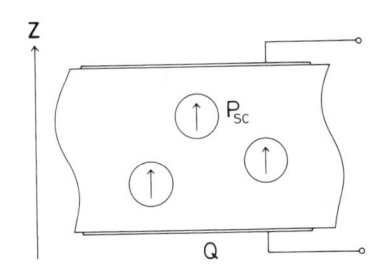

Figure 4.13 Model of semicrystalline polymer film. Crystallites are represented by spheres.

Since the observed piezoelectric constant of the film is the strain dependence of $Q = AP_s$, piezoelectricity can arise from four mechanisms:

1. Strain dependence of ϵ (electrostriction effect).
2. Strain dependence of l (dimensional effect).
3. Strain dependence of v.
4. Strain dependence of P_{sc} (intrinsic crystal piezoelectricity).

The pyroelectric effect is expressed by exchanging strain with temperature.

The observed e constant of the film defined by $e = (1/A)(\partial Q/\partial S)_{E=0}$ [see Eqs. (4.9) and (4.10)] is calculated from Eq. (4.51) as

$$e = P_s\left[\left(\frac{\epsilon_c}{2\epsilon_a + \epsilon_c}\right)\left(\frac{\kappa_a}{\epsilon_a} - \frac{\kappa_c}{\epsilon_c}g\right) - \frac{1}{l}\frac{\partial l}{\partial S}\right] \\ + \phi\left(\frac{3\epsilon_a}{2\epsilon_a + \epsilon_c}\right)\left(\frac{P_{sc}}{v}\frac{\partial v}{\partial S_c} + \frac{\partial P_{sc}}{\partial S_c}\right)g \quad (4.52)$$

where S_c is the strain in the crystal. For tensile strain (162),

$$g = \frac{\partial S_c}{\partial S} = \frac{5G_a}{3G_a + 2G_c} \quad (4.53)$$

The pyroelectric constant is similarly obtained as

$$p = P_s\left[\left(\frac{\epsilon_c}{2\epsilon_a + \epsilon_c}\right)\left(\frac{\lambda_a}{\epsilon_a} - \frac{\lambda_c}{\epsilon_c}\right) - \beta_3\right] + \phi\frac{3\epsilon_a}{2\epsilon_a + \epsilon_c}\left(\alpha_c P_{sc} + \frac{\partial P_{sc}}{\partial T}\right) \quad (4.54)$$

Comparing these equations with the general expressions derived for piezo- and pyroelectricity in polymers, Section 4.2, Eqs. (4.20–4.22), we note that $\partial P_{sc}/\partial S_c$ and $\partial P_{sc}/\partial T$ represent the intrinsic piezo- and pyroelectric effects of the crystal, respectively, mechanism (c). The other terms arise from the heterogeneities of the electrostriction and elastic constants which are intrinsic to the two phase system considered here, mechanism (b).

If space charges are trapped on the crystal surfaces and neutralize the polarization charge, P_{sc} in Eqs. (4.52) and (4.54) would have to be reduced, but $\partial P_{sc}/\partial S_c$ and $\partial P_{sc}/\partial T$ would remain unchanged because the motion of space charge occurs on a time scale long compared to the rate of change of S and T (131).

In the following, e_{31}, e_{33}, and p_3 of PVDF will be evaluated at 20°C from Eqs. (4.12) and (4.14), using the following assumptions:

1. $\phi = 0.5$.
2. The quantities relating to the amorphous phase, ϵ_a, κ_a, G_a, and λ_a, are approximated by those of the whole film, that is, we assume a single crystal embedded in a smeared-out continuum consisting of a mixture of crystal and amorphous phase.
3. P_{sc} is related to P_c, the spontaneous polarization of the ideal crystal ($P_c = 0.23$ C/m²) by $P_{sc} = \Phi P_c$. The factor Φ should depend on poling conditions. We assume $\Phi = 0.4$ for a standard poling condition, say, $E_p = 70$ MV/m and $T_p = 120$°C. This does not imply that the degree of dipole orientation in the crystal, $\Gamma = \langle \cos \varphi \rangle$, is 40% because Φ includes the effects of space charge and dipoles in the oriented amorphous phase.
4. The quantities relating to the crystal ϵ_c, κ_c, and λ_c at 20°C are assumed to be equal to those of the film at -100°C where the amorphous phase is frozen (Fig. 4.12). The value of G_c which is strongly anisotropic is taken from x-ray data (163) and calculated results (132).

Under the above assumptions, P_s in Eq. (4.51) is evaluated as 6.10×10^{-2} C/m² using $\epsilon_a/\epsilon_0 = 12.9$ and $\epsilon_c/\epsilon_0 = 3.4$ (144).

First, we consider the piezoelectricity for elongation along the x-axis. The elastic moduli for the crystal along the c-axis and for the amorphous phase are, respectively, $G_{c11} = 177 \times 10^9$ N/m² (132) and $G_{a11} = 2 \times 10^9$ N/m² (143). Therefore, from Eq. (4.53), $g = 2.8 \times 10^{-2}$, which is negligibly small. Equation (4.52) then reduces to

$$e_{31} = P_s \left[\left(\frac{\epsilon_c}{2\epsilon_a + \epsilon_c} \right) \frac{\kappa_a}{\epsilon_a} + m_{31} \right] \quad (4.55)$$

m_{31} for drawn PVDF has been reported as 0.7 (163). Sussner (164) estimated $m_{31} = 0.7$ from the data of Ohigashi (138). Richardson and Ward (165) observed $m_{31} = 0.8$ at room temperature for drawn low-density PE which has a morphology similar to that of PVDF. Assuming $m_{31} = 0.7$ and $\kappa_a/\epsilon_0 = 22.2$ (144), one obtains $e_{31} = 5.5 \times 10^{-2}$ C/m².

Since the contribution of m_{31} amounts to 78% of the total value of e_{31}, the temperature dependence of e_{31} (Figs. 4.11 and 4.12) should be largely determined by that of m_{31}. According to Richardson and Ward (165), m_{31} of drawn low-density PE decreases to 0.4 below T_g. Sussner (164) estimated $m_{31} = 0.2$ well below T_g. This decrease in m_{31} would account for the decrease in e_{31} observed with decreasing temperature (Figs. 4.11 and 4.12). The temperature dependence of d_{31} is enhanced by that of Y_1 ($d_{31} = e_{31}/Y_1$) which increases with decreasing temperature (Fig. 4.12). Similarly, the ratio e_{32}/e_{31} may be roughly equal to m_{32}/m_{31}, as pointed out by Sussner (164).

Next, we consider e_{33}. In this case, $(1/l)(\partial l/\partial S)$ in Eq. (4.52) is equal to unity, g is not small, and the electrostriction term is negligible compared with unity. Then Eq. (4.52) is approximated by

$$e_{33} = -P_s + \phi \frac{3\epsilon_a}{2\epsilon_a + \epsilon_c} e_{c33} g \qquad (4.56)$$

where e_{c33} is the contribution of the crystal; g amounts to 0.47 from $G_c = 10 \times 10^9$ N/m² and $G_a = 2.6 \times 10^9$ N/m² (157). If we ignore the contribution of space charge and amorphous dipoles and assume $\Gamma = \langle \cos \varphi \rangle = 0.2$, one obtains $e_{c33} = -0.13 \times 0.2$ C/m² (Section 4.5); and from Eq. (4.56), $e_{33} = -6.9 \times 10^{-2}$ C/m². If we take into account the contributions of amorphous dipoles and space charges, $|e_{33}|$ could be considerably larger. Since Poisson's ratio no longer appears in Eq. (4.56), one expects a much weaker temperature dependence for e_{33} than for e_{31}. This is in fact observed (Fig. 4.11).

Finally, p_3 will be evaluated from Eq. (4.54). According to Furukawa et al. (144), $\lambda/\epsilon = \partial \ln \epsilon / \partial T$ at 20°C is approximately equal to that at -100°C, and we may assume $\lambda_a/\epsilon_a \approx \lambda_c/\epsilon_c$. The intrinsic pyroelectric constant of the polar crystal $(\partial P_c/\partial T)$ has not been determined experimentally, but it may be approximately calculated using the rigid dipole approximation (131) for the crystal given in Eqs. (4.45) and (4.48). Using the numerical values for β_3, α_c, ϕ_0, and γ_G given by Broadhurst et al. (131), Eq. (4.54) yields $p_3 = -2.7 \times 10^{-5}$ C/m². The contribution of β_3 (dimensional effect) amounts to 47% of the total value of p_3.

The values calculated above agree quite well with observed values by Kepler and Anderson (157) for uniaxially oriented films, $e_{31} = 5.5 \times 10^{-2}$ C/m², $e_{32} = 0.6 \times 10^{-2}$ C/m², $e_{33} = -8.7 \times 10^{-2}$ C/m², and $p_3 = -2.7 \times 10^{-5}$ C/m² K. A better agreement between calculation and experiment for e_{33} is achieved by including an effect due to space charge and amorphous dipoles. For a complete agreement, e_{c33} should be taken as -8.4×10^{-2} C/m², in which case the contribution of e_{c33} to e_{33} would be 30%.

Table 4.1 summarizes the comparison between calculation and experiment. The agreement is satisfactory. Since the calculated values are proportional to P_{sc}, which was rather arbitrarily assumed as 0.4 times P_c, the physical significance of the agreement is that all the constants are reasonably well explained by a simple, unified model based on spheres with spontaneous polarization P_{sc} embedded in an amorphous matrix.

Table 4.1 Comparison Between Observed and Calculated Values of Piezoelectric and Pyroelectric Constants of PVDF Electret at 20°C

	Observed[a]	Calculated[b]	Fractional Contributions in Calculated Value (%)[c]			
			$\partial \epsilon / \partial X$	$\partial l / \partial X$	$\partial v / \partial X$	$\partial P_{sc} / \partial X$
e_{31} (10^{-2} C/m²)	5.5	5.5	22	78	0	0
e_{32} (10^{-2} C/m²)	0.6					
$-e_{33}$ (10^{-2} C/m²)	8.7	>6.9	0	70	<0	>30
$-p_3$ (10^{-5} C/m² K)	2.7	2.7	0	47	−38	91

[a] Reference 157.
[b] Parameters used: $P_{sc} = 9.2 \times 10^{-2}$ C/m², $\phi = 0.5$, $\epsilon_a/\epsilon_0 = 12.9$, $\kappa_{a31}/\epsilon_0 = 22.2$, $G_a = 2.6 \times 10^9$ N/m², $\epsilon_c/\epsilon_0 = 3.4$, $m_{31} = 0.7$, $\lambda_a/\epsilon_a = \lambda_c/\epsilon_c$, $G_{c11} = 177 \times 10^9$ N/m², $G_{c33} = 10 \times 10^9$ N/m², $e_{c33} = -2.6 \times 10^{-2}$ C/m², $\beta_3 = 2.1 \times 10^{-4}$ K⁻¹, $\alpha_c = 1.7 \times 10^{-4}$ K⁻¹, $\gamma_G = 5$, $\phi_0 = 16°$.
[c] X denotes strain and temperature for e and p, respectively.

An alternative approach based on the two-phase system of crystals embedded in an amorphous matrix has been adopted by Broadhurst, et al. (131). They assume the crystals to be thin lamellae with the large surface perpendicular to the film surface rather than the spherical approximation discussed here. Hence the factor $3\epsilon_a/(2\epsilon_a + \epsilon_c)$ in Eq. (4.51) reduces to unity. Furthermore, they approximated the e and p constants of the crystal by those of the polar single crystal given by Eqs. (4.44)–(4.48). The results are

$$e(\text{hydrostatic pressure}) = -\phi P_{sc}\left[\frac{\epsilon_c - \epsilon_0}{3\epsilon_0} + \frac{1}{2}\phi_0^2 \gamma_G + \frac{1}{l}\frac{\partial l}{\partial S_c}\right]\frac{K_c}{K} \quad (4.57)$$

$$p = -\phi \alpha_c P_{sc}\left[\frac{\epsilon_c - \epsilon_0}{3\epsilon_0} + \frac{1}{2}\phi_0^2\left\{\gamma_G + \frac{1}{2\alpha_c T}\right\} + \frac{1}{\alpha_c l}\frac{\partial l}{\partial T}\right] \quad (4.58)$$

where ϕ is the volume fraction of crystals; S_c is the volume strain of crystal ($S_c = d \ln v$), and K_c and K are, respectively, the compressibilities of crystal and film.

When the polarization charge on the crystal surface is completely neutralized by trapped charges, l in Eqs. (4.57) and (4.58) should be changed to l_c, the size of crystal along the z-axis. Numerical evaluation of each term in Eqs. (4.57) and (4.58) shows that thermal vibration of dipoles contributes 10 and 24% to the experimental value for e and p, respectively, when $\phi_0 = 16°$ and $\gamma_G = 5$ (131).

The calculations described in this section show that the dimensional effect is the main origin of both e and p. Kepler and Anderson (157) estimated the secondary pyroelectric effect from the phenomenologic relation $p(\text{secondary}) = \beta e$ and found that the secondary effect can account for about half the pyroelectricity observed.

Pyroelectricity of PVDF at low temperatures was measured by Glass and Negren above 6 K (166) and by Lang above 90 K (167).

In the above discussion, we assumed that the degree of crystallinity ϕ does not change with strain and temperature. Kepler and Anderson (149) proposed that reversible temperature-dependent crystallinity is responsible for part of pyroelectricity of PVDF.

Improvement of the activity of the PVDF electret was first made by increasing E_p and T_p. Corona and plasma charging are useful for increasing E_p without breakdown. Drawing and annealing processes before poling are also important. Tasaka et al. (139) showed that high-temperature annealing (up to 180°C) of the drawn film increases the crystallinity and degree of orientation of the crystals. Miyata et al. (168) employed annealing of drawn film at high temperature (240°C) and high pressure (2 kb). The resulting film exhibits a d_{31} after poling which is twice as large as that without high-pressure annealing. Matsushige and Takemura (86) also found the high-temperature, high-pressure annealing to be effective. Tasaka et al. noted that κ_{31} increases linearly with increasing degree of orientation in the amorphous phase (112). More recently, some workers have shown that the simultaneous drawing and poling can greatly increase the piezoelectricity of the films. According to Goho et al. (143), for example, a PVDF film with $d_{31} = 40 \times 10^{-12}$ C/N could be obtained by simultaneous drawing and corona charging.

Besides PVDF homopolymers, copolymers of vinylidene fluoride with vinylfluoride (50), trifluoroethylene (50, 156, 169–176), and tetrafluoroethylene (177– 179) have been extensively studied. These copolymers crystallize into form I crystal before drawing. It is to be stressed that the copolymer of VDF and TrFE (55:45 molar ratio) shows a sharp peak in the ϵ versus temperature curve at ~70°C, and the jump in the lattice constant at this Curie point was found to be proportional to P_s^2 (179). The coercive force vanishes at the Curie point. This co- polymer exhibits considerable piezoelectricity when appropriately poled (178). Blends of PVDF with PMMA and PVF have also been investigated (120, 180).

In addition to these PVDF-related polymers, various polymers have been studied in their piezo- and pyroelectricity after poling, including PVF (15, 16, 124, 153, 181, 182), odd nylon (15, 183, 184), PTFE (185–187), polyacrylonitrile and its copolymer (188–190), poly(ethylene terephthalate) (191), polypropylene (192), copolymer of vinylidene cyanide and vinyl acetate (193), polysulfone (194), poly(m-phenylene isophthalamide) (195), polycarbonate (15, 16), and thiourea formaldehyde polymers (196).

PVC, which is a noncrystalline polymer, shows relatively high piezo- and pyroelectricity when poled at a high temperature (15, 16, 115, 128, 181, 197, 198). Mopsik and Broadhurst (128) developed a dipole theory of piezo- and pyroelectricity of the glassy polymer electret and obtained a good agreement with the measured value for PVC. In Table 4.2, typical values of piezo- and pyroelectric constants at room temperature are listed for various polymers. Values depend greatly on the preparation conditions of the sample.

Table 4.2 Piezoelectric and Pyroelectric Constants of Polymers

	Piezoelectric Constant (10^{-13} C/N)				Pyroelectric Constant (10^{-5} C/m² K)
	d_{31}	d_{32}	d_{33}	$d_{14}{}^a$	p_3
Electret					
PVDF	200	20	−300		−4
PVF	10				−1.8
PVC	10			−7	−0.4
Nylon 11	3				−0.5
Polycarbonate	1				
Poly(*m*-phenylene isophthalamide)	4				
VDF–TFE copolymer	60				
VDF–VF copolymer	60				
VDF–TrFE copolymer	180	180			
PVDF–PVF blend					−1.5
Vinylidene cyanide–vinyl acetate copolymer	50				−0.9 (50°C)
Optically Active Polymer					
PBLG				−40	
PMLG				−30	
Cellulose triacetate				−1	
Cyanoethyl cellulose				−30	
Poly(D-propylene oxide)				−0.2	

*a*Coordinates in Fig. 4.1*a*.

4.7 PIEZOELECTRICITY AND PYROELECTRICITY OF POLYMER–CERAMIC COMPOSITES

A composite in which ferroelectric ceramic particles are dispersed in a polymer matrix exhibits piezo- and pyroelectricity when poled (199–202). Piezo- and pyroelectricity of the composite are expressed in general by Eqs. (4.52) and (4.54), respectively, in which P_{sc} stands for the spontaneous polarization of the ferroelectric ceramic (31). Two mechanisms are possible: (1) piezo- and pyroelectricity of the ferroelectric, and (2) coupling of the heterogeneity of the composite with polarization charge of ferroelectrics; that is, mechanisms (c) and (b), respectively, described in Section 4.2.

Furukawa et al. (203) studied the composite of PZT and epoxy resin (Fig. 4.14) and found that mechanism (1) is dominant in this case, because the epoxy resin at room temperature is glassy and the elastic modulus of the resin is not much lower than that of ferroelectrics. Equation (4.52) then yields

$$e = \phi \frac{3\epsilon_a}{2\epsilon_a + \epsilon_c} \frac{5G_a}{3G_a + 2G_c} e_c \qquad (4.59)$$

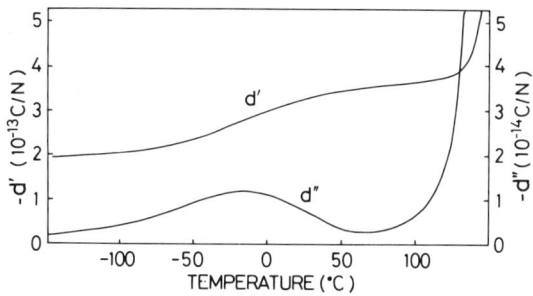

Figure 4.14 Real and imaginary parts of complex d constant for a composite of epoxy resin and PZT ceramics ($\phi = 0.131$, $E_p = 10$ MV/m, $T_p = 120°C$) measured for tensile stress in the film plane at 10 Hz. (From Furukawa et al. [203].)

When the matrix is rubbery, on the other hand, mechanism (2) dominates over mechanism (1), and from Eq. (4.52),

$$e = \phi \frac{3\epsilon_a}{2\epsilon_a + \epsilon_c} P_{sc} \left(\frac{\epsilon_c}{2\epsilon_a + \epsilon_c} \frac{\kappa_a}{\epsilon_a} - \frac{1}{l} \frac{\partial l}{\partial S} \right) \quad (4.60)$$

Kitayama and his co-workers (200–202) investigated piezo- and pyroelectricity of composite systems, using PVDF (form II crystal) and a blend of PVDF with fluorinated rubber (copolymer of hexafluoropropylene and TFE) as polymer matrix. In the latter case, the elastic modulus can be controlled by changing the blend ratio without changing the d constant significantly (202) (Fig. 4.15).

Model experiments with theoretical analysis were made by Furukawa et al. (203, 204) and Date (162).

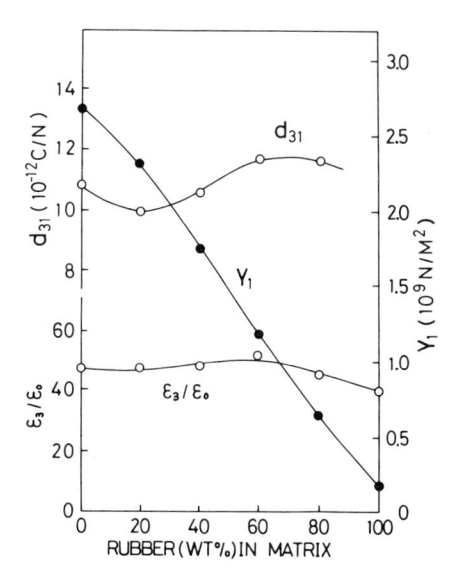

Figure 4.15 Permittivity (ϵ_3), piezoelectric constant (d_{31}), and Young's modulus (Y_1) of PZT ceramics (80 wt %)–polymer (blend of PVDF and fluorinated rubber) composition. (From Kitayama [202].)

4.8 PIEZOELECTRIC RELAXATION

Due to molecular motion and ionic conduction in the polymer, piezoelectricity exhibits relaxational behavior (13, 17, 34). The piezoelectric polarization possesses both in-phase and quadrature components with respect to sinusoidal excitation of strain and stress. The d and e constants are expressed as complex quantities:

$$e^*(i\omega) = e'(\omega) - ie''(\omega) \qquad (4.61)$$

$$d^*(i\omega) = d'(\omega) - id''(\omega) \qquad (4.62)$$

For a single relaxation system, the complex constants are

$$e^*(i\omega) = e(\infty) + \frac{\delta e}{1 + ix} \qquad x = \omega\tau \qquad (4.63)$$

$$d^*(i\omega) = d(\infty) + \frac{\delta d}{1 + ix} \qquad (4.64)$$

In contrast to dielectric and mechanical loss, e'' and d'' do not necessarily mean the energy loss (205) because the piezoelectricity is a combined effect of dielectric and elastic degrees of freedom. The relaxation behavior of pyroelectricity would be similar to that of piezoelectricity if it were not for the fact that the frequency dependence of p is usually dominated, on the high-frequency side by the interaction of the heterogeneous polarization and the thermal diffusivity of the material, and on the low-frequency side by the electrical and thermal time constants of the system (124, 206).

The Kramers–Kronig's relations hold for e^* and d^* (17):

$$e'(\omega) - e(\infty) = \frac{2}{\pi} P \int_0^\infty \frac{\zeta e''(\zeta)}{\zeta^2 - \omega^2} d\zeta \qquad (4.65)$$

$$e''(\omega) = -\frac{2\omega}{\pi} P \int_0^\infty \frac{e'(\zeta) - e(\infty)}{\zeta^2 - \omega^2} d\zeta \qquad (4.66)$$

The relaxation strength defined by $\delta e = e(0) - e(\infty)$ is

$$\delta e = \frac{2}{\pi} \int_{-\infty}^\infty e''(\zeta) \, d\ln \zeta \qquad (4.67)$$

Therefore, δe and e'' should have the same sign if e'' maintains its sign through the whole range of frequency. The same argument holds for d^*, too.

For a single-phase system with relaxational piezoelectricity, we can prove from irreversible thermodynamics (17) that

PIEZOELECTRIC RELAXATION

$$(\delta e)^2 \leq \delta\epsilon^S \, \delta G^E \tag{4.68}$$

where $\delta\epsilon^S = \epsilon^S(0) - \epsilon^S(\infty)$ and $\delta G^E = G^E(\infty) - G^E(0)$. The equality holds for a single relaxation system and the inequality for a system with multiple relaxations.

For a two-phase system in which a piezoelectric, nonrelaxational phase and a nonpiezoelectric, relaxational phase are mixed, the following inequality may be shown to hold (207):

$$\frac{d''}{d'} > 0 \tag{4.69}$$

For small dielectric loss tangent, one can further show that

$$\frac{e''}{e'} < 0 \tag{4.70}$$

The data shown in Fig. 4.14 satisfy the inequality Eq. (4.69).

A model theory for a composite in which piezoelectric, nonrelaxational spheres are dispersed in a nonpiezoelectric, relaxational matrix was developed by Fukada and his co-workers (45, 208–211) and was applied to piezoelectric relaxations of the systems epoxy resin–PZT (203) and PVDF (144).

As shown in the previous sections, the piezoelectric constant of a multiphase system, including polypeptide, PVDF, and artificial composites, may be described in terms of ϵ, G, and κ of the individual phases. Relaxational behavior of d and e is therefore closely related to these parameters. The complex permittivity $\epsilon^* = \epsilon' - i\epsilon''$ and complex elastic modulus $G^* = G' + iG''$ are, for a single relaxation system,

$$\epsilon^*(i\omega) = \epsilon(\infty) + \frac{\delta\epsilon}{1 + ix} \tag{4.71}$$

$$G^*(i\omega) = G(0) + \delta G \frac{ix}{1 + ix} \tag{4.72}$$

The complex electrostriction constant $\kappa^* = \kappa' - i\kappa''$ is obtained from Eq. (4.71) (207):

$$\kappa' = \frac{\partial\epsilon'}{\partial S} = \frac{\partial\epsilon(\infty)}{\partial S} + \frac{\partial\delta\epsilon}{\partial S}\frac{1}{1+x^2} - 2\delta\epsilon\, a_s\left(\frac{x}{1+x^2}\right)^2$$

$$\kappa'' = \frac{\partial\epsilon''}{\partial S} = \frac{\partial\delta\epsilon}{\partial S}\frac{x}{1+x^2} - \delta\epsilon a_s \frac{x(x^2-1)}{(1+x^2)^2} \tag{4.73}$$

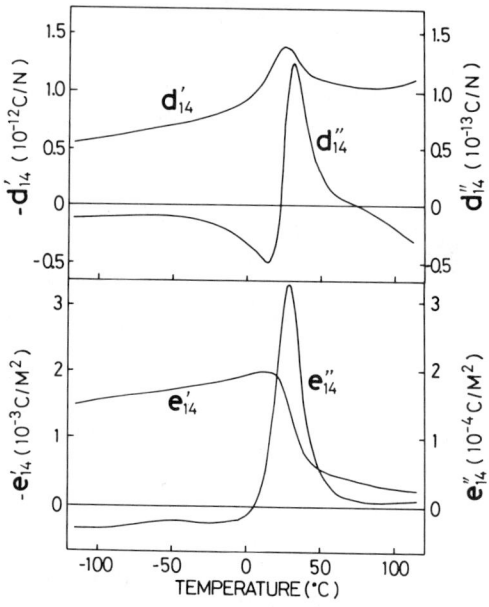

Figure 4.16 Temperature dependence of complex piezoelectric constants of uniaxially oriented PBLG (degree of orientation = 0.42) measured at 10Hz. (From Furukawa and Fukada [45].)

where $a_s = \partial \ln \tau / \partial S$. Usually the last terms on the right-hand side of Eqs. (4.73) are dominant for κ' and κ'', respectively, and a_s is negative for tensile strain. In this case, Eqs. (4.73) predict that κ' becomes a maximum at $x = 1$ similar to ϵ'', whereas κ'' reaches a minimum at $x = \sqrt{2} - 1$ and subsequently a maximum at $x = \sqrt{2} + 1$. This was experimentally verified for undrawn PVDF (207). The difference between the maximum and the minimum in κ'' gives $-\delta \epsilon a_s / 2$.

Figure 4.16 illustrates the temperature dependence of d' and d'' for PBLG (45), which arises from temperature dependence of both δd and τ. The relaxation around room temperature is ascribed to molecular motion of side chains surrounding piezoelectric backbones. Two relaxations are evident in Fig. 4.16; one manifested by a large peak of d'' and e'' at 30°C and the other by a small dip in d'' at 15°C. Signs of d''/d' and e''/e' do not follow the inequalities (4.69) and (4.70), respectively, indicating that the relaxations cannot be accounted for by a two-phase model.

Figure 4.17 illustrates the Cole–Cole plot of e^* for PMLG, indicating again two relaxations with positive and negative relaxation strengths similar to PBLG in Fig. 4.16. These double relaxations of polypeptides were interpreted by Furukawa and Fukada (45) in terms of a three-phase model since polypeptides possesses a grain structure. The first phase (the backbone) is embedded in the second phase (the side shain). Within each grain separated by the third phase (amorphous material), the backbones are aligned parallel.

MEASURING TECHNIQUES

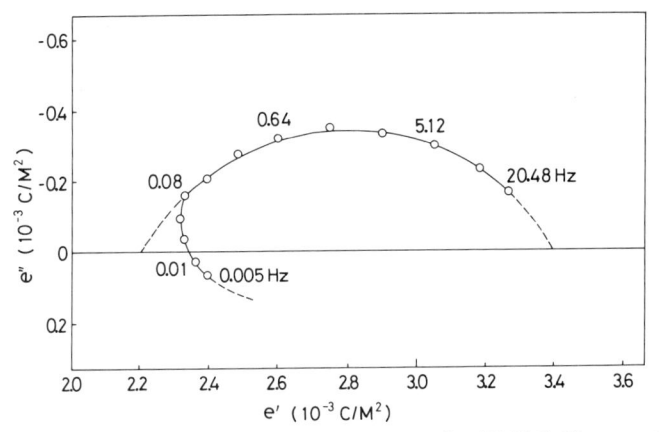

Figure 4.17 Cole–Cole plot of complex e constant for PMLG (draw ratio = 2) at 12°C measured for tensile stress along 45° direction against orientation axis. (From Hayakawa et al. [218].)

The effect of ionic conduction on the piezoelectric relaxation was discussed by Furukawa and Fukada (212). The difference between dielectric and piezoelectric relaxation was treated by a two-state model (213).

4.9 MEASURING TECHNIQUES

A variety of methods have been developed for measuring the piezoelectric constant of polymer films (214, 215). The most typical piezoelectric measurement is to apply sinusoidal stress or strain to the film and to observe the short-circuit current or open-circuit voltage at the same frequency across the electrodes that are deposited on to the film surfaces (16, 29, 40, 45, 184, 216, 217). The upper bound of frequency is limited by the mechanical resonance of the mechanical driver. The lower bound can be extended down to 10^{-3} Hz. The bias tension of the film should be kept constant during a series of measurement.

The complex piezoelectric constant is obtained by measuring both amplitude and phase of the response. Simultaneous measurement of frequency spectra of the complex constant was achieved by an apparatus constructed by Hayakawa and Wada (218, 219) for frequencies ranging from 10^{-3} to 30 Hz. An automatic recording system developed by Furukawa and Fukada (45) is capable of measuring d^* and e^* from 0.1 to 30 Hz at temperatures from -120 to 120°C by use of automatic control of the bias tension.

The inverse piezoelectricity is observed by applying a sinusoidal electric field to the sample and measuring the induced strain or stress of the same frequency (15, 28, 93, 100, 126).

The piezoelectric voltage is measured by application of a static stress (pressure) to the sample (175). Hydrostatic pressure can be used as the excitation by putting the sample into a gas cell (220).

The inverse piezoelectricity has been observed by application of a static electric field. The induced thickness change was measured with either a capacitance method (154) or a laser interferometer (157). X-ray analysis under a static electric field was used to obtain the d constant of the crystal (133, 134).

A piezoelectric film, cut in an appropriate form, may be excited into resonance vibration by apply a high frequency electric field. By measuring the frequency dependence of motional admittance of the film in the vicinity of resonance, the electromechanical coupling constant can be evaluated (138). Measurement of the coupling constant in a wide frequency range using divided electrodes was reported by Burgt et al. (114, 221).

The conventional method for measuring the pyroelectric constant is to observe the reversible pyroelectric current during heating or cooling the film at a constant rate (51, 147, 148, 151, 154).

The dynamic response of pyroelectricity is obtained by pulse excitation with a laser (146, 149, 152) or periodic excitation with chopped incident light (93, 185). Sussner et al. (222) reported a new method for pyroelectric measurement using dielectric heating which produces a uniform temperature distribution in a short time.

The inverse effect of pyroelectricity, the electrocaloric effect, is measured from the adiabatic temperature rise upon application of an electric field (21).

Higher-order constants, including κ and ζ, are important in discussing piezoelectricity. Since the change in strain due to the electrostriction is given by

$$\Delta S = \frac{\kappa E^2}{2G^D} \tag{4.74}$$

whereas ΔS due to linear piezoelectricity is proportional to E, κ may be calculated from a field reversal experiment. For a sinusoidal electric field at a frequency ω, the 2ω component of ΔS gives κ (15, 28, 93, 126). Similarly, the 2ω component of ΔP by application of E at ω gives ζ. In these measurements, care should be taken to keep the input signal free from second harmonics. The ζ value at optical frequency is estimated from the second harmonic generation by the intense incident light beam (98, 147).

These higher-order constants are also measured by applying two fields simultaneously to the sample—one static and the other sinusoidal, or both sinusoidal. The electrostriction constant is measured from the apparent piezoelectric constant under a dc bias field. The constant can also be measured by simultaneously applying an electric field at a frequency ω_1 and a strain at ω_2 and detecting the electrical response at $\omega_1 - \omega_2$ (113). The second-order susceptibility ζ is obtained analogously by simultaneously applying two sinusoidal electric fields at ω_1 and ω_2 and detecting the response at $\omega_1 - \omega_2$ (113). These techniques can avoid the possible errors due to distortion of input signals. The frequency dependence of the complex electrostriction constant in a wide frequency range of electric field was measured by Hayakawa et al. (207).

4.10 APPLICATIONS

Since the discovery of strong piezo- and pyroelectricity of PVDF electrets about 10 years ago, these materials have received great attention as new transducer materials. The piezo- and pyroelectric polymers offer new application opportunities not realizable with conventional crystalline substances.

Table 4.3 compares piezo- and pyroelectric constants of various materials. Values for polymers depend greatly on preparation conditions and are expected to increase still in the future. For illustration, Fig. 4.18 shows the increase in d_{31} of PVDF reported in the scientific literature during the past decade. This increase has been brought about by improvements in preparation techniques.

With respect to conventional piezo- and pyroelectric crystals, the polymer films offer the following advantages:

1 They are flexible and tough.
2 They can be made very thin (less than 10 μm).
3 They may be obtained in large areas.
4 They can be shaped into cylindrical, conical, and spherical forms.
5 They have a low mechanical impedance and thus exhibit good acoustic coupling to water, human body, and so forth.
6 They can be made locally or heterogeneously active by appropriate poling.
7 The mechanical properties of composite films of polymer and ceramics can be controlled over a wide range.

The relatively high mechanical and dielectric loss of polymers, however, can be a disadvantage in some cases.

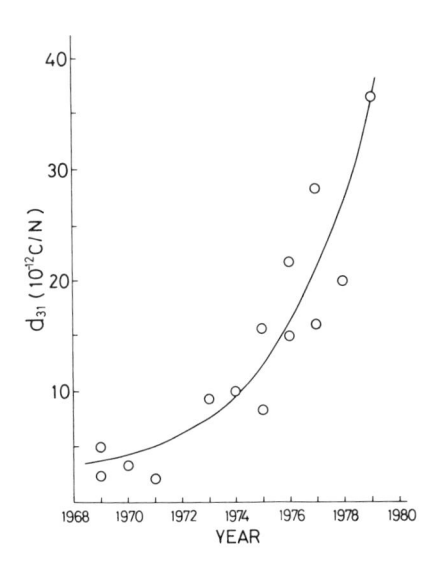

Figure 4.18 Values of d_{31} of PVDF electrets reported in scientific literature over the past decade.

Table 4.3 Comparison of Physical Properties of Piezoelectric and Pyroelectric Materials

Material	Density (10^3 kg/m^3)	Elastic Modulus (10^{10} N/m^2)	Dielectric Constant ϵ/ϵ_0	Piezo-electric Constant[a] (10^{-12} C/N)	Electro-mechanical Coupling Constant (%)	Acoustic Impedance (10^6 kg/m^2 s)	Pyro-electric Constant[a] (10^{-5} C/m^2 K)
Quartz (0°X)	2.65	7.7	4.5	2.3	10	14.3	
Rochekle salt (45°X)	1.77	1.8	350	275	73	5.7	
Triglycine sulfate	1.69	3	45	25			30
BaTiO$_3$ ceramic	5.7	11	1700	78	21	25	20
PZT ceramic	7.5	8.3	1200	110	30	25	27
PVDF	1.78	0.3	12	20	11	2.3	4
Polymer–ceramic composite	5.5	0.3	118	30	5	4	10
PBLG (45°Z)	1.3	0.3	3.8	2	2	2	

[a]Absolute value.

APPLICATIONS

Numerous applications of piezo- and pyroelectric polymers have been explored, and some of them have been commercialized (96, 223–232). (Compare also Chapter 3, Section 3.6.)

Acoustic devices such as high frequency speakers and headphones were the first commercial products utilizing piezoelectric PVDF film (233–236). Since then, various acoustic devices, including microphones and cartridges have been designed with piezoelectric polymers (237–241).

The low acoustic impedance of the polymer film makes it suitable for applications in hydrophones (242–247) and medical microphones for fetal phonocardiographs and heart rate monitors (248). The feasibility of using piezoelectric PVDF to vibrate the surface of hulls or marine installations has been explored, and the effect of vibration on fouling has been studied (249).

Piezoelectric polymer films can be used as pressure sensors in various devices: printing pressure monitors, blood pressure monitors, pressure sensors in the material testing devices, strain gauges to detect strains accompanying stress waves in the soil and rock (250), accelerometers, and so forth.

Such films are also widely used as ultrasonic transducers (251–266), especially in microwave ultrasonics (253), ultrasonic light modulation (252), medical ultrasonic equipment (254, 262, 264), nondestructive material testing (263), and acoustic emission devices. A concave ultrasonic transducer is helpful in ultrasonic microscopes (258, 260) in which focusing ultrasonic beams are necessary. Excitation of PVDF film into thickness shear mode was tried by Auld and Gagnepain (267).

In telephone communication systems, piezoelectric polymer films seem to have a wide potential applicability. Kitayama and his co-workers have used the piezoelectric composite film as pressure sensors in connection with electronic keyboards and input devices for handwritten figures and letters (268, 269, 202). Applications of PVDF to switchboards and vibration detectors for blind people (270) have also been reported. Induction of bone by implanting PVDF film has also been tried (271).

Bimorphs or multilayered piezoelectric polymer films show a very large amount of bending of up to $\pm 90°$ even when a relatively low voltage is applied (272–275). Possible applications of these films for visual display, large area display (276), electric fans (277–280), and light shutters (281) have been proposed. The deformation properties of an adaptive optical element made from a PVDF bimorph plate was analyzed by Kokorowski (282).

Pyroelectric polymer films have been utilized as an optical detector for infrared radiation (98, 146, 147, 283–288) and millimeter waves (289). Their small heat capacity permits rapid response (146, 147). In addition, large-device optical receiving areas become possible without substrates and windows.

Second harmonic generation due to nonlinear optical properties has been observed for PVDF (98, 147).

Calibration and analysis of response of pyroelectric detectors have been reported (284, 285, 287). Applications of pyroelectric films to absolute reflectivity measurements (290), copying machines (291), vidicon targets (292–296), fire alarm, and so on, have been proposed.

PVDF transducers can detect both acoustic and infrared radiation, which has possible applications for vehicle ranging, target classificiation (297, 298) and intrusion detectors.

ACKNOWLEDGMENTS

The author wishes to express his sincere thanks to Prof. A. Odajima at Hokkaido University, to Dr. D. K. Das-Gupta at University College of North Wales, and to Dr. R. Hayakawa at the University of Tokyo for carefully reading the manuscript and providing many valuable comments. Thanks are also due to Dr. M. G. Broadhurst, Dr. D. K. Das-Gupta, Dr. E. Fukada, Prof. J. B. Lando, Prof. S. Miyata, Prof. H. Tadokoro, Prof. T. Takemura, and Dr. M. Toda for kindly informing me of their recent results, including unpublished data. Permissions of the American Institute of Physics, the Publication Office of Japanese Journal of Applied Physics, and John Wiley and Sons, Inc., for reproducing figures in their publications are gratefully acknowledged.

REFERENCES

1 K. R. Brain, *Proc. Phys. Soc.*, **36**, 81 (1924).
2 I. S. Rez, *Sov. Phys.-Crystallogr.*, **6**, 521 (1962).
3 E. L. Kern and S. M. Skinner, *J. Appl. Polym. Sci.*, **6**, 404 (1962).
4 M. M. Kocharyan and Kh.B. Pachadzhyan, *Dokl. Akad. Nauk Arm. SSR*, **36**, 277 (1963).
5 G. E. Hauver, *J. Appl. Phys.*, **36**, 2113 (1965).
6 P. Harris, *J. Appl. Phys.*, **36**, 739 (1965).
7 F. E. Allison, *J. Appl. Phys.*, **36**, 2111 (1965).
8 S. N. Levine, *J. Appl. Polym. Sci.*, **9**, 3351 (1965).
9 N. M. Kocharyan, Kh.B. Pachadzhyan, and Sh.A. Mkhitaryan, *Izv. Akad. Nauk Arm. SSR*, **1**, 217 (1966).
10 E. Fukada, M. Date, and N. Hirai, *Nature*, **211**, 1079 (1966).
11 E. Fukada, M. Tamura, and I. Yamamoto, Reports of the 6th International Congress on Acoustics, 1968, Tokyo, D-69.
12 E. Fukada, *Ultrasonics*, 229 (Oct. 1968).
13 E. Fukada, *Adv. Biophys.*, **6**, 121 (1974).
14 H. Kawai, *Jpn. J. Appl. Phys.*, **8**, 975 (1969).
15 H. Kawai, *Oyobuturi*, **39**, 413 (1970).
16 H. Kawai, *Oyobuturi*, **38**, 1133 (1969).
17 R. Hayakawa and Y. Wada, *Adv. Polym. Sci.*, **11**, 1 (1973).
18 Y. Wada and R. Hayakawa, *Jpn. J. Appl. Phys.*, **15**, 2041 (1976).

REFERENCES

19 M. G. Broadhurst and G. T. Davis, *Topics in Applied Physics*, G. M. Sessler, Ed., Springer, Berlin, 1979, Vol. 33, p. 285.
20 W. G. Cady, *Piezoelectricity*, Dover, New York, 1964.
21 T. Brossat, G. Bichon, C. Lemonon, M. Royer, and F. Micheron, Thomson-CSF LCR 78-93/R/G. 5.
22 Sh.M. Kogan, *Sov. Phys. Solid State*, **5**, 2069 (1964).
23 T. Furukawa, Y. Uematsu, K. Asakawa, and Y. Wada, *J. Appl. Polym. Sci.*, **12**, 2675 (1968).
24 H. Kawai, *Oyobuturi*, **39**, 869 (1970).
25 T. Ibe, *Jpn. J. Appl. Phys.*, **13**, 197 (1974).
26 L. Breger, T. Furukawa, and E. Fukada, *Jpn. J. Appl. Phys.*, **15**, 2239 (1976).
27 R. W. Greaves and D. R. Lamb, *J. Mat. Sci.*, **6**, 74 (1971).
28 M. Oshiki and E. Fukada, *J. Mat. Sci.*, **10**, 1 (1975).
29 R. L. Zimmerman, C. Suchicital, and E. Fukada, *J. Appl. Polym. Sci.*, **19**, 1373 (1975).
30 R. Hayakawa and Y. Wada, *Rep. Prog. Polym. Phys. Japan*, **14**, 467 (1971).
31 R. Hayakawa and Y. Wada, *Rep. Prog. Polym. Phys. Japan*, **19**, 321 (1976).
32 R. W. Greaves, E. P. Fowler, and A. Goodings, *J. Mat. Sci.*, **9**, 1602 (1974).
33 R. E. Salomon, H. Lee, C. S. Bak, and M. M. Labes, *J. Appl. Phys.*, **47**, 4206 (1976).
34 E. Fukada, *Prog. Polym. Sci. Japan*, Vol. 2, Kodansha, Tokyo, 1971, p. 329.
35 T. Furukawa and E. Fukada, *Nature*, **221**, 1235 (1969).
36 E. Fukada, M. Date, and T. Emura, *J. Soc. Mat. Sci. Japan*, **17**, 335 (1968).
37 S. Sasaki and E. Fukada, *Rep. Prog. Polym. Phys. Japan*, **18**, 361 (1975).
38 S. Sasaki and E. Fukada, *J. Polym. Sci. Polym. Phys. Ed.*, **14**, 565 (1976).
39 T. Furukawa and E. Fukada, *Rep. Prog. Polym. Phys. Japan*, **19**, 309 (1976).
40 M. Date, S. Takashita, and E. Fukada, *J. Polym. Sci. A-2*, **8**, 61 (1970).
41 T. Konaga and E. Fukada, *J. Polym. Sci. A-2*, **9**, 2023 (1971).
42 E. Fukada and S. Takashita, *Jpn. J. Appl. Phys.*, **10**, 722 (1971).
43 E. Fukada, T. Furukawa, E. Baer, A. Hiltner, and J. M. Anderson, *J. Macromol. Sci. Phys.*, **B8**, 475 (1973).
44 T. Furukawa, K. Ogiwara and E. Fukada, *J. Polym. Sci. Polym. Phys. Ed.*, **18**, 1677 (1980).
45 T. Furukawa and E. Fukada, *J. Polym. Sci. Polym. Phys. Ed.*, **14**, 1979 (1976).
46 K. Namiki, R. Hayakawa, and Y. Wada, *J. Polym. Sci. Polym. Phys. Ed.*, **18**, 993 (1980).
47 G. M. Sessler, Ed., *Electrets, Topics in Applied Physics*, Springer, Berlin, Vol. 33, 1979.
48 J. M. Kenney and S. C. Roth, *J. Res. Natl. Bur. Stand.*, **84**, 447 (1979).
49 N. Murayama, *J. Polym. Sci. Polym. Phys. Ed.*, **13**, 929 (1975).
50 N. Murayama, T. Oikawa, T. Katto, and K. Nakamura, *J. Polym. Sci. Polym. Phys. Ed.*, **13**, 1033 (1975).

51 N. Murayama and H. Hashizume, *J. Polym. Sci. Polym. Phys. Ed.*, **14**, 989 (1976).
52 T. Furukawa, M. Date, and E. Fukada, *J. Appl. Phys.*, **51**, 1135 (1980).
53 J. E. McKinney, G. T. Davis and M. G. Broadhurst, *J. Appl. Phys.*, **51**, 1676 (1980).
54 W. M. Prest, Jr. and D. J. Luca, *J. Apply. Phys.*, **46**, 4136 (1975).
55 R. Hasegawa, M. Kobayashi, and H. Tadokoro, *Polym. J.*, **3**, 591 (1972).
56 R. Hasegawa, Y. Takahashi, Y. Chatani, and H. Tadokoro, *Polym. J.* **3**, 600 (1972).
57 D. K. Das-Gupta and D. B. Shier, *J. Appl. Phys.*, **49**, 5685 (1978).
58 K. Matsushige, K. Nagata, S. Imada, and T. Takemura, *Polymer*, **21**, 1391 (1980).
59 A. Odajima and A. Yoshida, *Oyobuturi*, **48**, 249 (1979).
60 B. P. Kosmynin, Y. L. Gal'perin, and D. Y. Tsvankin, *Polym. Sci. USSR*, **A12**, 1418 (1970).
61 B. P. Kosmynin, Y. L. Gal'perin, and D. Y. Tsvankin, *Polym. Sci. USSR*, **A14**, 1530 (1972).
62 J. B. Lando, H. L. Olf, and A. Peterlin, *J. Polym. Sci.* **A-1**, 941 (1966).
63 J. P. Luongo, *J. Polym. Sci. A-2*, **10**, 1119 (1972).
64 P. D. Southgate, *Appl. Phys. Lett.*, **28**, 250 (1976).
65 D. K. Das-Gupta and K. Doughty, *Appl. Phys. Lett.*, **31**, 585 (1977).
66 D. Naegle, D. Y. Yoon, and M. G. Broadhurst, *Macromolecules*, **11**, 1297 (1978).
67 D. K. Das-Gupta and K. Doughty, *J. Appl. Phys.*, **49**, 4601 (1978).
68 D. K. Das-Gupta and K. Doughty, *J. Phys. D*, **11**, 2415 (1978).
69 G. T. Davis, J. E. McKinney, M. G. Broadhurst, and S. C. Roth, *J. Appl. Phys.*, **49**, 4998 (1978).
70 G. R. Davis and H. Singh, *Polymer*, **20**, 772 (1979).
71 D. K. Das-Gupta, K. Doughty, and D. B. Shier, *J. Electrostat.* **7**, 267 (1979).
72 H. Dvey-Aharon, P. L. Taylor, and A. J. Hopfinger, *J. Appl. Phys.*, **51**, 5184 (1980).
73 S. Osaki and Y. Ishida, *J. Polym. Sci. Polym. Phys. Ed.*, **13**, 1071 (1975).
74 W. M. Prest, Jr., and D. J. Luca, *J. Appl. Phys.*, **49**, 5042 (1978).
75 N. C. Banik, F. F. Boyle, T. J. Sluckin, P. L. Taylor, S. K. Triphathy, and A. J. Hopfinger, *Phys. Rev. Lett.*, **43**, 456 (1979).
76 S. Weinhold, M. H. Litt, and J. B. Lando, *Polym. Lett.*, **17**, 585 (1979).
77 M. A. Bachman, W. L. Gordon, J. L. Koenig, and J. B. Lando, *J. Appl. Phys.*, **50**, 6106 (1979).
78 N. C. Banik, P. L. Taylor, S. K. Triphathy, and A. J. Hopfinger, *Macromolecules*, **12**, 1015 (1979).
79 S. Weinhold, M. H. Litt, and J. B. Lando, *Macromolecules*, 13, 1178 (1980).
80 Y. Takahashi and H. Tadokoro, *Macromolecules*, **13**, 1317 (1980).
81 G. J. Welch, *J. Polym. Sci. Polym. Phys. Ed.*, **14**, 1683 (1976).

REFERENCES

82. K. Nakagawa and Y. Ishida, *J. Polym. Sci. Polym. Phys. Ed.*, **11**, 2153 (1973).
83. K. Matsushige, K. Nagata, and T. Takemura, *Jpn. J. Appl. Phys.*, **17**, 467 (1978).
84. K. Matsushige and T. Takemura, *J. Polym. Sci. Polym. Phys. Ed.*, **16**, 921 (1978).
85. J. Scheinbeim, C. Nakafuku, B. A. Newman, and K. D. Pae, *J. Appl. Phys.*, **50**, 4399 (1979).
86. K. Matsushige and T. Takemura, *J. Cryst. Growth*, **48**, 343 (1980).
87. K. Matsushige and T. Takemura, *J. Polym. Sci. Polym. Phys. Ed.*, **18**, 1665 (1980).
88. G. Cessac and J. G. Curro, *J. Polym. Sci. Polym. Phys. Ed.*, **12**, 695 (1974).
89. R. G. Kepler, E. J. Graeber, and P. M. Beeson, *Bull. APS Ser. II*, **20**, 350 (1975).
90. N. Takahashi and A. Odajima, *Charge Storage, Charge Transport and Electrostatics with Their Applications*, Y. Wada et al., Eds., Elsevier, Amsterdam, 1979, p. 148.
91. N. Takahashi and A. Odajima, Ferroelectrics, **32**, 49 (1981).
92. R. G. Kepler and R. A. Anderson, *J. Appl. Phys.*, **49**, 1232 (1978).
93. M. Tamura, S. Hagiwara, S. Matsumoto, and N. Ono, *J. Appl. Phys.*, **48**, 513 (1977).
94. T. Takahashi, M. Date, and E. Fukada, *Appl. Phys. Lett.*, **37**, 791 (1980).
95. D. Naegele and D. Y. Yoon, *Appl. Phys. Lett.*, **33**, 132 (1978).
96. E. Fukada, *Proceedings of 2nd Meeting on Ferroelectric Materials and Their Applications*, 1979, I-1.
97. K. Makamura and Y. Wada, *J. Polym. Sci. A-2*, **9**, 161 (1971).
98. J. G. Bergman, Jr., J. H. McFee, and G. R. Crane, *Appl. Phys. Lett.*, **18**, 203 (1971).
99. P. Buchman, *Ferroelectrics*, **5**, 39 (1973).
100. M. Tamura, K. Ogasawara, N. Ono, and S. Hagiwara, *J. Appl. Phys.*, **45**, 3768 (1974).
101. R. G. Kepler, *Org. Coatings Plast. Chem.*, **38**, 278 (1978).
102. H. Dvey-Aharon, T. J. Slucken, P. L. Taylor, and A. J. Hopfinger, *Phys. Rev. B*, **21**, 3700 (1980).
103. Y. Wada and K. Tsuge, *Jpn. J. Appl. Phys.*, **1**, 64 (1962).
104. S. Suehiro, T. Yamada, H. Inagaki, T. Kyu, S. Nomura, and H. Kawai, *J. Polym. Sci. Polym. Phys. Ed.*, **17**, 763 (1979).
105. E. W. Aslaksen, *J. Chem. Phys.*, **57**, 2358 (1972).
106. M. G. Broadhurst and G. T. Davis, 1979 Annual Report, Conference on Electrical Insulation and Dielectric Phenomena, p. 447.
107. T. Furukawa and G. E. Johnson, *Appl. Phys. Lett.* 38, 1027 (1981).
108. K. Nakagawa and Y. Ishida, *J. Polym. Sci. A-2*, **11**, 1503 (1973).
109. G. Pfister and M. A. Abkowitz, *J. Appl. Phys.*, **45**, 1001 (1974).
110. M. Abkowitz and G. Pfister, *J. Appl. Phys.*, **46**, 2559 (1975).
111. J. B. Enns and R. Simha, *J. Macromol. Sci. Phys.*, **B13**, 11 (1977).

112 S. Tasaka, M. Yoshikawa, and S. Miyata, *Polym. Prepr., Japan,* **28,** 1778 (1979).
113 R. Hayakawa, J. Kusuhara, K. Hattori, and Y. Wada, *Rep. Prog. Polym. Phys. Japan,* **16,** 477 (1973).
114 P.T.A. Klaase and J. van Turnhout, *IEE Conf. Pub. No. 177,* 411 (1979).
115 M. Blukis, C. Lewa, S. Letowski, and A. Sliovinski, *Ultras. Intern. 1977, Brighton,* IPC Sci. & Technol. Press, Guildford, 1977, p. 474.
116 B. Stoll, *Colloid Polym. Sci.,* **256,** 521 (1978).
117 N. Murayama, *Charge Storage, Charge Transport and Electrostatics with Their Applications,* Y. Wada et al., Eds., Elsevier, Amsterdam, 1979, p. 118.
118 E. Fukada, M.Date, and T. Furukawa, *Org. Coatings Plast. Chem.,* **38,** 262 (1978).
119 K. Takahashi, H. Lee, R. E. Salomon, and M. M. Labes, *J. Appl. Phys.,* **48,** 4694 (1977).
120 K. Takahashi, H. Lee, R. E. Salomon, and M. M. Labes, *Org. Coatings Plast. Chem.,* **38,** 328 (1978).
121 H. Sussner, *Org. Coatings Plast. Chem.,* **38,** 331 (1978).
122 T. Ibe, Rep. 142 Committee (B Division) of Japan Society for the Promotion of Science, 1979, p. 50.
123 H. Sussner and K. Dransfeld, *J. Polym. Sci. Polym. Phys. Ed.,* **16,** 529 (1978).
124 R. J. Phelan, Jr., R. L. Peterson, C. A. Hamilton, and G. W. Day, *Ferroelectrics,* **7,** 375 (1974).
125 G. W. Day, C. A. Hamilton, R. L. Peterson, R. J. Phelan, Jr., and L. D. Mullen, *Appl. Phys. Lett.,* **24,** 456 (1974).
126 M. Oshiki and E. Fukada, *Jpn. J. Appl. Phys.,* **15,** 43 (1976).
127 W. R. Blevin, *Appl. Phys. Lett.,* **31,** 6 (1977).
128 F. I. Mopsik and M. G. Broadhurst, *J. Appl. Phys.,* **46,** 4204 (1975).
129 R. Hayakawa and Y. Wada, *Rep. Prog. Polym. Phys. Japan,* **15,** 377 (1972).
130 W. P. Mason, *Piezoelectric Crystals and Their Application to Ultrasonics,* Van Nostrand, New York, 1950.
131 M. G. Broadhurst, G. T. Davis, J. E. McKinney, and R. E. Collins, *J. Appl. Phys.,* **49,** 4992 (1978).
132 K. Tashiro, M. Kobayashi, H. Tadokoro and E. Fukada, *Macromolecules,* **13,** 691 (1980).
133 N. Takahashi, A. Odajima, K. Nakamura, and N. Murayama, *Polym. Prepr. Japan,* **27,** 364 (1978).
134 N. Takahashi, T. Ishibashi, and A. Odajima, *Polym. Prepr. Japan,* **27,** 1788 (1978).
135 E. Fukada and S. Takashita, *Jpn. J. Appl. Phys.,* **8,** 960 (1969).
136 E. Fukada and T. Sakurai, *Polym. J.,* **2,** 656 (1971).
137 R. J. Shuford, A. F. Wilde, J. J. Ricca, and G. R. Thomas, *Polym. Eng. Sci.,* **16,** 25 (1976).
138 H. Ohigashi, *J. Appl. Phys.,* **47,** 949 (1976).

REFERENCES

139 S. Tasaka, Y. Ando, R. Okano, S. Miyata, K. Sakaoku, and E. Fukada, *Polym. Prepr. Japan,* **26,** 1474 (1977).
140 K. Matsushige and T. Takemura, *Rep. Prog. Polym. Phys. Jap.,* **21,** 345 (1978).
141 M. Oie, K. Yamada, and M. Takayanagi, *Rep. Prog. Polym. Phys. Japan,* **23,** 509 (1980).
142 K. Matsushige, K. Tagashira, S. Imada, and T. Takemura, *Polymer,* **21,** 1391 (1980).
143 T. Furukawa, T. Goho, M. Date, T. Takamatsu, and E. Fukada, *Kobunshi Ronbunsenshu,* **36,** 685 (1979).
144 T. Furukawa, J. Aiba, and E. Fukada, *J. Appl. Phys.,* **50,** 3615 (1979).
145 D. K. Das-Gupta and K. Doughty, *J. Phys. D,* **13,** 95 (1980).
146 A. M. Glass, J. H. McFee, and J. G. Bergman, Jr., *J. Appl. Phys.,* **42,** 5219 (1971).
147 J. H. McFee, J. G. Bergman, Jr., and G. R. Crane, *Ferroelectrics,* **3,** 305 (1972).
148 G. Pfister, M. Abkowitz, and R. G. Crystal, *J. Appl. Phys.,* **44,** 2064 (1973).
149 R. G. Kepler and R. A. Anderson, *J. Appl. Phys.,* **49,** 4918 (1978).
150 J. I. Scheinbeim, K. T. Chung, K. D. Pae, and B. A. Newman, *J. Appl. Phys.,* **51,** 5106 (1980).
151 D. K. Das-Gupta and J. S. Duffy, *J. Appl. Phys.,* **50,** 561 (1979).
152 D. K. Das-Gupta and K. Doughty, *J. Appl. Phys.,* **51,** 1733 (1980).
153 J. Cohen, S. Edelman, and C. F. Vezzetti, *Electrets, Charge Storage and Transport in Dielectrics,* Electrochemical Society, N.J., 1973.
154 H. Burkard and G. Pfister, *J. Appl. Phys.,* 45, 3360 (1974).
155 K. Ogasawara, K. Shiratori, and M. Tamura, *Rep. Prog. Polym. Phys. Japan,* **19,** 313 (1976).
156 S. Tasaka, K. Murakami, Y. Ando, S. Miyata, K. Sakaoku, and E. Fukada, *Polym. Prepr. Japan,* **27,** 1784 (1978).
157 R. G. Kepler and R. A. Anderson, *J. Appl. Phys.,* **49,** 4490 (1978).
158 G. R. Davies, A. Killey, A. Rushworth, and H. Single, *Org. Coatings Plast. Chem.,* **38,** 257 (1978).
159 M. G. Broadhurst, W. P. Harris, F. I. Mopsik, and C. G. Malmberg, *ACS Polym. Prepr.,* **14,** 820 (1973).
160 F. Micheron, G. Bichon, C. Lemonon, H. Facoetti, and M. Royer, *Charge Storage, Charge Transport and Electrostatics with Their Applications,* Y. Wada et al., Eds., Elsevier, Amsterdam, 1979, p. 158.
161 Y. Wada and R. Hayakawa, *Ferroelectrics,* **32,** 115 (1981).
162 M. Date, *Polym. J.,* **8,** 60 (1975).
163 S. Tasaka and S. Miyata, *Ferroelectrics,* **32,** 17 (1981).
164 H. Sussner, *Phys. Lett.,* **58A,** 426 (1976).
165 I. D. Richardson and I. M. Ward, *J. Polym. Sci. Polym. Phys. Ed.,* 16, 667 (1978).
166 A. M. Glass and T. J. Negren, *J. Appl. Phys.,* **50,** 5557 (1979).
167 S. B. Lang, *J. Appl. Phys.,* **50,** 5554 (1979).

168 S. Miyata, S. Tasaka, Y. Ando, K. Sakouku, and E. Fukada, Extended Abstracts, International Workshop on Electric Charges in Dielectrics, Kyoto, 1978, p. 96.
169 G. T. Davis, *Proc. Piezoelectric and Pyroelectric Symposium*, M. G. Broadhurst, Ed., National Bureau of Standards Internal Report, 1975, p. 120.
170 A. I. Baise, H. Lee, B. Oh, R. E. Salomon, and M. M. Labes, *Appl. Phys. Lett.*, **26**, 428 (1975).
171 M. Latour, *Polymer*, **18**, 278 (1977).
172 M. Latour, *J. Electrostat.*, **2**, 241 (1977).
173 R. E. Collins, M. G. Broadhurst, G. T. Davis, and A. S. DeReggi, *Charge Storage, Charge Transport and Electrostatics with Their Applications*, Y. Wada et al., Eds., Elsevier, Amsterdam, 1979, p. 170.
174 J. C. Hicks, T. E. Jones, and J. C. Logan, *J. Appl. Phys.*, **49**, 6092 (1978).
175 H. Stefanou, *J. Appl. Phys.*, **50**, 1486 (1979).
176 K. Murakami, S. Tasaka, and E. Fukada, *Polym. Prepr. Japan*, **28**, 1794 (1979).
177 T. Furukawa, M. Date, E. Fukada, Y. Tajitsu, and A. Chiba, *Jap. J. Appl. Phys.*, **19**, L109 (1980).
178 Y. Higashihata, J. Sako, and T. Yagi, *Ferroelectrics*, **32**, 85 (1981).
179 Y. Tajitsu, A. Chiba, T. Furukawa, M. Date, and E. Fukada, *Appl. Phys. Lett.*, **36**, 286 (1980).
180 H. Lee, R. E. Salomon, and M. M. Labes, *Macromolecules*, **11**, 171 (1978).
181 J. Cohen and S. Edelman, *J. Appl. Phys.*, **42**, 3072 (1971).
182 E. Fukada and K. Nishiyama, *Jpn. J. Appl. Phys.*, **11**, 36 (1972).
183 B. A. Newman, P. Chen, K. D. Pae, and J. I. Scheinbeim, *J. Appl. Phys.*, **51**, 5161 (1980).
184 M. H. Litt, C. Hsu, and P. Basu, *J. Appl. Phys.*, **48**, 2208 (1977).
185 V. S. Likhovidov, V. V. Golovanov, and A. V. Vannikov, *Vysokomol. Soedin.* **A18**, 2058 (1976).
186 V. S. Likhovidov, V. V. Golovanov and A. V. Vannikov, *Sov. Phys. Solid State*, **17**, 1159 (1975).
187 S. Ikeda and K. Matsuda, *Rep. Prog. Polym. Phys. Japan*, **20**, 333 (1977).
188 H. Lee, R. E. Salomon, and M. M. Labes, *J. Appl. Phys.*, **50**, 3773 (1979).
189 V. S. Likhovidov, V. V. Golovanov, and A. V. Vannikov, *Vysokomol. Soedin.*, **A20**, 71 (1978).
190 S. I. Stupp and S. H. Carr, *J. Appl. Phys.*, **46**, 4120 (1975).
191 H. J. Wintle and J. Turlo, *J. Appl. Phys.*, **50**, 7128 (1979).
192 J. J. Crosnier, F. Micheron, G. Drefus, and J. Lewiner, *J. Appl. Phys.*, **47**, 4798 (1976).
193 S. Miyata, M. Yoshikawa, S. Tasaka, and M. Ko, *Polym. J.*, **12**, 857 (1980).
194 T. Ibe, *Jpn. J. Appl. Phys.*, **13**, 899 (1974).
195 K. Kato, T. Furukawa, E. Fukada, and H. Tadokoro, *Rep. Prog. Polym. Phys. Japan*, **22**, 343 (1979).
196 P. Vasudevan, H. S. Nalwa, and U. S. Tewari, *J. Appl. Phys.*, **50**, 4324 (1979).
197 J. Cohen and S. Edelman, *J. Appl. Phys.*, **42**, 893 (1971).

REFERENCES

198 S. Edelman, L. R. Grisham, S. C. Roth, and J. Cohen, *J. Acoust. Soc. Am.*, **48**, 1040 (1970).
199 L. A. Pauer, *IEEE Int. Conv. Rec.*, 49.2.1–49.2.5 (1973).
200 T. Kitayama, *Ceramics*, **14**, 209 (1979).
201 H. Yamasaki and T. Kitayama, *Proceedings of First Meeting on Ferroelectric Materials and Their Application*, Kyoto, 1977, p. 259.
202 T. Kitayama, Rep. 142 Committee (B Division) Japan Society for the Promotion of Science, 1979, p. 44.
203 T. Furukawa, K. Fujino, and E. Fukada, *Jpn. J. Appl. Phys.*, **15**, 2119 (1976).
204 T. Furukawa, K. Ishida, and E. Fukada, *J. Appl. Phys.*, **50**, 4904 (1979).
205 R. Holland, *IEEE Trans. SU-14-1*, 18 (1967).
206 L. F. Hu, K. Takahashi, R. E. Salomon, and M. M. Labes, *J. Appl. Phys.*, **50**, 2910 (1979).
207 R. Hayakawa, J. Kusuhara, and Y. Wada, *J. Macromol. Sci. Phys.*, **B8**, 483 (1973).
208 E. Fukada, *Proceedings of Fifth International Congress on Rheology*, S. Onogi, Ed., University of Tokyo Press, 1970, Vol. 3, p. 285.
209 E. Fukada and M. Date, *Polym. J.*, **1**, 410 (1970).
210 E. Fukada and M. Date, *J. Macromol. Sci. Phys.*, **B8**, 463 (1973).
211 H. Yamaguchi and E. Fukada, *Polym. J.*, **5**, 309 (1973).
212 T. Furukawa and E. Fukada, *Jpn. J. Appl. Phys.*, **16**, 453 (1977).
213 K. Nishinari and S. Koide, *J. Phys.*, **39**, 771 (1978).
214 I. M. Tsygel'nyi and V. L. Kus'menko, *Fiz.-Khim. Mekh. Mat.*, **11**, 119 (1975).
215 Kh. B. Pachadzhyan, A. K. Yagubyan, and A. P. Kodzhabashyan, *Izv. Akad. Nauk Arm. SSR, Fiz.*, **10**, 201 (1976).
216 R. Hayakawa, Y. Tanabe, N. Yamamoto, K. Yamakawa, and Y. Wada, *Rep. Prog. Polym. Phys. Japan*, **17**, 601 (1974).
217 K. Koga, T. Kajiyama, and M. Takayanagi, *J. Phys., E*, **8**, 299 (1975).
218 R. Hayakawa, K. Namiki, T. Sakurai, and Y. Wada, *Rep. Prog. Polym. Phys. Japan*, **19**, 317 (1976).
219 R. Hayakawa and Y. Wada, *IEE Conf. Pub. No. 177*, 396 (1979).
220 M. G. Broadhurst, C. C. Malmberg, F. I. Mopsik, and W. P. Harris, *Electret, Charge Storage and Transport in Dielectrics*, Electrochemical Society, Princeton, N.J., 1973, p. 492.
221 C. M. van der Burgt, G. A. van Maanen, and J. W. Vegt, Ultras. Int. 1977, Brighton, IPC Sci. & Technol. Press, Guildford, 1977, p. 500.
222 H. Sussner, D. E. Horne, and D. Y. Yoon, *Appl. Phys. Lett.*, **32**, 137 (1978).
223 H. R. Gallantree and R. M. Quilliam, *Marconi Rev.*, **39**, 189 (1976).
224 N. Murayama, K. Nakamura, H. Obara, and M. Segawa, *Ultrasonics*, 15. (Jan. 1976)
225 A. L. Robinson, *Science*, **200**, 1371 (1978).
226 F. Micheron, *Rev. Technique Thomson-CSF*, **10**, 445 (1978).
227 M. Jacobs, *Dimensions*, **62-2**, 2 (1978).

228 N. Chubachi and M. Tamura, *Oyobuturi*, **47**, 1162 (1978).
229 M. G. Broadhurst, G. T. Davis, and S. Edelman, Symposium on New Applications for Plastics, ACS, Houston 1980.
230 Y. Wada, *J. IECE Japan*, **61**, 17 (1978).
231 E. Fukada, *J. Soc. Electrostat. Japan*, **3**, 83 (1979).
232 H. Sussner and K. Dransfeld, *Colloid Polym. Sci.*, **257**, 591 (1979).
233 M. Tamura, T. Yamaguchi, T. Oyaba, and T. Yoshimi, 49th Convention Audio Engineers Society, Preprint No. 986, 1974.
234 M. Tamura, T. Yamaguchi, T. Oyaba, and T. Yoshimi, *J. Audio Eng. Soc.*, **23**, 21 (1975).
235 M. Tamura, K. Iwama, and T. Yoshimi, *J. Acoust. Soc. Japan*, **31**, 414 (1975).
236 M. Tamura, K. Ogasawara, and T. Yoshimi, *Ferroelectrics*, **10**, 125 (1976).
237 J. F. Sear and R. Carpenter, *Electron. Lett.*, **11**, 532 (1975).
238 H. Naono, T. Gotoh, M. Matsumoto, S. Ibaraki, and Y. Rikow, 58th Convention of Audio Eng. Soc., Preprint No. 1271, 1977.
239 M. Tamura and K. Iwama, *J. Acoust. Soc. Am.*, **64**, S56 (1978).
240 M. Tamura, *Oyobuturi*, **46**, 706 (1977).
241 F. Micheron and C. Lemonon, *J. Acoust. Soc. Am.*, **64**, 1720 (1978).
242 J. M. Powers and T. D. Sullivan, *J. Acoust. Soc. Am.*, **60**, S47 (1976).
243 B. Woodward, *Acustica*, **38**, 264 (1977).
244 J. M. Powers, A. T. Corcella, and R. E. Crooks, *J. Acoust. Soc. Am.*, **64**, S56 (1978).
245 J. F. Kilpatrick, *J. Acoust. Soc. Am.*, **64**, S56 (1978).
246 D. Ricketts, *J. Acoust. Soc. Am.*, **64**, S55 (1978).
247 A. S. DeReggi, S. Roth, J. M. Kenney, S. Edelman, and G. Harris, *J. Acoust. Soc. Am.*, **64**, S55 (1978).
248 K. Kobayashi and T. Yasuda, *Ferroelectrics*, **32**, 181 (1981).
249 M. Latour, O. Guelorget, and P. V. Murphy, *Charge Storage, Charge Transport and Electrostatics with Their Applications*, Y. Wada et al., Eds., Elsevier, Amsterdam, 1979, p. 175.
250 B. T. Brady, *Nature*, **260**, 108 (1976).
251 H. Sussner, D. Michas, A. Assfalgs, H. Hunklinger, and K. Dransfeld, *Phys. Lett.*, **45A**, 475 (1973).
252 H. Ohigashi, R. Shigenari, and M. Yokota, *Jpn. J. Appl. Phys.*, **14**, 1085 (1975).
253 C. Alquié, J. Lewiner, and C. Friedman, *Appl. Phys. Lett.*, **29**, 69 (1976).
254 H. Ohigashi, Abstracts, 4th World Congress on Ultrasonic Medicine, 1979, Miyazaki.
255 L. Bui, H. J. Shaw, and L. T. Zitelli, *Electron. Lett.*, **12**, 393 (1976).
256 A. S. DeReggi, S. Edelman, S. Roth, H. Warner, and J. Wynn, *J. Acoust. Soc. Am.*, **61**, S17 (1977).
257 S. D. Bennett and J. Chambers, *Electron. Lett.*, **13**, 110 (1977).
258 N. Chubachi and T. Sannomiya, *Proceedings of 1977 IEEE Ultrasonic Symposium*, IEEE Cat. No. 77 CH1264-1SU.

REFERENCES

259 L. N. Bui, H. J. Shaw, and L. T. Zitelli, *IEEE Trans.*, **SU-24**, 331 (1977).
260 N. Chubachi and T. Sannomiya, *Jpn. J. Appl. Phys.*, **16**, 2259 (1977).
261 E. F. Carome, G. R. Harris, and A. S. DeReggi, *J. Acoust. Soc. Am.*, **64**, S56 (1978).
262 D. T. Wilson, R. H. Tancrell, and J. Callerame, *Proceedings of 1979 IEEE Symposium on Ultrasonics*, IEEE Cat. No. 79, CH1482-9SU.
263 E. Carome, H. J. Shaw, D. Weinstein, and L. T. Zitelli, 1979 IEEE Ultras. Sym. Proc., IEEE Cat. No. 79, CH1482-9SU.
264 J. Callerame, R. H. Tancrell, and D. T. Wison, 1979 IEEE Ultras. Sym. Proc., IEEE Cat. No. 79, CH1482-9SU.
265 B. T. Khuri-Yukub and C. H. Chou, *Electron. Lett.*, **15**, 308 (1979).
266 O. Yano, K. Mizutani, S. Saito, Y. Sato, and Y. Wada, *Rep. Prog. Polym. Phys. Japan*, **23**, 425 (1980).
267 B. A. Auld and I. J. Gagnepain, *J. Appl. Phys.*, **50**, 5511 (1979).
268 T. Kitayama and T. Ueda, *Proceedings of First Meeting on Ferroelectric Materials and Their Applications*, 1977, p. 263.
269 T. Kitayama, *Oyobuturi*, **46**, 708 (1977).
270 J. G. Linvill, Stanford University Technical Rep. No. 4834-3, 1978.
271 H. Suzuki, *Shikai Tenbo*, **51**, 433 (1978).
272 M. Toda, *Ferroelectrics*, **32**, 127 (1981).
273 M. Toda and S. Osaka, *Trans. IECE Japan*, **E61**, 507 (1978).
274 M. Toda and S. Osaka, *Trans. IECE Japan*, **E61**, 513 (1978).
275 M. Toda and S. Osaka, *Proceedings of International Conference on Information Displays*, San Francisco, 1978.
276 M. Toda, S. Osaka, and S. Tosima, *Ferroelectrics*, **23**, 115 (1980).
277 M. Toda, *Ferroelectrics*, **22**, 911 (1979).
278 M. Toda, *Ferroelectrics*, **22**, 919 (1979).
279 M. Toda and S. Osaka, *Proceedings of Second Meeting on Ferroelectric Materials and Their Applications*, 1979, V-6, p. 85.
280 M. Toda and S. Osaka, *Proc. IEEE*, **67**, 1171 (1979).
281 M. Toda and S. Osaka, *Ferroelectrics*, **23**, 2301 (1980).
282 S. A. Kokorowski, *J. Opt. Soc. Am.*, **69**, 181 (1979).
283 R. J. Phelan, Jr., R. J. Mahler, and A. R. Cook, *Appl. Phys. Lett.*, **19**, 337 (1971).
284 R. J. Phelan, Jr., and A. R. Cook, *Appl. Opt.*, **12**, 2494 (1973).
285 R. L. Peterson, G. W. Day, P. M. Gruzensky, and R. J. Phelan, Jr., *J. Appl. Phys.*, **45**, 3296 (1974).
286 P. D. Southgate, *RCA Eng.*, **20** (2), 30 (1974).
287 G. W. Day, C. A. Hamilton, P. M. Gruzensky, and R. J. Phelan, Jr., *Ferroelectrics*, **10**, 99 (1976).
288 E. J. Sharp and L. E. Garn, *Appl. Phys. Lett.*, **29**, 480 (1976).
289 S. Iwasaki, T. Inoue, and T. Nemoto, *Bull. Electrotech. Lab. Japan*, **41**, 516 (1977).

290 W. R. Blevin and J. Geist, *Appl. Opt.*, **13,** 2212 (1974).
291 J. G. Bergman, G. R. Crane, A. A. Ballman, and H. O'Bryan, *Appl. Phys. Lett.,* **21,** 497 (1972).
292 A. W. Stephens, A. W. Levine, J. Fech, Jr., T. J. Zrebiec, A. V. Cafiero, and A. M. Garofalo, *Thin Solid Films,* **24,** 361 (1974).
293 L. E. Garn and E. J. Sharp, *IEEE Trans.* **PHP10,** 208 (1974).
294 Y. Hatanaka, S. Okamoto, and R. Nishida, *Proceedings of First Meeting on Ferroelectric Materials and Their Applications,* 1977, p. 251.
295 E. Yamaka and A. Teranishi, *Proceedings of First Meeting on Ferroelectric Materials and Their Applications,* 1977, p. 255.
296 Y. Hatanaka, S. Okamoto, and R. Nishida, *Preprint of Seventh Symposium on Photoelectric Image Devices,* London, 1978.
297 K. Shigiyama and K. Miura, *J. Acoust. Soc. Am.,* **64,** S56 (1978).
298 N. P. Huffnagle, *J. Acoust. Soc. Am.,* **64,** S57 (1978).

Chapter 5

ENERGY TRANSFER

Walter Klöpffer
Battelle-Institut e. V. D-6000 Frankfurt am Main, West Germany

5.1	Introduction	161
	5.1.1 Nature of Excited States in Polymers, 161	
	5.1.2 Scope of Chapter, 164	
5.2	Basic Concepts of Electronic Energy Transfer	164
	5.2.1 Excitons, 164	
	5.2.2 Single-Step Singlet Transfer, 169	
	5.2.3 Single-Step Triplet Transfer, 172	
	5.2.4 Excimers and Exciplexes, 174	
	5.2.5 Deactivation of Excited Electronic States, 177	
5.3	Measuring Techniques	180
	5.3.1 General, 180	
	5.3.2 UV Absorption, 181	
	5.3.3 Fluorescence, 181	
	5.3.4 Phosphorescence and Delayed Fluorescence, 182	
	5.3.5 Electron Spin Resonance, 182	
	5.3.6 Measurement of Formation and Decay of Excited States, 183	
	5.3.7 Chemical Techniques, 186	
5.4	Results and Interpretation	187
	5.4.1 Introduction, 187	
	5.4.2 Polymers with Carbazole Groups, 187	
	5.4.3 Polymers with Naphthalene Groups, 196	
	5.4.4 Polystyrene, 198	
	5.4.5 Polymeric Aromatic Ketones and Imides, 202	
5.5	Applications	205
5.6	Summary	206
	References	208

5.1 INTRODUCTION

5.1.1 Nature of Excited States in Polymers

The excitation of solids using radiation of the order of several electron volt quantum energy creates electronically excited states of widely different nature depending on the chemical and physical properties of the material. In nonmetallic solids, which can be classified as semiconductors and insulators according to their electrical properties, or in covalent, ionic, and molecular solids depending on their type of binding, we distinguish the following groups of excited electronic states:

1. Localized states.
2. Frenkel excitons.
3. Wannier excitons.
4. Charge carriers (electrons and holes).

Localized states are preferably formed at chemical impurities or lattice defects and in amorphous solids. If the concentration of these defects and/or the interaction between them is very small, the excited state remains at the same site during its lifetime which may range, depending on the efficiency of the deactivation processes, from picoseconds to seconds. In contrast to this behavior, Frenkel excitons (1) are not confined to the lattice site where they have been created by absorption of radiation or as an intermediate step in a deactivation cascade from higher excited states. Here the electronic excitation energy may be spread out over several lattice sites (ideally over the whole crystal) in a wave-like form, controlled by the lattice symmetry and the law of diffraction (exciton band model). Alternatively (2), the Frenkel exciton can perform incoherent motion if the lattice is disordered or the thermal energy induces strong lattice vibrations, breaking the coherence of the exciton waves (exciton hopping model).

Wannier excitons (3), too, are mobile and neutral excited states. However, unlike for the Frenkel exciton where the electron–hole pair can formally be considered to occupy the same site, the electron and hole of the Wannier exciton are located at different positions of the lattice, often showing a hydrogen-like absorption spectrum with the energy gap as the dissociation limit. The properties of the resulting free charge carriers are treated in chapters 6 and 7 of this book and are discussed here only from the perspective of energy transfer. Electrons and holes, when moving independently in a lattice, clearly constitute an electronically excited state of the lattice (4), and their movement can therefore be considered as one form of electronic energy transfer. In contrast to energy transfer by the electrically neutral exciton, this type of energy transfer can be influenced by electric fields. For most inorganic solids, this is by far the most important way of transferring energy, excitons playing only a marginal role.

The reverse is true, however, for molecular solids. For these, the exciton is the basic excited state, which is, most often, of the Frenkel type. The fundamental reason for this behavior of organic, molecular solids is the weak van der Waals interaction between the individual molecules occupying the lattice sites.

Organic polymers, the main topic of this book, are molecular solids with weak intermolecular bonding and small dielectric constants. The intramolecular, covalent bonding, however, is strong and can be described by one-dimensional (1D) energy bands (5) similar to the 3D energy bands of most other solids. This

INTRODUCTION

applies to the backbone of polymers, especially for polyethylene and similar linear polymers without side groups. Many polymers absorbing in the near UV are composed of an aliphatic backbone and aromatic side groups, for example, aromatic vinyl polymers:

$$\left[\begin{array}{c} H \\ | \\ C \\ | \\ R \end{array} \text{———} CH_2 \right]_n \quad (5.1)$$

The characteristic low-lying excited states in these polymers are those of R, more or less disturbed by the polymeric environment to which the side groups are chemically linked.

Most polymers are amorphous or only partly crystallized. Exciton waves can therefore not be expected in these compounds, unless the samples are highly crystallized and investigated at low temperatures. So far, no experimental evidence, such as Davydov splitting (6), is available for exciton waves in organic polymers. The characteristic energy transfer process in these polymers is therefore (Frenkel) exciton hopping (7). Competing with this "pure" form of energy transfer are several single-step transfer processes which occur between localized states deriving, for instance, from polymer degradation, structural defects, or additive molecules. The transition between the two transfer processes is not distinct as could be shown by experiments using polymers doped with aromatics (8, 9), and it has led to the somewhat misleading use of the term "localized state" in connection with exciton hopping.

Single-step energy transfer shows few, if any, features that are specific for polymers. The solid polymer acts as an essentially inert matrix, comparable to the glassy solvents frequently used in studies of energy transfer between isolated molecules. Actually, amorphous polymers have often been used for this purpose.

In addition to the fundamental role excitons play in excited-state processes, they are essential in various related photophysical and photochemical phenomena. (a) In photoconductivity, discussed in Chapter 6, Frenkel excitons can liberate trapped charge carriers (10, 11), and Wannier charge transfer excitons (CT) may act as intermediates in the charge generation process (12), for instance, the Onsager process (Chapter 6, Section 6.4). Furthermore, exciton traps may also constitute traps for charge carriers (Section 5.2.4) or may be involved in charge separation of trapped excitons leading to the formation of polar exciplexes or excited charge transfer complexes. (b) In plastic scintillators, exciton energy transfer is the basic mechanism of operation (see Section 5.5) (13, 14). (c) Polymeric sensitizers in photochemical processes may also function on the basis of energy transfer.

Energy transfer in polymers is preferably studied under three experimental conditions: (a) in the solid state, mostly in form of thin films; (b) in dilute, rigid solutions where the polymer coils form quasi one-dimensional systems (15); and (c) in dilute, liquid solution where the superposition of molecular and excitonic dynamics severely complicates interpretation but can lead to information on polymer dynamics.

5.1.2 Scope of Chapter

The term "energy transfer" is most frequently used synonymously with "transfer of electronic excitation energy," and so it will be used in this chapter. In a strict sense, it also encompasses transfer of vibrational energy (phonons) and vibrational relaxation following radiationless transitions or radiative transitions to the vibrationally excited electronic ground state. These processes will be discussed insofar as they are connected with the main theme. This involves the role of excitons as the characteristic excited states of polymers and the role of specific traps or localized states which dominate the energy transfer in the solid state to such an extent that the observation of excitons is mostly an indirect one.

In accord with the main line of this book, the emphasis in this chapter is on the solid-state behavior of polymers, whereas results from other studies will be referred to as appropriate. The role of traps is important under all conditions, but especially so in solid-state studies. The nature of these traps is therefore discussed. In order to provide a basis for interpretation of the experiments presented in Section 5.4, we first briefly consider the different concepts of energy transfer (Section 5.2), including a short discussion of excimers and deactivation processes of excited states. Section 5.3 is devoted to experimental techniques, including spectroscopic and chemical techniques. The emphasis in this section is on polymer specific problems rather than on the details of the mostly well-known methods. The main part of this chapter is Section 5.4, dealing with experimental results and their interpretation. The order in this section is the chemical class of the optically dominant groups. Some actual and future applications of energy transfer are discussed in Section 5.5, which is followed by a summary (Section 5.6).

5.2 BASIC CONCEPTS OF ELECTRONIC ENERGY TRANSFER

5.2.1 Excitons

As mentioned in Section 5.1.1, there are two basically different types of excitons: Wannier excitons and Frenkel excitons. We first consider the Wannier excitons in which pairs of electrons and holes occupy different lattice sites (which in the lowest energy state are neighboring ones). Pope (12) suggested that these excited charge pairs may constitute potential intermediates for intrinsic charge carrier formation in organic aromatic crystals. In these solids, the extremely narrow valence band is separated from the equally narrow conduction band by an energy gap E_g, which depends on the molecular ionization energy I, the electron affinity A, and the polarization energies of the hole, P_+, and the electron, P_-, which are assumed to be nearly equal (16):

$$E_g = I - A - P_+ - P_- \approx I - A - 2P \qquad (5.2)$$

In organic crystals, the dielectric constant is in general so small that the charge carriers feel their mutual Coulomb attraction up to distances of more than 10 nm, so that Eq. (5.2) is only valid for charge carriers with large average distances. If they come closer together, the possibility of forming a Wannier exciton with E_g of Eq. (5.2) as dissociation limit clearly exists. The lowest energy level of this exciton, corresponding to an electron–hole pair at neighboring lattice sites, is reported to lie not deeper than 0.5 eV below E_g (17). This exciton could be an intermediate in intrinsic photoconductivity (dissociation) and in electrochemiluminescence (17), that is, recombination of positive and negative charge carriers.

There is no reason to believe that aromatic polymers will behave completely different from crystals, the main difference being the absence of energy bands in the amorphous polymers (18), since P in Eq. (5.2) is somewhat different for each side group, thus favoring the hopping model of charge carrier transport. In this case, the holes and electrons are localized on the aromatic side groups and thus may be considered as radical cations and radical anions, respectively (19). Wannier excitons in organic solids, usually referred to as "CT excitons," may indeed be restricted to nearest-neighbor pairs of opposite charges. Although they are likely to occur in organic crystals and polymers, it turns out to be extremely difficult to prove their existence directly (74). The main reason for this difficulty is the energetically lower-lying Frenkel excitons, which in case of the spin-allowed singlets are characterized by a strong absorption below E_g. In N-isoproplycarbazole (NIPCA), a monomeric model compound of the photoconductive polymer poly(N-vinylcarbazole) (PVCA), E_g amounts to about 4.4 eV (19, 20), whereas S_1, the energy level of the singlet Frenkel exciton, is 3.55 eV (21). Between these energies, the energy levels of the Wannier exciton would be expected, both in the crystal and in the polymer. The absorption spectrum in this spectral region (280–350 nm), however, is due to allowed "intramolecular" transitions, not involving intermolecular charge separation. Direct evidence for a Wannier exciton showing a dipole moment of 23 debyes has been found by Hanson (22) by means of electro reflectance measurements in crystals of 9,10-dichloroanthracene.

Frenkel excitons do not require full separation of charges. They can be identified with neutral electronically excited states of the molecules forming the crystal or of side groups of the polymer. The first excited singlet (S_1) and triplet (T_1) states alone will be considered here, since only these have lifetimes sufficiently long for real multistep energy transfer processes. Furthermore, only the hopping aspect of the Frenkel exciton will be discussed as the relevant one for amorphous or partly crystalline polymers, since exciton waves require unperturbed crystal lattices.

The exciton hopping model considers the exciton to be an excited state of the individual molecules rather than of the crystal as a whole. The excited state spends some time at one molecule, jumps at random to another neighboring one, and so on, until it is finally deactivated (see Section 5.2.5) by some radiative or radiationless process. These include bimolecular processes such as trapping by guest molecules and mutual annihilation.

The main evidence for Frenkel excitons in the hopping limit is energy transfer, as revealed by the spectra, lifetimes, and intensities of the exciton emission of sensitized guest molecules or traps. The most important processes used as indicators for exciton hopping are the following:

1. For singlet (S_1) excitons:
   Sensitized fluorescence of guest (G) molecules (23, 24, 25): $S_1(G) < S_1(H)$, where H is the host crystal or polymer.
   Fluorescence quenching by nonfluorescent guest molecules (14, 7): $S_1(G) < S_1(H)$.
   Singlet–singlet and singlet–triplet annihilation (17).
2. For triplet (T_1) excitons:
   Sensitized phosphorescence of guest molecules which are unable to trap singlet excitons for energetical reasons (26, 27): $S_1(G) > S_1(H)$, $T_1(G) < T_1(H)$.
   Triplet–triplet annihilation (28, 17).
   Sensitized chemiluminescence (29, 30).

Although all these processes can be used for studying excitons in organic solids, including polymers, it is often difficult to distinguish between (multistep) energy transfer by exciton hopping and single-step transfer (see Sections 5.2.2 and 5.2.3). It frequently requires a detailed kinetic analysis and estimates of the relative probabilities of the different competing processes.

The exciton hopping model in its simplest form assumes a constant hopping time (τ_h), that is, the average time spent by an exciton at a particular molecule or side group:

$$\tau_h = \frac{1}{k_h} = \frac{\tau_0}{n} \tag{5.3}$$

The hopping time is the reciprocal of the hopping rate (k_h) and is given by the lifetime of the exciton in the absence of energy transfer (τ_0) and the number of jumps it performs during its lifetime (n). As can be seen from Table 5.1, typical values for n in aromatic crystals are in the range 10^4–10^7, depending on the state forming the exciton. Triplet excitons are in general slower than singlet excitons, but their decay times are orders of magnitude longer. They therefore cover more

BASIC CONCEPTS OF ELECTRONIC ENERGY TRANSFER

Table 5.1 Typical Exciton Hopping Parameters for Singlet and Triplet Excitons

	τ_0(s)	n	τ_h(s)	$\bar{r}^a$(nm)	D^a(cm^2/s)
S_1	10^{-8}	10^4	10^{-12}	50	8×10^{-4}
T_1	10^{-3}	10^7	10^{-10}	1600	8×10^{-6}

^{a}Using Eq. (5.4) with $a = 0.5$ nm

lattice sites than singlets do. Another illustrative magnitude is the average end-to-end distance ($\bar{r}$) of the exciton path, which can be considered as a random walk between nearest neighbors of distance $a \approx (M/N_L d)^{1/3}$ (M is molar mass, d is density N_L is Loschmidt's number):

$$\bar{r} = a\sqrt{n} = \sqrt{3D\tau_0} \tag{5.4}$$

where D is the exciton diffusion constant (31). The number of steps, n, can be determined from luminescence quenching and decay time experiments using a Stern–Volmer formula (2, 21, 24, 31):

$$Q = \frac{I_o - I}{I} = \frac{\tau_0 - \tau}{\tau} = K c_G \tag{5.5}$$

where Q is the quenching factor, as described by Northrop and Simpson (24). I_0 and τ_0 are the values of the fluorescence or phosphorescence intensity and decay time without added quencher molecule (guest), and I and τ are the values with added guest in the molar ratio (guest to host) c_G. It is assumed in Eq. (5.5) that the intensities and lifetimes measured are those of the mobile state, that is, the exciton. As will be seen in Section 5.4, this is in general not the case in polymers (7).

Now, $\tau_0 = (k_e + k_i)^{-1}$ and $\tau = (k_e + k_i + k_{tr})^{-1}$, where k_e, k_i, and k_{tr} are the probabilities per second for radiative decay, radiationless decay, and exciton trapping by the guest molecule, respectively. Assuming $k_{tr} = k_h C_G$, where k_h is the hopping rate defined in Eq. (5.3), one finds from Eq. (5.5) that $K = n$; that is, a plot of Q versus c_G yields n (21).

Sometimes a factor 0.66 has been introduced in Eq. (5.5) which takes into account, according to three-dimensional random walk theory (33), the probability of returning of the exciton to its starting point (21, 32). This modification has not found broad application since the true dimensionality of the random walk is often not exactly known and since other approximations in the interpretation of the experiments as well as experimental difficulties exclude precise determinations of n anyway. Equation (5.5) predicts a slope of 1 in a log–log plot of Q versus c_G (Fig. 5.1) which can be used to distinguish between exciton hopping

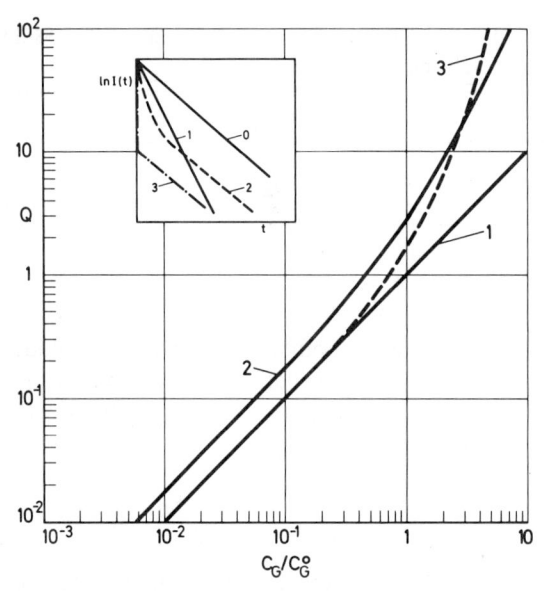

Figure 5.1 Dependence of quenching factor $Q = (I_0 - I)/I$ on energy acceptor (guest) concentration: 1, exciton hopping model; 2, Förster's dipole–dipole resonance theory (44, 45); 3, Perrin's model (49). Insert: schematic decay behavior 0, exponential decay of pure donor or host exciton luminescence; 1, 2, 3, as in figure.

and single-step transfer processes (see Sections 5.2.2 and 5.2.3). Deviations from a slope of 1 to lower values have frequently been observed in crystals and may be due to an inhomogeneous distribution of the guest molecules, especially in disordered regions of the host.

In general, Eq. (5.5) successfully accounts for the host luminescence quenching. However, discrepancies with that equation are often observed when analyzing the kinetics of sensitized guest fluorescence (34). The main reason for this discrepancy may be the assumption that the exciton jump rate from the host to guest molecule is equal to k_h, the host–host jump rate [Eq. (5.3)]. Powell (34) improved Eq. (5.5) by assuming extended trapping zones around each guest molecule.

Biexcitonic annihilation processes follow the rate equation (28)

$$\frac{dN}{dt} = N_0 - \beta N - \gamma N^2 \qquad (5.6)$$

where N_0 = number of excitons created $(cm^{-3} s^{-1})$
N = concentration of excitons (cm^{-3})
$\beta = 1/\tau_0$ = sum of first-order decay rates (s^{-1})
γ = second order annihilation rate $(cm^3 s^{-1})$

Annihilation occurs between singlet as well as triplet excitons, with the latter very important process leading to delayed fluorescence (28). This is the favorite method for investigating triplet excitons (see Section 5.4). Annihilation processes offer the opportunity for studying exciton kinetics without introduction of guest molecules, which always—except for isotopic substitution—disturb the host lattice. A detailed discussion of exciton annihilation processes has been given by Swenberg and Geacintov (17). Very recently, Bässler and coworkers (35) suggested that f_h may not be constant in amorphous solids (in close similarity with the hopping of charge carriers, Chapter 6). A numerical (Monte Carlo) simulation of exciton diffusion shows that for a Gaussian distribution of exciton energy levels exciton hopping is faster immediately following their creation and decreases until thermalization is completed (35). Clearly, the time dependence of exciton diffusion is less important for the intrinsically long-lived triplet excitons as compared to singlet excitons. The experimental verification of this new concept, which eventually could remove some inconsistencies of the simple hopping model, is a challenge to picosecond spectroscopy (Section 5.3.6). There are two extreme types of "self-trapped" Frenkel excitons whose hopping rate increases with temperature:

Excimer excitons, observed in crystals allowing for good overlap of the molecular planes, for example, pyrene, perylene, 9-cyanoanthracene (36–39)

CT excitons in donor–acceptor complexes (40–43)

Both excitons derive from excited complexes (see Section 5.2.4) which are energetically below the monomer states of the constituent molecule(s). CT excitons may play a role in heavily doped donor polymers, for example, photoconductive PVCA/TNF, although direct evidence is still missing. Excimer excitons are likely to be identified in polymers of suitable structure, either crystallizing with sandwiched or stapled aromatic rings or, hypothetically, showing cyclophane groups in the basic units.

5.2.2 Single-Step Singlet Transfer

If two fluorescent substances, the absorption spectrum of the one overlapping with the fluorescence spectrum of the second, are molecularly dissolved in a transparent, rigid solvent, for example, a polymer, sensitized fluorescence can be observed upon excitation of the compound absorbing at shorter wavelengths. The intensity of the longer-wavelength fluorescence increases with concentration of the energy-accepting species, denoted G in order to stress the analogy to host-sensitized fluorescence in solids. The relative guest fluorescence intensity is independent of the concentration of the energy-donating, initially excited molecules (D), at least up to the concentration of beginning D–D transfer, the transition region to exciton hopping:

$$D \xrightarrow{h\nu_A} D^* + G \longrightarrow D + G^* \qquad (5.7)$$
$$\phantom{D \xrightarrow{h\nu_A}} \downarrow \phantom{{} + G \longrightarrow D + {}} \downarrow$$
$$\phantom{D \xrightarrow{h\nu_A}} D + h\nu_F \phantom{{} \longrightarrow {}} G + h\nu'_F$$

The fluorescence of G* facilitates detection of energy transfer but is not vital to the process which otherwise can be classified as fluorescence quenching. According to Förster's theory (44, 45) of dipole–dipole resonance [multipole interactions are included in the more comprehensive treatment by Dexter (46)], the rate constant of energy transfer k_{DG} can be calculated from spectroscopic data according to

$$k_{DG}(R) = \frac{9000 \ln 10}{128 \pi^5 N_L} \frac{\kappa^2 k_e}{n_s^4 R^6} \Omega \qquad (5.8)$$

where κ^2 is the orientation factor ≈ 0.6 in solid solutions with random orientation of D and G molecules, k_e is the radiative rate of the donor fluorescence, N_L is Loschmidt's number, n_s is the refractive index of the solvent (polymer), R is the distance between D and G, and Ω is the spectral overlap, defined as

$$\Omega = \int_0^\infty \epsilon_G(\nu') f_D(\nu') \frac{d\nu'}{\nu'^4} \approx \sum_{\nu'_1}^{\nu'_2} \epsilon_G(\nu') f_D(\nu') \frac{\Delta \nu'}{\bar{\nu}'^4}$$

where f_D is the spectral distribution of the fluorescence of the donor normalized according to $\int_0^\infty f_D(\nu') d\nu' = 1$, ϵ_G is the molar decadic extinction coefficient of the acceptor, ν'_1 and ν'_2 are the limits in wavenumbers of fluorescence (D) and absorption (G) overlap, $\bar{\nu}'$ is the average wavenumber of fluorescence (D) and absorption (G), and $\Delta \nu'$ is the wavenumber increment chosen for summation.

From Eq. (5.8), the energy transfer rate is a strong function of the intersite distance R and depends on the spectral overlap Ω and the intrinsic strength of the transitions involved, as expressed by k_e for D and ϵ_G for G. Equation (5.8) can be transformed into a more familiar form introducing the critical radius of energy transfer, R_0, where the rate of energy transfer equals the rates of radiative (k_e) and nonradiative (k_i) deactivation of D:

$$k_{DG}(R_0) = (k_e + k_i)_D \qquad (5.9)$$

From Eqs. (5.8) and (5.9) and introducing the true quantum efficiency of fluorescence [$\phi_F = k_e/(k_e + k_i)$] of the donor molecule D, one obtains

$$R_0^6 = \frac{9000 \ln 10 \, \kappa^2}{128 \pi^5 N_L n_s^4} \phi_F \Omega \qquad (5.10)$$

which relates R_0 and observable quantities.

BASIC CONCEPTS OF ELECTRONIC ENERGY TRANSFER

The corresponding critical concentration of the guest (acceptor) molecules, C_G° (mol/liter), is usually defined as

$$C_G^\circ = \frac{3000}{N_L 2\pi^{3/2} R_0^3} \qquad (5.11)$$

which roughly corresponds to one guest molecule per sphere of radius R_0.

In polymers single step singlet energy transfer is often weak due to small spectral overlap Ω such that exciton hopping may become a competitive process. Experimentally one can distinguish between the two transfer processes by analyzing the stationary quenching efficiency $Q(C_G)$ and the decay curves $I(t)$ of the fluorescence following an excitation pulse. In the presence of G molecules, $I(t)$ is given by

$$I(t) = I_0 \exp\left(-\frac{t}{\tau_0}\right) \exp\left[-\left(\frac{t}{\tau_0}\right)^{1/2} \frac{2C_G}{C_G^0}\right] \qquad (5.12)$$

where the first term describes the first order deactivation and the second term accounts for the quenching efficiency by the G molecules which is strongly concentration dependent (compare R^{-6} dependence, Eq. (5.8)). Integrating Eq. (5.12) yields the relative intensity of the donor fluorescence as a function of guest concentration:

$$\frac{I}{I_0} = 1 - \pi^{1/2} \frac{C_G}{C_G^0} \exp\left(\frac{C_G}{C_G^0}\right)^2 (1 - \delta) \qquad (5.13)$$

where δ is the error function

$$\delta = \frac{2}{\pi^{1/2}} \int_0^{C_G/C_G^0} \exp(-x^2)\, dx$$

For small concentrations, $C_G < 0.2\, C_G^0$, Eq. (5.13) reduces to

$$Q \approx \frac{\pi^{1/2}}{C_G^0} C_G \qquad (5.14)$$

where Q is the quenching factor defined in Eq. (5.5). Hence for small C_G, exciton hopping (Eq. 5.5) and single-step transfer exhibit the same concentration dependence. For large concentrations, $C_G > 3C_G^0$, Eq. (5.13) reduces to

$$Q = \frac{2}{C_G^{0\,2}} C_G^2 \qquad (5.15)$$

Thus, in contrast to the exciton hopping process, the Förster mechanism shows

Table 5.2 Annihilation (fusion) Processes as Special Cases of Energy Transfer

Annihilation	Transition in Energy Donor	Transition in Energy Acceptor
S–$S^{(a)}$	$S_1 \rightarrow S_0$	$S_1 \rightarrow S_n$
S–$T^{(a)}$	$S_1 \rightarrow S_0$	$T_1 \rightarrow T_n$
T–T	$T_1 \rightarrow S_0$	$T_1 \rightarrow S_n^{(b)}$ and $T_1 \rightarrow T_n$
T–$S^{(c)}$	$T_1 \rightarrow S_0$	$S_1 \rightarrow S_n$

[a] Förster-allowed processes.
[b] This event, the singlet channel of T–T annihilation, leads to delayed fluorescence.
[c] Förster-allowed only if the slow decay of T_1 compensates the inherently weak transition in the energy donor; this may not be the case in fusion experiments.

a quadratic concentration dependence at high C_G doping levels. An example is shown in Fig. 5.1, curve 2. In practice, however, it is often difficult to measure quenching factors accurately at high acceptor concentrations, that is, under strong quenching conditions.

In general, dipole–dipole resonance is the main driving force of singlet exciton hopping, although in case of a very small (S_0–S_1) transition moment, and hence unfavorable k_{DD} (analogous to k_{DG} in Eq. 5.8), multipole and exchange terms may become crucial. Dipole–dipole resonance is always possible provided an appropriate spectral overlap exists between the D emission and G absorption spectra and provided the energy-accepting transition is allowed. Therefore $T_1 \rightarrow S_0$ in D and $S_0 \rightarrow S_1$ in G give reasonable transfer efficiencies, at least at low temperature (ϕ_F in Eq. (5.10) has to be replaced by ϕ_P, the true quantum efficiency of phosphorescence). This and similar cases have been treated by Bennet and co-workers in a series of articles (47). Equally Förster-allowed are singlet–singlet and singlet–triplet annihilation where the energy acceptor is an excited singlet or triplet state.

Annihilation or fusion processes, which can occur if at least one partner is mobile (as an excited molecule or exciton), are basically energy transfer processes, as can be seen in Table 5.2. It should be noted that here, and in all true energy transfer processes, the transitions in the energy-donating and energy-accepting molecules are coupled and no real emission and reabsorption of photons takes place. This latter process is commonly referred to as the "trivial process" and is strongly dependent on sample geometry.

5.2.3 Single-Step Triplet Transfer

Triplet transfer according to Eq. (5.7), replacing $h\nu_F$ (fluorescence) by $h\nu_P$ (phosphorescence), involves multiplicity changes in D and G. It is therefore a

BASIC CONCEPTS OF ELECTRONIC ENERGY TRANSFER

spin-forbidden process and hence Förster's dipole resonance mechanism cannot operate. Triplet transfer is nevertheless an efficient process, if D and G molecules are molecularly dissolved in rigid solvents, as discovered by Ermolaev (48). The interaction responsible for T—T transfer is electron exchange (46), which in contrast to dipole resonance requires overlap of electronic wave functions. Unfortunately, no simple theory relating transfer efficiency with spectroscopic parameters of D and G is available in this case. Some general rules can nevertheless be deduced from experiment and theory:

1 Triplet energy transfer occurs, if $T_1(G) < T_1(D)$. In order to distinguish it from singlet transfer, $S_1(G) > S_1(D)$ should furthermore be fulfilled.

2 Overlap of electronic wave functions decays exponentially with distance. It is therefore to be expected that k_{DG} decreases even more rapidly with R than observed in the case of singlet transfer (R^{-6} law):

$$k_{DG}(R) = \frac{1}{\tau_o'} \exp\left[\gamma\left(1 - \frac{R}{R_0}\right)\right] \qquad (5.16)$$

where $\tau_o' = 1/(k_e^T + k_i^T)$ and is phosphorescence decay time of D; R_0 is the critical radius, as defined in Eq. (5.9), and $\gamma = 2R_0/L$, where L is called the effective, average Bohr's radius, which cannot be calculated (in the order of 0.1 to 0.2 nm).

3 F. Perrin (49) has developed an approximate, convenient model to describe the steep decline of k_{DG} with increasing R in order to account quantitatively for some observations on singlet transfer made earlier by his father (50). The model works however much better for triplet transfer and is now used for this case exclusively. The basic assumption of the model is shown in Fig. 5.2. The critical radius, r_0, is defined differently from Eq. (5.9) as the radius of a sphere of immediate energy transfer. The decay of donor phosphorecence $I(t)$

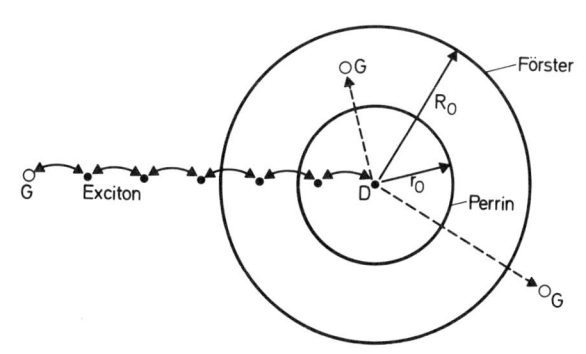

Figure 5.2 Schematic presentation of energy transfer models. Förster: $k_{DG} = k_e + \Sigma k_i$ at $R = R_0$; Perrin: $k_{DG} = 0$ if $r > r_0$; $k_{DG} = \infty$ if $r < r_0$. The exciton model is characterized by the hopping rate k_h and the number of host or donor molecules, n, visited during the exciton lifetime. Here a 1D-exciton is symbolized.

cannot be adequately reproduced by this model, which allows only two kinds of donor molecules: one kind decaying at an infinite rate and a second one decaying according to their intrinsic life time τ_o' (see insert in Fig. 5.1). Inokuti and Hirayama (51) have developed a numerical method for calculating $I(t)$ during triplet transfer. Perrin's model, although incapable of describing the decay behavior, is quite successful in describing the relative phosphorescence intensity (quantum efficiency) as a function of guest (acceptor) concentration:

$$\frac{I}{I_0} = \exp\left(-\frac{C_G}{C_G^0}\right) \tag{5.17}$$

where the critical concentration is defined as

$$C_G^0 = \frac{3000}{4\pi N_L r_0^3} \tag{5.18}$$

corresponding to exactly one guest molecule per sphere with radius r_0.

In order to quantify the efficiency of single-step triplet transfer, most frequently Perrin's r_0 value according to Eq. (5.17) and (5.18) is given. It should be stressed that r_0 is defined in a different way as R_0 and that the case $k_{DG} = k_e + k_i$ does, in a strict sence, not exist in this model; r_0 is often found to be about 1–1.5 nm (48). The steep increase in the quenching factor Q with increasing C_G is shown in Fig. 5.1.

Electron exchange, as the interaction leading to single-step triplet transfer, also causes multistep transfer (triplet excitons), as discussed in Section 5.2.1. In completely disordered solids, triplet excitons may be considered to perform a three-dimensional random walk; this is likely the case in amorphous polymers. In crystals however, and the same may be true for ordered regions of polymers, electron exchange interaction is much stronger in those crystal axes or planes showing strong overlap of molecular planes. Hence apparent two- or one-dimensional diffusion is frequently observed for triplet excitons as a direct consequence of their transfer mechanism, for example, two-dimensional in anthracene (ab-plane) and one-dimensional in 1,4-dibromonaphthalene (c-axis) (52, 53).

5.2.4 Excimers and Exciplexes

The formation of complexes which in general exist only in their excited state is an important exciton trapping mechanism in aromatic polymers. Since these complexes have no bonding ground state (there are however borderline cases between exciplexes and weak charge transfer (CT) complexes), they cannot act as energy acceptors in the single-step processes discussed in Sections 5.2.2 and 5.2.3 and hence are suitable probes in distinguishing between the different transfer mechanisms. Excimers are formed between identical entities, for exam-

BASIC CONCEPTS OF ELECTRONIC ENERGY TRANSFER

ple, adjacent side groups in a polymer, if one of the groups is excited by absorption of a photon or if by chance it is visited by an exciton. These conformations have been termed "excimer-forming sites" by the present author and will be abbreviated as EFS in the following (7). Excimer fluorescence, which is broad and red-shifted with respect to the monomer emission, has first been identified by Förster (54) in solutions of pyrene and investigated in detail by the same group (55), by Birks (56), and by many others. Excimer formation is a very common phenomenon in aromatic molecules, although not all complexes are stable enough to be detected in solution at ambient temperature. The binding force (see Fig. 5.3) is provided by mixing charge and excitation energy transfer states:

$$D^* + D \rightleftharpoons (D \cdots D)^* = D^*D \rightleftharpoons DD^* \rightleftharpoons D^+D^- \rightleftharpoons D^-D^+ \quad (5.18)$$

as first proposed by Konijnenberg (57). This type of binding is supposed to be

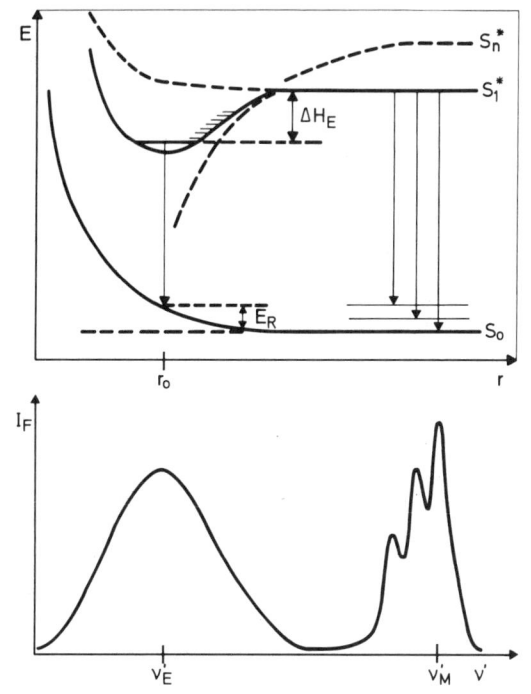

Figure 5.3 Schematic energy level diagram (E) and fluorescence spectrum (I_F) of monomer and excimer: S_0, S_1^*, S_n^* = ground and excited singlet states of separated excimer forming molecules; ΔH_E = enthalpy of excimer formation; r_o = equilibrium distance in excimer state; ν'_E, ν'_M = maximum wavenumber of excimer and monomer fluorescence; E_R = repulsion energy in ground state at r_o (r = interplanar distance).

significantly weaker for phosphorescent states. Reports on triplet excimers have been contradictory, but recently ESR and phosphorescence spectroscopy measurements on paracyclophane could unambiguously establish the existence of triplet state excimers in this material (58).

Exiplexes are formed between different molecules, mostly weak electron donors and weak electron acceptors (59, 60). Electronic excitation decreases the ionization energy of the donor and increases the electron affinity of the acceptor, thus allowing the formation of an excited charge transfer complex even in the absence of measurable ground-state interactions.

There is a second, less polar type of exciplex which resembles more closely the excimer and is termed "mixed excimer." Exciplexes are formed in donor polymers mostly by adding weak, monomeric acceptors, the sites thus created acting in the same way as the EFS in pure polymers as exciton traps. They also act as centers for charge carrier generation (61, 62) due to the charge separation which is characteristic for this complex. The same is true for the formation of real donor–acceptor complexes which are recognized by their characteristic absorption (CT) bands. These complexes are readily formed in aromatic polymers due to the relatively low ionization energy of the side groups (Table 5.3) (63). The potential energy diagram of exciplexes is similar to that of excimers (Fig. 5.3) in that their fluorescence is in general broad, structureless, and red-shifted with regard to the fluorescence of the components. The energy of the exciplex triplet state is slightly below that of the corresponding singlet (especially in polar exciplexes, due to the separation of the electron and hole on different molecules). It can only be observed if no "local" triplet state, belonging to one of the components, has a lower energy than the exciplex (67).

Excimer and exciplex formation will frequently be considered as important exciton trapping process in Section 5.4. Here we should briefly discuss the possible role of EFS and corresponding exciplex-forming centers (involving suitable guest molecules, mostly acceptors in donor polymers) in charge carrier trapping. There can be no doubt that weak acceptors act as electron traps in donor polymers. However, electrons seem to be trapped rapidly in polymers (68) anyhow, so that additional exciplex-forming guest molecules cause small changes. The suspicion remains that traces of weak acceptors, not easily detected analytically, are the reason for this strange behavior of polymers (68) (see also Chapter 6, Section 6.3).

Excimer formation is possible in chemically pure polymers, and there is no obvious answer with regard to trapping of electrons and holes. The question is whether the ionization energy (I) of an EFS is lower and/or its electron affinity (A) is higher in the solid polymer than the I and A of the basic units not capable of excimer formation. Photoelectron spectroscopy of cyclophanes, having a similar closely overlapping sandwich structure as EFS, shows that the ionization energy of free cyclophanes is lower than that of open analogues (64). Some examples are given in Table 5.3. The electron affinities of [2,2-] and of [3,3-] paracyclophane are higher than that of p-xylene, as evidenced by Gerson (65)

Table 5.3 Ionization Energy I and Electron Affinity A of Basic Units of Aromatic Polymers[a]

Polymer	Model of Basic Unit	I (eV)	A (eV)
Polystyrene	Toluene	8.82	<0 (66)
	[2.2] Paracyclophane[b]	8.1 (64)	[c]
Poly(vinylnaphthalene)[d]	Naphthalene	8.12	0.15
	1-Methylnaphthalene	7.96	
	2-Methylnaphthalene	7.95	
	syn-[2.2](1,4) Naphthalenophane[b]	7.2 (64)	
Poly(N-vinylcarbazole)	N-Isopropylcarbazole	7.4	0
Poly(vinylpyrene)[d]	Pyrene	7.72 (66)	0.58
Poly(vinylphenyl ketone)	Acetophenone	9.65 (66)	0.33 (66)

[a] Reference 19 if not quoted otherwise.
[b] Model for excimer-forming site.
[c] Affinity larger than in open-chain analogues (65).
[d] And related polymers.

by ESR in solution. Although there can be little doubt about the trapping potential of free cyclophanes—and therefore probably of EFS—this effect may be counterbalanced by decreasing polarization energy [P in Eq. (5.2)] due to the delocalization of the electron or the hole over the two groups of EFS.

Convincing evidence in favor of trapping by EFS however is offered by charge carrier mobility studies in pyrene (69) and perylene (70), both excimer-forming crystals and both showing thermally activated electron *and* hole transport. It should be pointed out that this behavior shows a close resemblance to exciton transport in the same crystals (36, 71).

5.2.5 Deactivation of Excited Electronic States

Up to this point, deactivation of excited electronic states has been considered from the viewpoint of energy transfer to guest molecules, EFS, or exciplex-forming centers, whereby the original state vanishes and a new excited state of lower energy is created. This state is eventually deactivated either by an intramolecular (radiative or radiationless) process, or again by energy transfer or some other bimolecular quenching process. The deactivation processes are qualitatively the same in the original state and in the trap (guest or EFS), the only difference being that traps, if present in small concentration, cannot form excitons.

The intramolecular deactivation processes are conveniently displayed in the form of a Jablonski diagram (72) (Fig. 5.4). These processes are now well known (73) and can therefore be treated concisely.

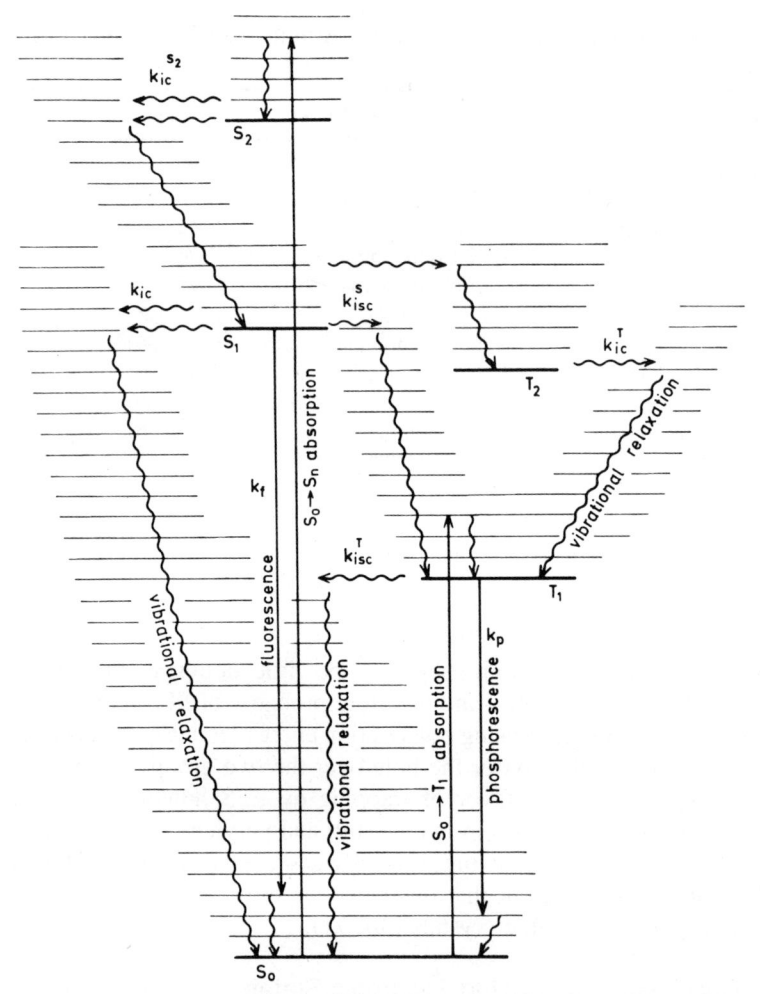

Figure 5.4 Jablonski diagram: $S_0, S_1, \ldots S_n$ = singlet states; $T_1, T_2, \ldots T_n$ = triplet states; k_{ic}, k_{isc} = rates of internal conversion and intersystem crossing; k_f, k_p = rates of fluorescence and phosphorescence.

Absorption

Absorption of photons of a few electron volts (visible and near-UV) creates excited singlet states which in aromatic polymers are essentially those of the side groups or other aromatic parts of the macromolecules. The absorption spectra in this spectral region are therefore similar to those of monomeric model compounds, representing isolated basic units, although some broadening of the bands is usually observed. The excited states differ from the ground state with respect to energy content, lifetime, and electron distribution. The transitions

BASIC CONCEPTS OF ELECTRONIC ENERGY TRANSFER

$S_0 \to S_n$ are quantified by their absorption coefficient, for example, the molar decadic extinction coefficient (ϵ):

$$\epsilon(\nu') = \frac{E(\nu')}{Cd} \tag{5.19}$$

where ν' is the wavenumber (cm^{-1}), E is the absorbance (log I_0/I), C is the concentration of light absorbing groups, for example, the basic unit of the polymer (mol/liter); and d is the thickness of sample, for example polymer film or solution (cm). Equation (5.19) is the usual form of expressing Lambert–Beer's law in chemical spectroscopy.

Fluorescence

Fluorescence is the radiative transition from the lowest excited singlet (S_1) state to the ground state (S_0). The transition is spin-allowed, but in some cases symmetry-forbidden; its decay time τ_o varies in most cases between 1 and 100 ns. The natural fluorescence rate k_e^S is roughly proportional to the maximal molar absorption coefficient of the $S_0 \to S_1$ absorption band.

The fluorescence is quantified by decay time and (true) quantum efficiency ϕ_F:

$$\tau_o = \frac{1}{k_e^S + \Sigma k_i^S} \tag{5.20}$$

$$\phi_F = k_e^S \tau_o \tag{5.21}$$

In the absence of bimolecular quenching processes, Σk_i^S is the sum of the rate of internal conversion to the vibrationally excited ground state and the rate of intersystem crossing to the triplet system.

Radiative S–T Processes

The transitions $S_0 \to T_n$ are spin-forbidden and can therefore not be observed by absorption spectroscopy, unless an enhancement procedure is applied, for example, oxygen perturbation or the heavy atom effect. The emission from T_1 to S_0 is called phosphorescence and decays, owing to its forbidden character, very slowly, $\tau_o' \approx 10^{-3}$–10s. Its decay time and true quantum efficiency ϕ_P are defined as

$$\tau_o' = \frac{1}{k_e^T + \Sigma k_i^T} \tag{5.22}$$

$$\phi_P = k_e^T \tau_o' \tag{5.23}$$

The long lifetime makes T_1 very susceptible to quenching so that phosphorescence can be observed only at low temperature, where bimolecular quenching is reduced, unless extreme care is taken to prevent quenching processes. At low temperature, the dominant intramolecular deactivation process is intersystem crossing back to S_0 (see Fig. 5.4). The most important diffusional quenching process is due to paramagnetic oxygen, yielding excited singlet molecular oxygen. This process can be suppressed by cooling with liquid nitrogen, except for bimolecular processes involving excitons ($T-T$ annihilation).

Internal Conversion (IC) and Intersystem Crossing (ISC)

Radiationless transitions are classified as IC when occurring within the singlet or triplet system and ISC if the spin state is changed. The latter processes are made possible, as in the case of radiative $T-S$ processes, by some mixing of the states with different multiplicity, most importantly by spin-orbit coupling. Radiationless transitions are always isoenergetic processes (see Fig. 5.4) leading to vibrationally excited lower electronic states, followed by rapid vibrational relaxation in condensed media. This relaxation constitutes, of course, a form of irreversible energy transfer to the matrix where rapid dissipation occurs.

Intermolecular quenching processes involving molecular diffusion are in general less important in solids than in liquids. Amorphous polymers do allow slow diffusion even below the glass transition temperature. This is especially true for small molecules such as oxygen, whose triplet deactivation ability has been mentioned. Singlet oxygen is the only known form of molecular diffusional energy transfer in solid polymers.

5.3 MEASURING TECHNIQUES

5.3.1 General

The experimental techniques used for studying energy transfer in polymers are mostly spectroscopic methods, although chemical methods are used in addition (Section 5.3.7). Since all polymer spectroscopic techniques are widely used for other research work, there is no need to discuss them in detail, except for some advanced instrumentation not yet in general use and where special problems exist in studying energy transfer in polymers. One important feature of polymer spectroscopy is that it is not merely the spectroscopy of polymers, since it involves the comparative investigation of monomeric, dimeric, and oligomeric model compounds, measurements on molecular crystals, glassy and liquid solutions, and so on. The deviations from the more or less undisturbed behavior of models point to specific polymer effects. The techniques applied have to be sensitive enough to detect these differences or to measure effects which can only be observed in polymers. Sensitivity, however, is often found to be much less important as a limiting factor than sample purity and preparation.

5.3.2 UV Absorption

Absorption spectra in the near UV (200–400 nm) and visible range of spectrum (400–700 nm) are conveniently measured with double-beam spectrophotometers recording the absorbance (log I_0/I) or transmission (I/I_0) as a function of wavelength or wavenumber. From this, the molar extinction coefficient (Eq. 5.19) can be deduced if the concentration of chromophores is known. In polymer films, especially in partly crystalline ones, light scattering can cause problems since it is in general not possible to find a reference sample showing exactly the same intensity and wavelength dependence of scattering. The films to be measured therefore need to be as thin as possible, consistent with homogeneity and absorbance. Most films are cast from solution and allowed to form slowly, for example, by evaporation in a nearly solvent-saturated atmosphere (7) or by centrifugation (75).

The preparation of small-size films of uniform thickness is a nontrivial problem which still needs improvement. The average sample thickness can be determined from area and weight, or by interferometric and microscopic techniques. Typical film thicknesses used for weak absorptions and luminescence studies are of the order of 10 μm. Strong absorptions have to be studied in much thinner layers, which can be prepared by dipcoating from solutions of different concentrations; the dipping of the substrate has to be performed slowly by means of a continuously working motor. Insoluble polymers can be prepared from the melt by pressing between inert plates. Unavoidable light scattering can be corrected for by plotting log absorbance versus log wave-number and extrapolating the linear part observed at longer wavelengths into the region of absorption (76).

5.3.3 Fluorescence

The measurement of prompt fluorescence involves recording the fluorescence spectrum (if necessary as a function of excitation wavelength) and determination of the relative quantum efficiency and of the decay behavior (Section 5.3.6). Modern fluorescence spectrophotometers consist of a source of "white" radiation (e.g., a xenon lamp), a monochromator selecting the excitation wavelength, a sample compartment, a second monochromator for analyzing the emitted light, a photomultiplier, an amplifier, and a recorder. More advanced spectrometers have built-in facilities for correction of the spectra to equal quantum intensity; otherwise this correction has to be applied by calibrating the lamp (for correcting excitation spectra) and the photomultiplier (for correcting the emission spectra).

Excitation spectra are in general not very useful in solid polymers, except in extremely thin films, since very often excitation maxima actually correspond to absorption minima, and vice versa (surface quenching). Emission spectra are often disturbed by reabsorption of fluorescence. In this case, very thin films have to be used and the geometry of the film in the optical light path has to be optimized. The most convenient part of the UV–visible spectrum for studying the fluorescence is between 300 and 600 nm due to the combined effects of light source intensity and detector sensitivity.

The determination of relative quantum efficiencies is very simple in principle (77) but more difficult in practice. The film to be measured has to be compared with a sample of known quantum efficiency and showing the same absorbance (preferably $E > 2$) at the excitation wavelength. The quantum efficiency is calculated from the ratio of the areas under the quantum-corrected fluorescence curves (7, 21, 77). In practice it is often difficult to find a suitable standard showing not only a well-defined fluorescence of known quantum efficiency but also similar optical characteristics, as refractive index, light scattering, and so on, compared to the sample. Since polymers of defined quantum efficiency are not available, monomers, for example, 9, 10-diphenylanthracene, have to be used. If the quantum efficiency of the standard is unity, reabsorption only shifts the emission toward the red without changing its intensity. More precise determinations are possible using an integrating (Ulbricht's) sphere for measuring sample and reference.

Fluorescence depolarization in rigid systems allows in principle the demonstration of energy transfer, as already shown by T. Perrin (50), but proves difficult in polymer films, probably because of light scattering.

5.3.4 Phosphorescence and Delayed Fluorescence

Delayed luminescence is usually studied using fluorescence spectrometers equipped with a mechanical device (phosphoroscope), such as a rotating cylinder that prevents the prompt emissions from entering into the second monochromator. Since phosphorescence and delayed fluorescence are in general extremely weak at ambient temperature, cooling of the sample by means of an appropriate cryostat is necessary. Most investigations are performed by cooling with boiling nitrogen at 77 K, or, more rarely with liquid helium. Low-temperature measurements omitting the phosphoroscope give the total luminescence spectrum, including prompt fluorescence, phosphorescence, and delayed fluorescence. Delayed excitation spectra can in favorable cases be used to measure S_0–T_n absorption.

5.3.5 Electron Spin Resonance (ESR) Spectroscopy

ESR measures unpaired electrons via the magnetic field effect on the magnetic moment associated with electron spin. It is therefore well suited for studying doublet and triplet states. Whereas doublet states have been extensively studied in polymers (78), mostly in connection with degradation processes, triplet work on polymers is still in its infancy (79). ESR spectrometers essentially consist of a strong, variable-field electromagnet, a microwave source (klystron), a wave guide, a cavity with sample holder, and a detection/recording system. Low-

MEASURING TECHNIQUES

temperature work is accomplished, as in the case of phosphorescence, by cryostats mostly using liquid nitrogen.

The information obtained from ESR allows one to identify the nature of paramagnetic species, for example, triplet states and their approximate concentration. More advanced techniques use microwave double resonance and optical detection via the phosphorescence emitted from sublevels of the triplet state.

5.3.6 Measurement of Formation and Decay of Excited States

The complexity of experiments on the kinetics of the formation and deactivation of excited states depends on the characteristic time scale, which may span 14 orders of magnitude from 100 s to about 1 ps. The decay of excited states is usually monitored following a short excitation pulse or rapid termination of continuous excitation. For decay measurements of reasonably intense phosphorescences, mechanical termination of the exciting radiation, for instance, by means of a camera shutter, is often sufficient. The decay curve is then observed using fast recorders ($\tau_o' > 0.5$s) or oscilloscopes. For very weak phosphorescences, computerized sampling/averaging techniques have recently been used (80).

The microsecond range, very important in solution work, is scarcely used in solid-state work. Strong flash lamps are available for this time range which can be used in conjunction with oscilloscopes. The same technique can be used down to the important nanosecond range, although the intensity of conventional spark lamps is small and sampling of pulses becomes imperative. However, severe noise problems hinder the decay measurement over more than one decade of intensity using this relatively simple technique.

Significant progress has been made by the time-correlated single-photon counting technique (81) the application of which to polymer fluorescence has recently been described (82). The most advanced nanosecond device includes 0.2-ns laser excitation and time gating of the emitted photons which pass through a scanning monochromator (82, 83). The multichannel analyzer and computer provide either the decay behavior at a selected wavelength or the emission spectra in time intervals after the excitation pulse. This system therefore is able to elucidate complex decay behavior involving more than one species, as often found in polymers.

A further extension to shorter time intervals is offered by picosecond techniques, made possible by ultrashort, intense pulse trains produced by mode-locked lasers (84). These pulses can be isolated and the frequency doubled or tripled, if necessary, and used for kinetic measurements in the picosecond–nanosecond range. In most experimental setups the laser beam is split by means of a semitransparent mirror and one of the beams is delayed by increasing its optical path, thus creating a time difference between the pulses (10 ps correspond to 3 mm in air). Thus one pulse can be used as zero-time or trigger pulse for the detection electronics.

Kinetic measurements in the picosecond range can be performed using the rapid buildup and decay of emissions or, alternatively, the transient absorption spectra of excited species. The picosecond measurement of emissions has recently been improved by use of streak cameras (85, 86, 81), where special phototubes convert the fast light pulse into an electron beam which is deviated by a fast, variable electric field and made visible on a phosphorescent screen. The duration of the original pulse is thus transformed into proportional distance on the screen, the intensity of the luminescence being a measure for the intensity of the light pulse. This technique seems to be extremely useful in studying energy transfer processes in aggregates of chlorophyll (86), which have some similarities with analogous processes in polymers. When studying energy transfer by means of picosecond techniques, it is critical to control the intensity of the exciting pulses. When using high intensities, S–S exciton annihilation can be observed, whereas at lower intensities, first-order quenching by exciton trapping is observed (86). In principle, the signals obtained from the streak camera can be photographed and evaluated by densitometry. A better way, however, seems to be the use of a vidicon combined with electronic averaging and an optical multichannel analyzer (81, 86). The resulting signal indicates the reference pulse followed by the buildup and decay of the emission selected by an appropriate filter.

▲₁ SINGLE PULSE 1060nm
▲₂ SINGLE PULSE 530nm
▲₃ SINGLE PULSE 353nm

(a)

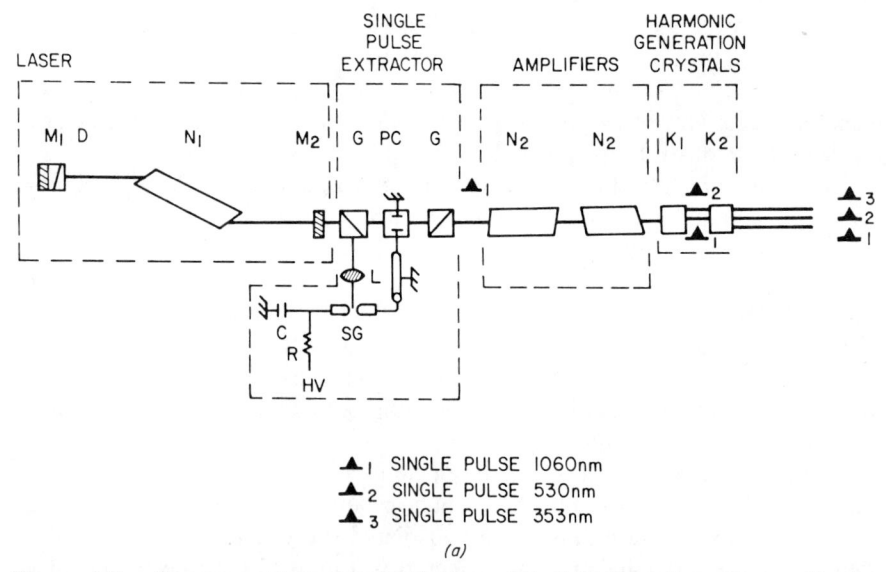

Figure 5.5a Picosecond laser block diagram. M_1, 100% cavity reflector; D, dye cell for mode locking solution; N_1, Laser oscillater rod (Nd^{+3}/glass); M_2, 50% cavity output coupling reflector; G, Glan laser prism polarizer; PC, Pockels cell; C, coaxial capacitor; R, charging resistor; HV, high-voltage power supply; SG, N_2 pressurized spark gap; L, lens; N_2, laser amplifier rod; K_1, type I KD*P doubling crystal; K_2, type II KD*P tripling crystal.

Figure 5.5 shows an apparatus used for picosecond transient absorption studies at the Xerox Webster Research Center (87). In this apparatus, single pulses of 1064 nm, 35 mJ per pulse with 7 ± 2 ps are generated and isolated as shown in Fig. 5.5a. The third harmonic, 1 mJ per pulse, is used as the excitation pulse in the near UV. After passing a time delay device, the fundamental creates "white" radiation in a cell filled with carbon tetrachloride (visible + near IR) or ethanol (near UV). This radiation is divided into a reference and a probe pulse. The sample is exposed to the excitation pulse and to the probe pulse through the same aperture. By appropriate adjustment of the length of the different light paths, the excitation pulse excites the sample first followed by the probe pulse. After passing through the sample the probe pulse is analyzed by a combined spectrograph and multichannel analyzer. Delaying the probe pulse with respect to the excitation pulse allows the transient absorption spectra to be recorded as a function of time.

▲ 1 SINGLE PULSE 1060nm
▲ 2 SINGLE PULSE 530nm
▲ 3 SINGLE PULSE 353
▲ 4 CONTINUUM PULSE

(b)

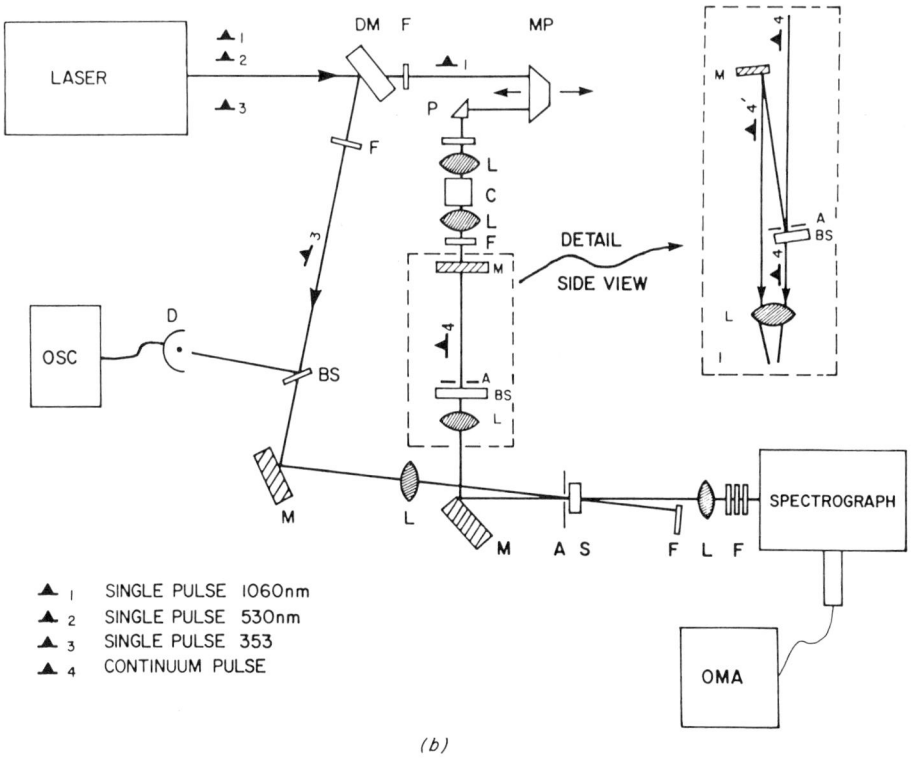

Figure 5.5b Laser photolysis and time-resolved spectra apparatus. DM, dichroic mirror; F, filter; MP, moveable prism; L, lens; C, continuum cell (2 cm path); A, aperture; BS, beam splitter; D, calibrated photodiode; OSC, 7904 Tektronix oscilloscope; S, sample cell. The "Laser" box is shown in detail in Fig. 5.5a.

5.3.7 Chemical Techniques

Energy transfer in polymers, although in most cases induced and detected physically, may also be studied in part by chemical means. Excited states can be created thermally, for example, by dioxetanes forming ketones in excited singlet and triplet states (29):

$$R_1-\underset{R_2}{\underset{|}{\overset{O-O}{\overset{|}{C}}}}-\underset{R_2}{\underset{|}{\overset{|}{C}}}-R_1 \xrightarrow{\Delta} R_1-\overset{O^*}{\overset{\|}{C}}-R_2(S_1 \text{ or } T_1) + R_1-\overset{O}{\overset{\|}{C}}-R_2 (S_0) \quad (5.24)$$

When molecules of this type and suitable guest molecules are added to a polymer at low temperature, energy transfer can be studied (29, 30) if reaction (5.24) is accelerated by warming the polymer sample and using very sensitive chemiluminescence techniques (88). Of course, only polymers with energy levels below or slightly above the ones created according to Eq. (5.24) can be investigated.

Energy transfer can also be detected chemically if the energy donor or acceptor groups are photochemically active. The first case is realized when using polymers containing ketone groups giving rise to main chain splitting. The reduction in the number-average molecular mass as a function of irradiation time and guest concentration is a measure of energy transfer. Stern–Volmer kinetics can be used for quantitative analysis (89). Another type of photochemical detection of energy transfer efficiency uses photochromic acceptor molecules (90, 91).

Finally it should be pointed out that numerous chemical as well as physical methods used in polymer research have an auxiliary function in energy transfer studies, such as nuclear magnetic resonance, gel permeation chromatography, and all other techniques that can be used to produce pure and well-defined polymers. Purification of polymers is achieved by multiple dissolution/precipitation from suitable solvent–nonsolvent systems. This procedure is very effective in removing physically dissolved impurities, including nonreacted monomer and additives. It is advisable to use carefully polymerized laboratory samples as starting materials instead of commercial polymers, although a general statement is not possible in this case. Since most solid polymer samples are studied as films produced from solution, the purity of the solvent used is of key importance.

It should be stressed that polymers can be obtained in a high degree of purity, although—as always in solid-state research—the search for impurities and their removal form an important part of the experimetal work.

5.4 RESULTS AND INTERPRETATION

5.4.1 Introduction

The experimental work devoted to energy transfer in solid polymers has been increasing for the last years so that quite a number of articles have now to be reviewed. The ordering of the material follows the chemical structure of the active groups, so that families of spectroscopically similar polymers are discussed together. Results obtained with monomeric and dimeric model compounds are mentioned, where appropriate, but not discussed exhaustively. We start with polymers containing carbazole groups, since these polymers, especially poly(N-vinylcarbazole) (PVCA),* have been most intensively investigated and some basic observations have been made with them. Energy transfer was observed earlier in polystyrene (PS), although no clear mechanism could be deduced from this work (92). Other families of polymers frequently investigated contain naphthalene, phthalimide, and aromatic and aliphatic ketones.

5.4.2 Polymers with Carbazole Groups

Poly (N-vinylcarbazole)

Carbazole polymers and their complexes with acceptor molecules have been extensively discussed for their photoconductive properties, as described in Chapter 6. Many of these properties are related to the phenomena observed in energy transfer discussed in the following, that is, transport and trapping sites for excitons and charge carriers.

The discussion of energy transfer in solid PVCA is conveniently divided into measurements at ambient temperature, which can be explained with a simple model and low-temperature studies revealing a quite complex behavior due to the increasing influence of shallow exciton traps and triplet processes not observed at higher temperatures. Furthermore, puzzling relationships between the

---

*PVCa, PVCz, or PVK, from obsolete German Karbazol, to distinguish it from PVC; trade name Luvican, BASF.

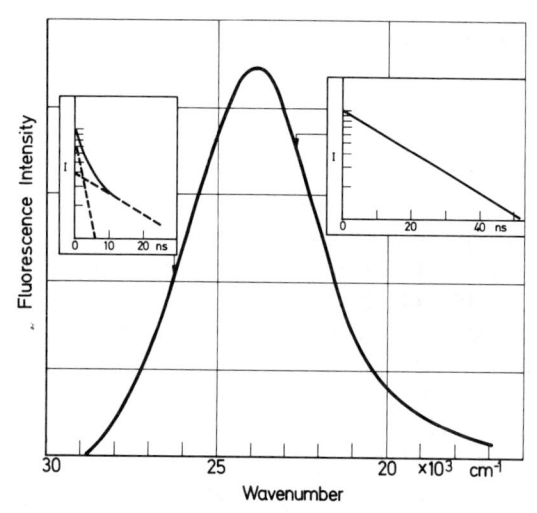

Figure 5.6 Fluorescence spectrum of a PVCA film at room temperature (excitation wavelength 320 nm) and fluorescence decay curves (inserts), measured at the wavenumbers indicated by arrows. (From Klöpffer and Bauser [97].)

mode of polymerization and emission properties appear in the low-temperature spectra.

Energy Transfer at Room Temperature. Solid PVCA shows a broad, structureless fluorescence spectrum (Fig. 5.6) which has been identified as excimer fluorescence by comparison with monomeric model compounds (7, 100). Since the fluorescence typical for N-alkylated carbazole groups (21), showing vibrational structure in the near UV, is absent even in very thin films where reabsorption effects can be ignored (101), it is concluded that excimer formation is very efficient in solid PVCA. Molecular mobility being minimal in this polymer far below the glass transition temperature of about 210°C (102), excimer formation can occur only at preformed sites (excimer-forming sites, EFS) (7).

For sterical reasons it is not possible for all side groups to form the sandwich-like structure required for (singlet) excimers. Hence exclusive excimer fluorescence in itself gives excellent evidence of energy transfer in solid PVCA and similar rigid aromatic polymers whose intrinsic monomer fluorescence efficiency is high.

The side groups forming the EFS show ordinary carbazole-type absorption since the ground state is nonbonding (Fig. 5.3). Therefore they cannot act as energy acceptors in a Förster energy transfer process, which points to exciton diffusion as the only alternative process. In order to quantify the range of the excitons, as controlled by the concentration of EFS, the simple exciton hopping model discussed in Section 5.2.1 has been modified and checked for different guest substances (7, 95, 98). The model predicts a linear increase of the quenching factor with guest concentration (7),

RESULTS AND INTERPRETATION

$$Q = \frac{c_G}{c_E} \quad (5.25)$$

and allows to estimate the concentration of EFS (c_E). It is assumed in this model that no significant long-range transfer from excimer to guest occurs, that k(exciton to EFS) = k(exciton to guest) = k_h, and that energy transfer is much more efficient than monomer fluorescence. Direct excitation of EFS is considered to be unimportant.

The simple model describes the experimental results very well (7, 98) and gives an estimate of $c_E \approx 2 \times 10^{-3}$ mol/mol basic unit (see Table 5.4). Most convincingly, guest molecules without any absorption below the carbazole S_0–S_1 transition (345 nm) and quenching by weak CT or exciplex interaction (HCX, DMTP, Table 5.4) act as efficient exciton traps. This excludes any long-range transfer that may offer an additional channel if strongly absorbing guests are used (61, 96). The quenching factor in Eq. (5.25) can be measured by fluorescence intensity decrease (I, I_0) as in Eq. (5.5) or by the ratio of guest-to-excimer fluorescence using

$$Q = \frac{I_G}{I_E} \frac{\phi_E}{\phi_G} \quad (5.26)$$

Table 5.4 Energy Transfer and Excimer Decay in PVCA at Ambient Temperature

Guest Molecules	Concentration of EFS ($c_E \approx c_I$) (mol/mol basic unit)	Decay Time (ns)	Reference
Hexachloro-p-xylene (HCX); perylene; 2,4,7-Trinitrofluorenone (TNF)	10^{-3}		7
		$\tau_I = 43$ (1 atm); 39 (30 kbar)	93
Perylene	a)	$\tau_I = 20$; $\tau_{II} = 10$	94
Dimethyl terephthalate (DMTP)[b]; perylene	$(1.5\text{–}3.3) \times 10^{-3}$		95
		$\tau_I = 23 \pm 1$; $\tau_{II} = 2 \pm 1$	97
DMTP	$(2.2\text{–}2.7) \times 10^{-3}$	$\tau_I = 22 \pm 2$; $\tau_{II} = 6 \pm 1$	98
		$\tau_I = 26$ (8 kbar); $\tau_{II} = 6$ (8 kbar)	99

[a] Deviation from simple hopping model claimed according to time dependence of guest fluorescence.
[b] Quenching of exciplex fluorescence by an external electric field has been reported for this guest (61, 96) as well as extrinsic photocharge carrier generation.

where I_G and I_E are the intensities of guest and excimer fluorescence in the presence of guest molecules, and ϕ_G and ϕ_E are the quantum efficiencies of guest (dissolved in polymer) and excimer fluorescence.

It is not possible, however, to determine Q from excimer decay times—as indicated in Eq. (5.5) for exciton fluorescence—since excimers do not actively take part in energy transfer and hence their decay time is independent of the presence of guest molecules. If, on the other hand, long-range transfer excimer-to-guest occurs, the excimer decay should follow Eq. (5.12).

Energy Transfer in PVCA at 77 K The binding energy of carbazole excimers in PVCA (singlet exciton trap depth) is not known exactly, although a rough value of $E \approx 1000$ cm$^{-1} \approx 5$ kT, at room temperature, can be estimated from the spectra. At low temperature, shallow traps can become effective.

Johnson and Offen (93) discovered a second fluorescence band in PVCA at 77 K which is due to a second intrinsic singlet exciton trap (101), most probably another excimer forming site. It should be noted that a weak shoulder at the short-wavelength side of the excimer fluorescence (94) and a second component in the fluorescence decay (97) have occasionally been observed even at room temperature (Fig. 5.6). As a consequence of the appearance of the new "trap II," singlet energy transfer to guest molecules should be less efficient at low temperature. At low temperature, triplet-state processes which are quenched at ordinary temperatures often become observable. In PVCA, phosphorescence (101, 103) as well as delayed fluorescence (80, 97, 104, 115) can be observed at 77 K. As in the case of fluorescence, polymer phosphorescence is red-shifted with respect to monomer (carbazole) phosphorescence (97, 101, 103) and has been interpreted as excimer phosphorescence (101).

Itaya, Okamoto, and Kusabayashi (104, 105) have observed that the fluorescence of PVCA depends on the method of polymerization: cationic polymerization favors "true" excimer fluorescence whereas in radically polymerized samples, "trap II" fluorescence dominates. There is obviously a relationship between polymerization and the ability of the resulting polymer to develop the different excimer-forming sites in solution (105) or, frozen in during film formation, in the solid state (104). It has been claimed by the same group (106) that the differences observed are due to predominant syndiotactic configuration of radicalic PVCA, whereas the cationic polymers are said to contain up to 50% isotactic sequences. From molecular models it can be deduced that the 1,3-sandwich conformation responsible for the long-wavelength excimer can easily be formed at isotactic sequences. Similarly, for trap II, a weakly overlapping 1,3-structure has been proposed (105) which can be formed without difficulty in syndiotactic regions.

Time-resolved spectroscopy in solutions of PVCA has recently shown that even the "true" excimer may be preceded by a species similar to trap II (107, 108). Unfortunately, PVCA samples of well-defined and homogeneous tacticity are not yet available. It is nevertheless important to compare polymers prepared by cationic and radicalic initiation due to the differences in forming the intrinsic traps.

RESULTS AND INTERPRETATION

The results of recent work in our laboratory (109, 110) using both cationic and radicalic PVCA samples can be summarized as follows:

In accordance with the work by Itaya et al. (104, 105), trap II emission dominates in radicalic, presumably syndiotactic, PVCA, whereas in cationic PVCA "true" excimer fluorescence is stronger (Fig. 5.7)

The delayed emission consists of phosphorescence (P) and delayed fluorescence (DF) in both types of polymers.

The relative importance of the two emissions is a function of excitation intensity, as can be seen in Fig. 5.8 for weak, medium, and high excitation intensity: the intensity of the delayed fluorescence increases relative to that of phosphorescence with increasing excitation intensity.

Quantitatively, the following can be shown:

The phosphorescence is proportional to excitation intensity up to about 10^{14} photons s^{-1} cm^{-2} and tends to saturate above this level.

Delayed fluorescence depends quadratically on radiation intensity below 10^{14} photons s^{-1} cm^{-2} and only linearly above. Fluorescence, on the other hand, is linear over the whole range investigated (10^{10}–10^{17} photons s^{-1} cm^{-2}).

The quantum efficiency of P and DF is much smaller than that of prompt fluorescence. The behavior of the radicalic and cationic PVCA is qualitatively similar.

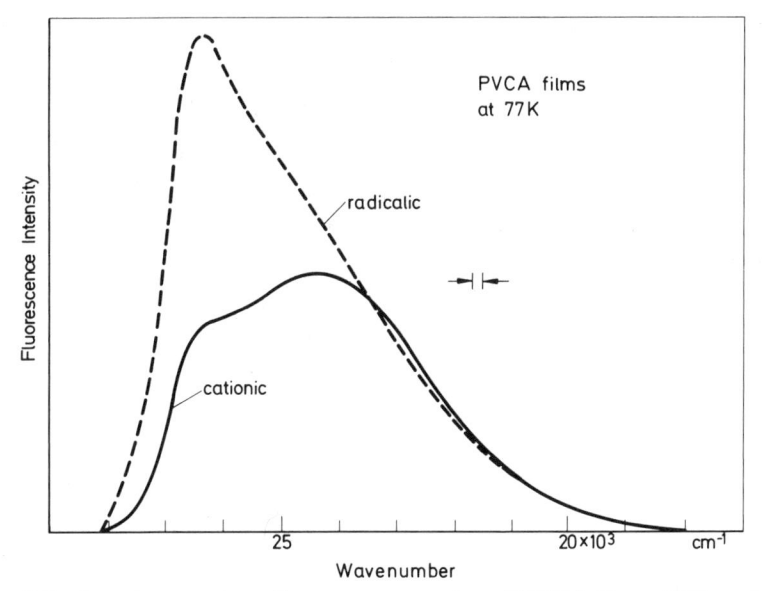

Figure 5.7 Low-temperature fluorescence spectra of PVCA films. Different ordinate scales. (From Rippen et al. [110].)

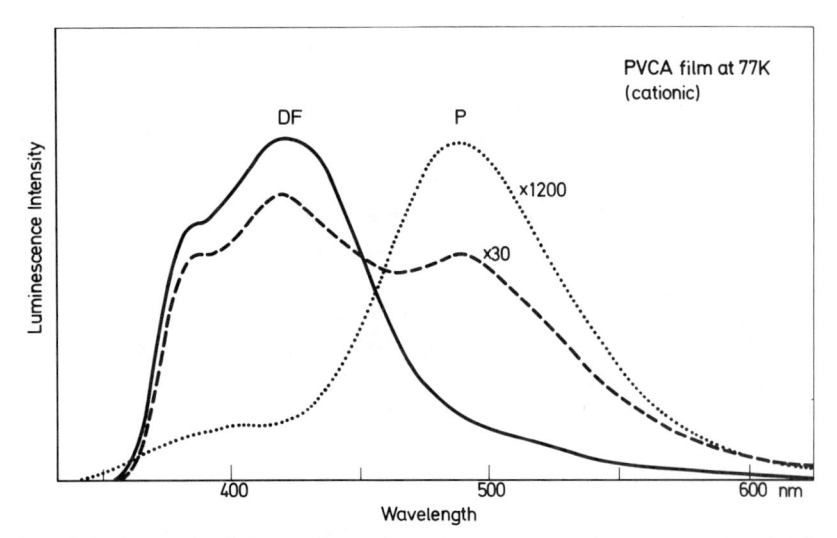

Figure 5.8 Delayed fluorescence (DF) and phosphorescence (P) of PVCA at 3×10^{13} (. . . .), 3.8×10^{14} (– – – –), and 1.0×10^{16} (——) photons s^{-1} cm^{-2} excitation intensity. Ordinate scale has been multiplied by the factors indicated. (From Rippen et al. [110].)

There are two exponential phosphorescence decay times of about 10 and 2 s. The decay of delayed fluorescence is much faster (40–50 ms) and seems to be independent of polymer preparation.

The delayed fluorescence is red-shifted with regard to prompt fluorescence.

A kinetic model has been tailored (111) specifically for PVCA at 77 K, making the following assumptions:

Energy transfer can be described with a simple hopping model (2, 7), the probability of trapping (or annihilation) being given by $k_h \times c_E$, where k_h is the triplet or singlet exciton hopping rate and c_E is the concentration of one of at least three different excimer-forming sites.

The differences between radicalic and cationic polymers consist only in different concentrations of the excimer-forming sites, not in their nature. All hopping rates are assumed to be equal in both types of polymers.

There are at least three different intrinsic traps indicated by roman numbers: I, "true excimer-forming sites"; II, traps II (or high-energy excimer-forming sites); III, phosphorescent excimer-forming sites (according to decay measurements most likely *two* different sites, but only the long-lived one has been considered in the model for the sake of simplicity).

Unless an unreasonably long radiative phosphorescence lifetime (III) in the order of hours is accepted (caused by small quantum efficiency and slow decay), an efficient nonradiative triplet trap has to be present in the PVCA samples. This has tentatively been ascribed to the triplet state of excimer I. Trap II, on the other hand, is unlikely to provide a binding triplet excimer state, since neither additional quenching of triplet exciton diffusion nor saturation of fluorescence is observed in radicalic PVCA, which is rich in trap II.

In this model, delayed fluorescence is due to T-T annihilation between migrating triplet excitons and immobile triplet excimers (III). In the low intensity regime, increasing the excitation intensity increases both the triplet exciton density (by intersystem crossing at EFS II or at the monomer level) and the probability of triplet excimer formation (both via singlet and triplet excitons diffusing to EFS III). Thus delayed fluorescence depends quadratically on exciton intensity in this range. Under high excitation intensity, the long-lived triplet excimer and phosphorescence intensity saturate, whereas the delayed fluorescence change to linear behavior, since only the triplet exciton production increases further with intensity.

The "delayed singlets" formed during T-T annihilation in part remain in the trap, causing the red-shifted DF, and in part migrate away until they are trapped at EFS I and II. S–T annihilation is quantitatively insignificant, due to the small concentration of EFS III. Therefore no saturation of prompt fluorescence is observed which should be observed if one of the main EFS was the phosphorescent excimer.

This model (111) describes the experimental facts quantitatively if the different emissions are formulated by suitable rate equations and solved for stationary conditions. The parameters used for calculating the intensities of phosphorescence and delayed fluorescence are summarized in Table 5.5. Although the data have to be considered as preliminary, it seems that EFS II and III are present in much higher concentration in radicalic PVCA and thus presumably related to syndiotactic sequences. The triplet hopping rate is much smaller than in molecular crystals (Table 5.7) or even in amorphous aromatic layers, for example, tetracene (112). This behavior points to the presence of additional shallow triplet traps in PVCA. Evidence for such traps are often found in rigid solutions of aromatic polymers at low temperature (15) where excimer formation is in general not observed.

An important result of the kinetic analysis of the experiments is the existence of separate phosphorescent excimers which are not related to one of the main singlet excimers. Otherwise, the linearity of fluorescence versus excitation intensity could not be explained: fluorescence should saturate at high intensities due to singlet–triplet (S–T) annihilation, a Förster-type, spin-allowed process which is bound to occur if saturated triplet traps are present in the polymer in a high concentration. A direct experimental proof for this fact in addition to the linear dependence of I_F versus I_A is the red-shifted delayed fluorescence, indicating that those singlets created in T-T annihilation and staying at the site of their formation have a higher excimer binding energy. They are therefore different from the states causing the prompt fluorescence.

Increasing the temperature leads to detrapping of singlets as well as triplets, thus effectively increasing the range of the respective excitons. A study made by the chemiluminescence technique slightly above room temperature (30) showed that triplet excitons are more mobile than at 77 K. Burkhart and Avilés (80, 113, 114) studied detrapping of triplets in PVCA over a large temperature range and found different behavior of the two phosphorescences showing different decay times. They assigned the faster decaying phosphorescence with a deep trap, whereas the long-lived triplet excimer, considered in the above kinetic treatment, was said to be a shallow trap. They also observed weak phosphorescence at room temperature when oxygen was carefully excluded (80).

Poly(2- N-carbazolylethyl vinyl ether)

PCEVE

PCEVE is a more open-structured analogue of PVCA which has been obtained as a low-molecular-weight product (104). It shows residual monomer (CA) fluorescence in films at 77 K in addition to the two excimer fluorescences already known from PVCA. The concentration of EFS seems to be lower in this polymer (98):

$$c_I + c_{II} = 4.5 \times 10^{-4} \text{ mol/mol basic unit}$$

so that the range of singlet excitons is higher than in PVCA. Phosphorescence is broad and centered at 495 nm; delayed fluorescence is weak and has the same spectral composition as prompt fluorescence (maxima at 370–371 and 388 nm, shoulder at 410 nm). In rigid solution, for comparison, these are maxima at 360 and 376 nm, indicating monomer fluorescence.

The fluorescence decays with two components, $\tau'_o = 3.5 \pm 0.2$ and 0.7 ± 0.05 s in solid films (6.6 ± 0.1 s in rigid solution). Delayed fluorescence again decays much faster ($\tau_{DF} = 45$ ms) and increases with the square of the excitation intensity (104). Triplet excitons in PCEVE have also been observed by sensitized phosphorescence, using naphthalene as guest. The phos-

Table 5.5 Data Used for Calculating Delayed Fluorescence and Phosphorescence of PVCA at 77 K.

Parameter[a]	PVCA (cationic)	PVCA (radicalic)	Reference
$\tau_{(P)}^{III}$ [b]		9.6 (± 0.9) s	110
k_h^S		2×10^{12} s^{-1}	111
$\tau_{(F)}^{I}$		28 ± 2 ns	97
$\tau_{(F)}^{II}$		4 ± 1 ns	97
$c_I \approx c_E$		2×10^{-3} mol/mol basic unit	7, 95, 98
ϕ_F^I		0.09	111
ϕ_F^{II}		0.5	111
$c_I / \Sigma c$	0.9	0.25	111
$c_{II} / \Sigma c$	0.1	0.75	111
c_{II}	2×10^{-4} mol/mol basic unit	6×10^{-3} mol/mol basic unit	111
ϕ_P^{III}		0.2	111
k_h^T		10^4 s^{-1}	111
c_{III} [c]	1.2×10^{-5} mol/mol basic unit	2.6×10^{-4} mol/mol basic unit	111

[a] τ = Decay times according to $I = I_0 \exp(t/\tau)$; ϕ = true quantum efficiency, defined as ratio of radiative rate/sum of all deaction rates; k_h = hopping rate; P = phosphorescence; F = fluorescence; S = singlet; T = triplet; Roman numerals indicate different EFS discussed in text.
[b] Long-lived triplet excimer.
[c] Estimated using a radiative lifetime of the triplet excimer of τ_e^{III} = 50 s. c_{III}/τ_e^{III} = 2.5×10^{-7} s^{-1} (cat.) and 5.3×10^{-6} s^{-1} (rad.).

concentration, thus indicating that the phosphorescent centers are traps, most probably triplet excimers.

Other Poly(vinylcarbazoles)

The structure of PVCA has been modified by introducing electron-withdrawing groups, for instance [brominated PVCA (104), poly(N-vinyl-3,6-dibromocarbazole) (116), or poly(N-acryloylcarbazole) (104)], or by linking the N-ethylcarbazole groups in positions 2 to 4 with the backbone (117, 118).

These polymers have not been investigated as films in detail. Bromination increases intersystem crossing rates and hence favors triplet state processes. Excimer phosphorescence with a maximum between 450 and 500 nm has been observed in films (104) as well as in rigid solutions (116). Low-molecular-weight systems with properties similar to films of PVCA consist of amorphous layers of 1,3-bis(N-carbazolyl)propane (119, 120) and polystyrene, highly doped with N-isopropylcarbazole (121).

5.4.3 Polymers with Naphthalene Groups

Poly(2-vinylnaphthalene)

[structure of P2VN: naphthalene group attached to —CH—CH$_2$— repeat unit, subscript n]

P2VN

Films of P2VN are characterized by broad excimer fluorescence with a maximum near 400 nm (122), whereas the monomer S_1 level, observable as weak transition in solution and in monomer fluorescence, is situated near 320 nm. Complete absence of monomer fluorescence indicates energy transfer via singlet excitons.

Alternating copolymers have been prepared from 2VN with methyl methacrylate and styrene (123). These copolymers do not show excimer fluorescence in solution or film, thus pointing to intramolecular excimer formation between neighboring side groups in the homopolymer. Kim and Webber (124) observed a molar mass dependence of triplet energy transfer efficiency similar to that observed in dilute, rigid solutions (15). This may indicate predominant intramolecular exciton migration, due to better overlap of electronic wave functions of naphthalene groups within the individual coils compared to interchain overlap and, related to it, increased transfer probability (see Section 5.2.3).

Delayed fluorescence in solid P2VN at 77 K is red-shifted by 1800 cm^{-1} with regard to prompt fluorescence (124), showing that triplet exciton–triplet excimer annihilation is the mechanism of delayed fluorescence, provided the broad phosphorescence band between 500 and 600 nm is due to excimer phosphorescence. The nonequivalence of triplet and singlet EFS implied in this interpretation is in agreement with the experiment of Chandross (125) on naphthalene sandwich pairs in rigid solution. The red-shifted delayed fluorescence had earlier been observed by Fox and co-workers (126) and interpreted as (delayed) exciplex fluorescence. This interpretation rests on the assumption that the phosphorescence observed is due to some acceptor-type impurity molecule or its complex formed with naphthalene side groups. Beardslee and Offen, however, observed a similar red shift in delayed fluorescence of naphthalene crystals (127) due to high pressure-induced lattice defects, pointing to intrinsic traps (EFS) rather than to impurities as cause of the red shift.

Finally, it should be mentioned that P2VN has been used as a probe in compatibility studies in polymer blends by Frank (128). Strong excimer fluorescence indicates crowding of naphthalene groups and beginning segregation of the guest polymer (P2VN), whereas monomer fluorescence indicates

RESULTS AND INTERPRETATION

unperturbed solution. This effect is of course enhanced by "down chain" singlet energy transfer to suitable EFS.

Poly(1-vinylnaphthalene)

P1VN

This polymer has first been investigated in liquid solution, showing excimer fluorescence (129), and in rigid solution at 77 K, showing monomer phosphorescence, monomer fluorescence (129), and delayed fluorescence (130), the latter being due to pseudo-one-dimensional triplet–triplet annihilation. The solid-state behavior has first been investigated by David, Demarteau, und Geuskens (131). Again excimer fluorescence (λ_{max} = 400 nm) only is observed, at room temperature and at 77 K. A second maximum in the low-temperature spectrum, similar to trap II in PVCA, may be due to an impurity (122).

Singlet energy transfer has been studied quantitatively (131) using benzophenone as guest. A reinterpretation of the data using the exciton hopping model [Eq. (5.25)] yields $c_E \approx 5 \times 10^{-3}$ mol/mol basic unit. The absence of monomer fluorescence has been correctly interpreted as evidence for "energy migration" to suitably oriented side groups (131); a quantitative formulation, however, has been attempted using Förster's theory. A similar confusion of the two models had earlier obscured the results obtained with polystyrene (132). The quantum efficiency of excimer fluorescence was later determined (133, 134) by the same group (Table 5.6).

Table 5.6 Data Related to Energy Transfer in Films of Naphthalene Polymers

	Temperature	P2VN	P1VN	PACN
$\phi_E{}^a$	RT		4×10^{-2}	1.5×10^{-2}
	77 K		0.13	8.2×10^{-2}
c_E(mol/mol basic unit)	RT and 77 K		5×10^{-3}	6×10^{-3}
λ_{max}(fluorescence) (nm)	RT	400	400	390
	77 K	390	400 (131)	
			360 (122)	
λ_{max}(delayed fluorescence) (nm)	77 K	413 (126)		
		417 (124)		

[a]Measured relative to PS (0.04) and PVCA (0.03) at RT (133).

Delayed luminescence spectra recorded by Fox and co-workers (126) strongly depend on sample purity. The intensity of delayed fluorescence has been found to depend on the 1.7th power of excitation intensity. Doubtlessly, triplet excitons do occur in films of P1VN at 77 K as well as in the other naphthalene polymers.

Polyacenaphthylene

$$\left[\begin{array}{c} \text{naphthalene} \\ -CH-CH- \end{array} \right]_n$$
PACN

Due to its rigid structure, this polymer is characterized by its inability to form next-neighbor excimers. Consequently, excimer fluorescence is weaker in solution relative to poly(vinylnaphthalenes) (133). In films, however, only excimer fluorescence is observed, indicating either next-nearest-neighbor- or interchain interactions. Singlet energy transfer has been investigated quantitatively (133); the concentration of EFS which can be estimated from the published data is about the same as found in P2VN (Table 5.6).

5.4.4 Polystyrene

$$\left[\begin{array}{c} \text{phenyl} \\ -CH-CH_2- \end{array} \right]_n$$
PS

Historically, PS was the first polymer to be studied from the viewpoint of energy transfer efficiency. Soon after the discovery of plastic scintillators (135), prepared by bulk polymerization of styrene containing small amounts of fluorescent guest molecules ("sensitizers"), the first mechanistic interpretation—emission of polymer fluorescence and reabsorption by guest molecules (136)—was shown to be only partly correct (137, 138). In a plastic scintillator, each high energy particle creates highly excited states of the matrix which quickly return to the metastable states observed by fluorescence spectroscopy. The guest molecules added have to shift the emission into the spectral region of highest sensitivity of the photomultiplier used as a detector. It soon turned out that the dramatic increase in sensitivity due to the presence of guest molecules can only be explained by nonradiative energy transfer, since guest molecules with high quantum efficiency of fluorescence can increase the total light output by pre-

RESULTS AND INTERPRETATION

venting excited states of the polymer from radiationless decay. Emission–reabsorption, on the other hand, can increase the sensitivity only by a bathochromic shift of the emitted radiation, the number of photons emitted being equal in the most favorable case. The guest molecules, for example, *p*-terphenyl (135), anthracene, *trans*-stilbene (137), and several oxadiazoles known from liquid scintillation counting, have been shown to be most effective if they showed intense absorption in the region of PS fluorescence and efficient fluorescence around 400 nm, conditions clearly favoring both Förster transfer and emission–reabsorption.

Three nonradiative mechanisms have been identified and discussed by Brown, Furst, and Kallmann (14):

Energy migration (exciton hopping).
Single-step (Förster) transfer.
Molecular diffusion of excited species.

The last mechanism can be disregaded in solid polymers, with the exception of singlet molecular oxygen which is not relevant here. A clear-cut distinction between the first two mechanisms, still difficult today, was impossible before Basile's milestone report on PS fluorescence (141). He showed that the fluorescence of bulk-polymerized PS invariably is due to emission of residual monomer (Fig. 5.9), whereas the pure PS shows a broad and structureless fluorescence band, later on to be identified as excimer fluorescence by Hirayama

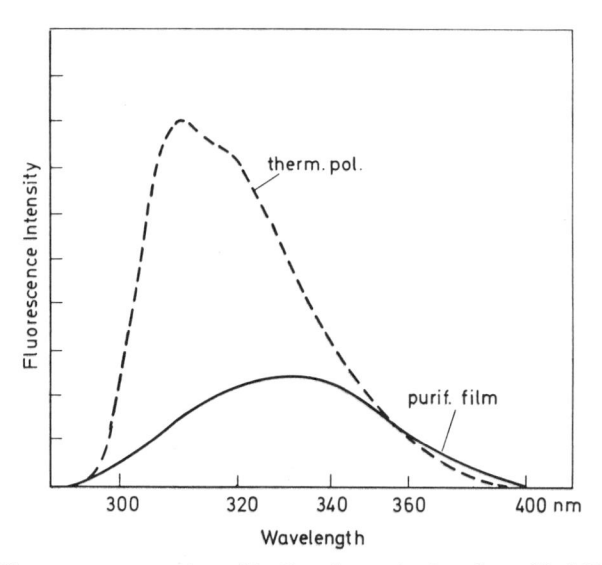

Figure 5.9 Fluorescence spectra of bulk-polymerized and purified PS in the solid state. (From Basile [141].)

(140) and Vala, Haebig, and Rice (129). Of course, excimer fluorescence indicates nonradiative energy transfer, most likely by excitons.

Exciton migration has been postulated by Hirayama et al. (132) but questioned by Heisel and Laustriat (142). The difficulties in distinguishing between the exciton and Förster mechanisms seem to be due to the high concentration of EFS in PS (Table 5.7), favored by the relatively small phenyl rings. Förster transfer from the excimer to suitable guest molecules is therefore dominating, the excimer itself being formed by trapping of excitons. Direct Förster transfer from excited side groups has been claimed (132), but this is incorrect (142, 143) due to the nondetectable phenyl fluorescence [see Eq. (5.10)].

The energy transfer mechanism involving Förster single-step transfer from excimer to guest molecules has clearly been recognized by Heisel (142). The agreement between calculated and experimental critical radii is quite satisfactory except for p-terphenyl, which is a "bad" acceptor due to its absorption band at short wavelengths (Table 5.7). In this case, additional exciton transfer can safely be assumed; however, other explanations have been given (142). Convincing evidence for energy transfer by exciton migration has been offered by Geuskens

Table 5.7 Energy Transfer Data for Different PS/Guest Systems

Guest	R_0 (experimental) (nm)	R_0 (calculated) (nm)	c_E (mol/mol basic unit)	Reference
p-Terphenyl	1.53	1.24	$(4.3 \times 10^{-3})^a$	142
	1.8	1.3		143
PBD[b]	1.87	1.75		142
α-NND[c]	2.08	1.85		142
Naphthalene-d_8	1.3	0.75		143
Pyrene-d_{10}	2.2	1.83		143
Cumene hydroperoxide	(1.12)	$\ll 1^e$	1.1×10^{-2d}	144
Tetraphenylbutadiene	1.95 ± 0.1			145
	2.18			146
Oxygen		$\ll 1^e$	$(2 \times 10^{-3})^f$	149

[a]Reinterpretation of Fig. 5.6, TP (142) taking $Q = 1$ at transfer efficiency = 0.5. The relatively low value of c_E shows that Förster transfer is at least partly involved.
[b]Phenyl(2-p-biphenyl-5-oxadiazole-1,3,4).
[c]Di(1-naphthyl)-2,5-oxadiazole-1,3,5).
[d]Reinterpretation, see text.
[e]Estimated.
[f]Reinterpretation of Fig. 5.3 in Ref. 149 using 1 bar $O_2 \triangleq 3.1 \times 10^{-3}$ mol/l^{-1} and assuming that Henry's law is valid. Oxygen forms a weak CT complex with the phenyl groups of PS (149); $I_0/I = Q + 1$ is linear up to about 15 bar.

RESULTS AND INTERPRETATION

and David (144), using weakly absorbing cumene hydroperoxide as guest, although the results have been interpreted in terms of a critical radius rather than using the hopping model. A reinterpretation of these results using Eq. (5.25) yields $c_E \approx 10^{-2}$ mol/mol basic unit (Table 5.7). We therefore understand that Förster transfer dominates in the case of guest molecules whose absorption bands are intense in the region of PS excimer fluorescence.

Similar although more complicated transfer pathways are likely to cause sensitization in bulk-polymerized plastic scintillators, monomeric styrene acting as intermediate trap in addition to the always present EFS. Exciton migration is to be expected as a competitive mechanism and as the only mechanism possible if the absorption by the guest is weak above, say, 290 nm.

David, Piens, and Geuskens (145) observed partly monomeric (side group) fluorescence at 77 K in atactic, amorphous PS, possibly indicating incomplete energy transfer at low temperature. The same group, however, observed purely excimeric fluorescence in 30% crystalline, isotactic PS at 77 K (147). This may indicate more efficient energy transfer in isotactic PS or, alternatively, higher c_E. It should be noted that in the case of incomplete energy trapping by EFS, as evidenced by side group fluorescence, the above criticized mechanism of Förster transfer from excited, monomeric side groups to guest molecules can operate, and that Eq. 5.25 is not applicable.

The origin of excimer fluorescence in PS, which in technical grades is always obscured by fluorescence bands originating from degradation products (148), seems to be essentially an intramolecular 1,3-interaction. In alternating copolymers of styrene and methyl methacrylate, monomeric phenyl fluorescence is exclusively observed (123) with a maximum near 290 nm and with some indication of vibrational structure. Using low-molecular-weight model compounds, Hirayama showed beyond any doubt the preferred role of 1,3-interaction in excimer formation of phenyl-substituted alkanes (140).

Using the chemiluminescence technique, Turro et al. (29, 150) studied different energy transfer processes in systems where PS served as solid matrix. Most processes could be explained by single-step transfer, except for the transfer from acetone T_1 (from thermally dissociated tetramethyl-1,2-dioxetane) to 1,4-dibromonaphthalene, a well-known triplet acceptor. The unusually large critical radius ($r_0 = 2.1$ nm) points to PS triplet excitons participating in energy transfer, although the triplet level of phenyl groups ($T_1 = 343$ kJ/mol) is somewhat higher relative to acetone T_1 (326 kJ/mol) (29) so that its population requires some thermal activation. The range of the triplet exciton may furthermore be restricted by triplet EFS. The existence of phenyl group excimers could clearly be established in cyclophane model compounds (58).

Excimer phosphorescence in PS has not yet been identified with certainty, since in the expected spectral region (400–450 nm) a strong impurity phosphorescence due to acetophenone-type end groups (148) is observed. It remains to be shown whether extremely careful preparation of PS films will succeed in showing the structureless excimer phosphorescence similar to that of [3,3-]paracyclophane. Indirect evidence for excimer formation is offered by the absence of any phenyl-type phosphorescence in films of PS, which can clearly be observed in rigid solutions (129).

Finally, a report by Guillet and co-workers (151) should be mentioned, describing energy transfer and emission spectra in copolymers of styrene and phenylacetylene. This copolymerization introduces isolated and conjugated double bonds into the main chain of PS molecules. The fluorescence spectra of these products resemble those of several technical PS samples investigated, whose blue fluorescence has been ascribed by partial degradation (148).

A few publications deal with energy transfer in the structurally related polymers poly(vinyltoluene) (138, 152, 153, 154) and poly(α-methylstyrene) (155) but are not discussed here for reasons of space.

5.4.5 Polymeric Aromatic Ketones and Imides

Poly(vinyl phenyl ketone)

PVPK

The chromophoric group of PVPK is structurally related to acetophenone and similar model compounds which have been intensively investigated photochemically (156). These compounds are characterized by lowest $n\pi^*$ excited states and high intersystem crossing (ISC) and phosphorescence yields. Energy transfer as well as photochemical reactions are therefore primarily triplet-state processes and are studied conveniently by phosphorescence and by combined photochemical and polymer chemical methods, for example, photochemical chain splitting, as measured by the decrease of molar mass. Energy transfer competes with photochemistry and therefore has a stabilizing effect (157–159, 144).

At low temperature, sensitized phosphorescence is the preferred tool of studying energy transfer at the triplet level (160), naphthalene being a suitable guest molecule due to its high-lying S_1 level (no singlet acceptor) and low T_1. A purely formal evaluation using Perrin's model yields $r_0 = 2.6$ nm, too high for single-step transfer and thus pointing to energy transfer by triplet excitons (160). The phosphorescence spectrum of PVPK shows distinct vibrational structure with maxima at 410, 430, 470, and 505 nm (160). This structure is due to the C=O stretching vibration, taken as evidence for the predominantly carbonyl-localized nature of the phosphorescent state (156). The shape of this spectrum is very similar to that of isolated acetophenone-type impurity endgroups in PS (148, 144) but is red-shifted by about 700 cm^{-1}. A somewhat smaller red shift of about 650 cm^{-1} has been observed in polymers prepared by Hrdlovič and co-workers (159).

The question therefore arises whether the 77 K phosphorescence of PVPK is due to traps, the actual triplet exciton level lying above the observed O—O band at 24, 500 cm^{-1}. Alternatively, the crowding of chromophores in the polymer could decrease the T_1 level of the acetophenone chromophore. There is no excimer phosphorescence in PVPK. The efficiency of energy transfer in PVPK, doped with naphthalene, amounts to 50% at $c_G = 1.9 \times 10^{-3}$ mol/mol basic unit (160) in the low-temperature spectroscopic experiment. The chemical technique has been applied using various technically important stabilizers and aromatic hydrocarbons as guest molecules. This method allows measurements at room temperature where phosphorescence is very weak. Evaluation of the results using Stern–Volmer (SV) kinetics typically yields quenching constants of $K_q = 21$ liters/mol for naphthalene. In terms of the exciton hopping model, which is just another form of SV, this result can be expressed as $Q = 208 c_G$. This means that the triplet exciton performs about 200 jumps during its lifetime [Eq. (5.5), $K = n$] or, alternatively, that 1/200 deep exciton traps are present [Eq. (5.25)]. The spectral shift observed at low temperature, however, does not seem to indicate deep traps at room temperature ($kT \approx 200$ cm^{-1}).

Slow exciton diffusion by shallow trapping therefore seems to be a more reasonable explanation for the relatively inefficient energy transfer observed. Single-step transfer, on the other hand, should be even less efficient and $Q(c_G)$ should not be a linear function up to $Q = 15$, as observed experimentally (159) (see also Fig. 5.7). Furthermore, quenching by oxygen decreases the triplet lifetime at room temperature. Chemical linking of a guest molecule by copolymerization (VPK–VN) does not increase the transfer efficiency, compared to naphthalene dissolved physically in PVPK at the same concentration. Hence "down chain" T energy transfer does not seem to be preferred in films of PVPK. The quantum efficiency of chain scissions in PVPK and in copolymers of VPK with styrene increases dramatically near the glass transition temperature (161), indicating a strong influence of molecular mobility on Norrish splitting in polymers.

Derivatives of PVPK have been prepared and investigated spectroscopically and photochemically by the Bratislava group (162, 163). A large bathochromic shift and loss of vibrational structure has been observed in p-methoxy-PVPK phosphorescence. This behavior may indicate the formation of T excimers in this and related polymers (163). The red shift observed in radicalic PVPK (160, 159), without loss of vibrational structure, seems to be less pronounced (350 cm^{-1}) in anionic PVPK (163) which, however, has a low molar mass so that a direct comparison of the differently prepared polymers is not yet possible.

Poly(vinylbenzophenone)

[Structure of PVB: a repeating unit with –CH–CH$_2$– backbone, where CH bears a phenyl group connected via C=O to another phenyl ring]

PVB

The UV absorption spectrum of PVB indicates a weak $n\pi^*$ transition near 350 nm ($\epsilon \approx 100$ liters mol^{-1} cm^{-1}) and a stronger $\pi\pi^*$ band around 260 nm ($\epsilon \approx 10^4$ liters mol^{-1} cm^{-1}) (164). The phosphorescence spectrum of a film of PVB is vibrationally structured ($\Delta\nu'_v \approx 1600-1700$ cm^{-1}) and red-shifted (634 cm^{-1} for O—O peak) with regard to benzophenone in nonpolar glass (165).

Naphthalene turns out to be an efficient triplet trap in films of PVB at 77 K, $r_0 = 3.6$ nm (compared to $r_0 = 1.1$ nm for benzophenone–naphthalene in glass!), indicating that a single-step electron exchange model is not applicable. Energy migration therefore has been postulated and quantified (164) using a somewhat simplified Voltz theory (166). The triplet hopping rate, $k_h^T = 7 \times 10^4$ s^{-1}, has been calculated on the assumption that the observed phosphorescence decay time ($\tau_0' = 12$ ms) corresponds to the triplet exciton lifetime rather than to an excited trap. The hopping rate is very small compared to that observed in benzophenone crystals (167), $k_h^T \approx 10^{10}$ s^{-1}, but much closer to that of amorphous layers of benzophenone (168), $k_h^T \approx 10^7$ s^{-1}

An interesting dependence of the spectral position of λ_{00} in phosphorescence has been observed in a series of styrene copolymers containing various amounts of vinylbenzophenone (169). The wavelength increases from 417 to 429 nm in films of copolymers ranging from 0.7 to 77 mol% VB. The energy transfer efficiency increases in the same order.

Aliphatic polymeric ketones, for example, copolymers of CO with ethylene and poly(methyl vinyl ketone) (PMVK) have been studied occasionally (170, 171). Energy transfer to suitable guest molecules has been observed, but no evidence for excitons has been detected in these systems. For instance, in PMVK, energy transfer to naphthalene yields $r_0 = 1.1$ nm, in accord with expectations on single-step transfer (171).

Poly(N-vinylphthalimide)

$$\left[\begin{array}{c} \text{O}=\text{C} \quad \text{C}=\text{O} \\ \diagdown \text{N} \diagup \\ | \\ -\text{CH}-\text{CH}_2- \end{array} \right]_n$$

PVPI

PVPI is the first polymer for which energy transfer by triplet excitons in the solid state has been claimed by Lashkov and Ermolaev (172, 173). This work, which was often overlooked in previous reviews, uses the fact that ISC $S_1 \rightarrow T_1$ is very efficient in the basic unit of PVPI (174) and the triplet lifetime is long, τ_o' (phthalimide) = 1.25 s (174). The first triplet state is at $T_1 = 24,000$ cm^{-1}, the phosphorescence maximum near 20,000 cm^{-1}. The structureless spectrum of the polymer is somewhat red-shifted compared to N-methylphthalimide, but there is no clear evidence for triplet excimers in PVPI. Energy transfer has been observed using different guests, including photochromic spiropyranes. It has been observed (173) that purification increases significantly the "migration length" of the excitons.

There is nearly no work related to energy transfer in nonvinyl polymers, for example, polyamides, and polyesters. An exception is the work by Cheung et al. (175) on energy transfer in poly(ethylene phthalate) containing 2,6-naphthalene dicarboxylic acid groups.

5.5 APPLICATIONS

This section is relatively brief since clearly there are no "excitonics" comparable in any way with electronics. Energy transfer, however, is important in several technical applications and has been suggested for several uses, not yet exploited.

The use of energy transfer in plastic scintillators was the starting point of research in this field and has been discussed in Section 5.4.4. Photoconductivity in organic polymers is treated in Chapters 6 and 7 of this book; here, too, energy transfer plays a major role in creating charge carriers.

In a more general way, energy transfer can be used in sensitizing and inhibiting photochemical processes in polymers. Sensitized photodegradation of polymers has been proposed as a means of controlling the lifetime of plastics. This development has reached the stage of industrial testing of materials and small scale production but the role of energy transfer processes in these products is not clear at the moment.

Sensitization of photochemical processes due to excitons has been proposed by Lashkov (173) and Irie et al. (176). Excitons arriving at the surface of polymer films can induce photochemical reactions at the polymer–solution interface (177); this process can be controlled by the penetration depth of the exciting light and thus is of potential interest for image formation. Dilute, rigid solutions of aromatic polymers are capable of sensitization within the polymer coils (178) and are thus in principle capable of forming a very dense "raster."

Inhibition of photochemical processes in polymers is the main aim of polymer stabilization, and energy transfer contributes to this aim (158, 159), although it is now recognized that energy transfer alone is not sufficient for good stabilizing efficiency (179).

Finally, a still further long-range outlook may concern the possible use of polymers in artificial solar energy conversion, for example, hydrogen producing systems. This research field is in a rapid development phase, and polymers are likely to be used at least as passive elements once a technical solution has been found. The use of polymers as photoactive elements, in analogy to the "light harvesting" role of chlorophyll (180)—a singlet exciton process—is of course far in the future but certainly conceivable.

5.6 SUMMARY

I hope to have shown in this chapter that many observations on energy transfer in solid polymers can be understood today, at least in a semiquantitative way, and that the main features of these processes are fairly well understood. It has been shown that transfer by excitons in the framework of the exciton hopping picture is the most important intrinsic process, at least in aromatic polymers excited in their characteristic absorption bands. It should be noted that equally interesting studies have been performed in liquid and rigid solutions, which, however, are outside the scope of this book.

Various mechanisms are found to be dominant in different polymer groups discussed in Section 5.5. The main features are summarized in the following:

Carbazole-containing polymers are characterized by efficient excitonic energy transfer. Pure polymer films show excimer fluorescence and phosphorescence, indicating EFS as exciton traps. Triplet excitons have been detected at low temperature by delayed fluorescence and at ordinary temperature using sensitized chemiluminescence and triplet detrapping experiments. The concentration of EFS is in the order of 10^{-3} mol/mol basic unit in PVCA and seems to be lower in related, but more open structured polymers.

SUMMARY

Energy transfer in naphthalene polymers is very similar to that observed in PVCA, except that the trap spectrum seems to be somewhat simpler, at least in very pure polymers. As in the case of PVCA, however, the triplet excimer originates from another EFS than the singlet excimer. Singlet EFS are present in concentrations of 10^{-3}–10^{-2} mol/mol basic unit. Triplet excitons have been detected using T–T annihilation as probe. The fluorescence of rigid PACN shows the existence of nonnearest-neighbor excimer formation.

Polystyrene is the first polymer in which nonradiative energy transfer has been observed due to its usefulness as plastic scintillator. The concentration of EFS is higher in this polymer than in the previously described ones, so that exciton diffusion to guest molecules is efficient only if the spectral overlap between excimer fluorescence and guest absorption is small. The EFS are filled, however, by singlet excitons as in other aromatic polymers. Triplet excitons contribute to energy transfer in PS; the existence of triplet excimers in PS and their possible role in energy transfer has not yet been demonstrated.

Polymeric aromatic ketones and imides are characterized by high intersystem crossing efficiency and phosphorescence yield. Energy transfer is due to triplet excitons. The phosphorescence in PVPK and PVB is red-shifted compared to monomeric models, but triplet excimers are likely only in polymers substituted with electron-donating groups. PVPI is the first polymer for which T excitons have been shown to occur in the solid state.

The following gaps in our present knowledge should be filled with further research work:

The trapping centers controlling energy transfer in polymers are only insufficiently characterized structurally, for example, triplet excimers and shallow traps.

Several details of exciton transport, for example, dispersive exciton motion, have been treated only theoretically and require experimental confirmation using picosecond spectroscopy.

Sterically well-defined (purely isotactic or syndiotactic) polymers have been used only exceptionally for energy transfer studies and are still not available for most aromatic polymers with large pendant groups. The same is true for highly crystalline polymers with regard to the possible role of exciton bands.

The role of intramolecular energy bands of the backbone has so far completely been neglected in aromatic polymers, assuming that only the side groups contribute to energy transfer and trapping.

Charge transfer excitons, discussed for aromatic crystals and complexes, have not been investigated in polymers. The same is true for the connections between these excitons and charge carrier generation in polymers, especially in highly doped ones.

Last, but not least, the development of practical applications based on the theoretical knowledge acquired should not be neglected. Hints to possible uses have been given in Section 5.5. There is no reason to believe that the use of excitons —the intrinsic and mobile excited states of organic matter—will remain forever a dream of polymer spectroscopists and photochemists.

REFERENCES

1. J. Frenkel, *Phys. Rev.*, **37,** 17 (1931).
2. H. C. Wolf, in *Festkörperprobleme,* F. Sauter, Ed., Vieweg, Braunschweig, 1964, Vol. 4, p. 57.
3. G. H. Wannier, *Phys. Rev.*, **52,** 191 (1937).
4. A. Terenin, E. Putzeiko, and I. Akimov, *Disc. Faraday Soc.*, **27,** 83 (1959).
5. J. Delhalle, J. M. André, S. Delhalle, J. J. Pireaux, R. Caudano, and J. J. Verbist, *J. Chem. Phys.*, **60,** 595 (1974).
6. A. S. Davydov, *Theory of Molecular Excitons,* 2nd ed., Plenum New York, 1971; translation of the original Kiev, 1951, by M. Kasha and M. Oppenheimer, McGraw-Hill, New York, 1962.
7. W. Klöpffer, *J. Chem. Phys.* **50,** 2337 (1969).
8. G. E. Johnson, *Macromolecules,* **13,** 145 (1980).
9. R. D. Burkhart, *Chem. Phys.*, **46,** 11 (1980).
10. H. Bauser and H. H. Ruf, *Phys. Status Solidi,* **32,** 135 (1969).
11. H. Bauser and W. Klöpffer, *Chem. Phys. Lett.*, **7,** 137 (1970).
12. M. Pope, *J. Polym. Sci. Part C,* **17,** 233 (1967).
13. J. B. Birks and K. N. Kuchela, *Disc. Faraday Soc.*, **27,** 57 (1959).
14. F. H. Brown, M. Furst, and H. Kallmann, *Disc. Faraday Soc.*, **27,** 43 (1959).
15. W. Klöpffer, *Spectrosc. Lett.*, **11,** 863 (1978).
16. F. Gutmann and L. E. Lyons, *Organic Semiconductors,* Wiley, New York, 1967.
17. C. E. Swenberg and N. E. Geacintov, in *Organic Molecular Photophysics,* J. B. Birks, Ed., Wiley, London, 1973, Vol. 1, p. 489, and 1975, Vol. 2, p. 395.
18. G. Pfister and H. Scher, *Adv. Phys.*, **27,** 747 (1978).
19. W. Klöpffer, *Z. Naturforsch.* **24a,** 1923 (1969).
20. J. H. Sharp, *J. Phys. Chem.*, **71,** 2587 (1967).
21. W. Klöpffer, *J. Chem. Phys.*, **50,** 1689 (1969).
22. D. M. Hanson, VIth Molecular Crystal Symposium, Elmau, 1973; S. C. Abbi and D. M. Hanson, *J. Chem. Phys.*, **60,** 319 (1974).
23. E. J. Bowen, E. Mikiewicz, and F. W. Smith, *Proc. Phys. Soc.* (London), **A62,** 26 (1949).
24. D. C. Northrop and O. Simpson, *Proc. R. Soc. Lond.*, **A234,** 136 (1956).
25. M. W. Windsor, in *Physics and Chemistry of the Organic Solid State,* D. Fox, M. M. Labes, and A. Weissberger, Eds., Interscience, New York, 1965, Vol. 2, p. 343.

REFERENCES

26. T. F. Hunter, R. D. Mc Alpine, and R. M. Hochstrasser, *J. Chem. Phys.*, **50**, 1140 (1969).
27. G. I. Lashkov and V. L. Ermolaev, Opt. Spectr. (Engl. Transl.), **22**, 462 (1967).
28. P. Avakian and R. E. Merrifield, *Mol. Crystals*, **5**, 37 (1968).
29. N. J. Turro, I. E. Kochevar, Y. Noguchi, and M.-F. Chow, *J. Am. Chem. Soc.*, **100**, 3170 (1978).
30. W. Klöpffer, N. J. Turro, M.-F. Chow, and Y. Noguchi, *Chem. Phys. Lett.*, **54**, 457 (1978).
31. H. C. Wolf, *Advances in Atomic Molecular Physics*, Academic, New York, 1967, Vol. 3, p. 119.
32. H. B. Rosenstock, *J. Chem. Phys.*, **48**, 532 (1968).
33. M. Rudemo, *SIAM J. Appl. Math.*, **14**, 1293 (1966).
34. R. C. Powell and Z. G. Soos, *J. Luminescence*, **11**, 1 (1975).
35. G. Schönherr, R. Eiermann, H. Bässler, and M. Silver, *Chem. Phys.*, **52**, 287 (1980).
36. W. Klöpffer, H. Bauser, F. Dolezalek, and G. Naundorf, *Mol. Cryst., Liq. Cryst.*, **16**, 229 (1972).
37. D. Fischer, G. Naundorf, and W. Klöpffer, *Z. Naturforsch.*, **28a**, 973 (1973).
38. Z. Ludmer, *Chem. Phys.*, **26**, 113 (1977).
39. V. Yakhot, M. D. Cohen, and Z. Ludmer, *Advances in Photochemistry*, Wiley, New York, 1979, Vol. 11, p. 489.
40. D. Haarer and N. Karl, *Chem. Phys. Lett.*, **21**, 49 (1973).
41. D. Haarer, *Chem. Phys. Lett.*, **27**, 91 (1974).
42. C. P. Keijzers and D. Haarer, *J. Chem. Phys.*, **67**, 925 (1977).
43. M. R. Philpott and A. Brillante, *Mol. Cryst. Liq. Cryst.*, **50**, 163 (1979).
44. Th. Förster, *Fluoreszenz organischer Verbindungen*, Vandenhoeck und Ruprecht, Göttingen, 1951.
45. Th. Förster, *Disc. Faraday Soc.*, **27**, 7 (1959).
46. D. C. Dexter, *J. Chem. Phys.*, **21**, 836 (1953).
47. R. G. Bennet et al., *J. Chem. Phys.*, **41**, 3037, 3040, 3042, 3046, and 3048 (1964).
48. V. L. Ermolaev, *Sov. Phys. Uspekhi*, **80**, 333 (1963) (Review, Engl. Transl.).
49. F. Perrin, *C. R. Acad. Sci. (Paris)*, **178**, 1978 (1924).
50. J. Perrin, *C. R. Acad. Sci. (Paris)*, **177**, 469 (1923).
51. M. Inokuti and F. Hirayama, *J. Chem. Phys.*, **43**, 1978 (1965).
52. V. Ern, H. Bouchriha, M. Schott, and G. Castro, *Chem. Phys. Lett.*, **29**, 453 (1974).
53. H. Bouchriha, V. Ern, J. L. Fave, C. Guthmann, and M. Schott, *Chem. Phys. Lett.*, **53**, 288 (1978).
54. Th. Förster and K. Kasper, *Z. Phys. Chem.*, **NF1**, 275 (1954).
55. Th. Förster, *Angew. Chem.*, **81**, 364 (1975).
56. J. B. Birks, *Rep. Prog. Phys.*, **38**, 903 (1975).
57. E. Konijnenberg, Thesis, Amsterdam, 1963.

58 G. Melzer, D. Schweitzer, K. H. Hausser, J. P. Colpa, and M. W. Haenel, *Chem. Phys.*, **39**, 229 (1979).
59 M. Gordon, W. R. Ware, Eds., *The Exciplex*, Academic, New York, 1975.
60 H. Leonhardt and A. Weller, *Ber. Bunsenges. Phys. Chem.*, **67**, 791 (1963).
61 M. Yokoyama, Y. Endo, and H. Mikawa, *Bull. Chem. Soc. Japan*, **49**, 1538 (1976).
62 A. Itaya, K.-I. Okamoto, and S. Kusabayashi, *Bull. Chem. Soc. Japan*, **52**, 2218 (1979).
63 J. M. Pearson, S. R. Turner, and A. Ledwith in *Molecular Association*, R. Foster, Ed., Academic, New York, Vol. 2, 1979, p. 79.
64 B. Kovač et al., *J. Electron Spectr. Rel. Phen.*, **19**, 167 (1980).
65 F. Gerson and W. B. Martin, Jr., *J. Am. Chem. Soc.*, **91**, 1883 (1969).
66 R. P. Blaunstein and L. G. Christophorou, *Rad. Res. Rev.*, **3**, 69 (1971).
67 H. Beens and A. Weller, in *Organic Molecular Photophysics*, J. B. Birks, Ed., Wiley, London, 1975, Vol. 2, p. 159.
68 J. H. Pearson, in *Photochemical Processes in Polymer Chemistry*, G. Smets, Ed., Pergamon, Oxford, 1977, Vol. 2, p. 463.
69 A. Suzuki, H. Inokuchi, and Y. Maruyama, *Bull. Chem. Soc. Jap.*, **49**, 3347 (1976).
70 Y. Maruyama, T. Kobayashi, and H. Inokuchi, *Mol. Cryst. Liq. Cryst.*, **20**, 373 (1973).
71 M. D. Cohen, R. Haberkorn, E. Huler, Z. Ludmer, M. E. Michel-Beyerle, D. Rabinovich, R. Sharon, A. Warshel, and V. Yakhot, *Chem. Phys.*, **27**, 211 (1978).
72 A. Jablonski, *Z. Phys.*, **94**, 38 (1935).
73 J. B. Birks, *Photophysics of Aromatic Molecules*, Wiley, London, 1970.
74 D. M. Hanson, *Crit. Rev. Solid State Sci.*, **3**, 243 (1973).
75 K. P. Hofmann and G. Zundel, *Rev. Sci. Instr.*, **42**, 1727 (1971).
76 E. Schauenstein and W. Klöpffer, *Acta Histochim.*, *Suppl. VI*, 227 (1965).
77 C. A. Parker, *Photoluminescence of Solutions* Elsevier, Amsterdam, 1968.
78 B. Rånby and J. F. Rabek, *ESR Spectroscopy in Polymer Research*, Springer, Berlin, 1977.
79 G. Rippen, Thesis, Göttingen, 1976.
80 R. D. Burkhart and R. G. Avilés, *J. Phys. Chem.*, **83**, 1897 (1979).
81 F. Heisel, J. A. Miehe, and B. Sipp, *Ann. Phys. (France)*, **4**, 331 (1979).
82 S. W. Beavan, J. S. Hargreaves, and D. Phillips, *Adv. Photochem.*, **11**, 207 (1979).
83 K. P. Ghiggino, R. D. Wright, and D. Phillips, *J. Polym. Sci. Polym. Phys. Ed.*, **16**, 1499 (1978).
84 A. J. DeMaria, D. A. Stetser, and H. Heynan, *Appl. Phys. Lett.*, **8**, 174 (1966).
85 D. J. Bradley, in *Ultrashort Light Pulses, Picosecond Techniques and Applications*, S. C. Shapiro, Ed., Springer, Berlin, 1977, p. 17.
86 A. J. Campillo and S. L. Shapiro, *Photochem. Photobiol.*, **28**, 975 (1978).

REFERENCES

87 R. W. Anderson, Jr., D. E. Damschen, G. W. Scott, and L. D. Talley, *J. Chem. Phys.*, **71**, 1134 (1979).
88 G. D. Mendenhall, *Angew. Chem.*, **89**, 220 (1977).
89 J. Danecek, P. Hrdlovič, and I. Lukáč, *Eur. Polym. J.*, **12**, 513 (1976).
90 R. D. Burkhart, *Macromolecules*, **9**, 234 (1976).
91 G. I. Lashkov, L. S. Shatseva, and E. N. Bodunov, *Opt.-Spectrosc.* (Engl. Transl.), **34**, 57 (1973).
92 W. Klöpffer, *Ann. N. Y. Acad. Sci.*, Vol. 366, 373 (1981).
93 P. C. Johnson and H. W. Offen, *J. Chem. Phys.*, **55**, 2945 (1971).
94 R. C. Powell and Q. Kim, *J. Luminesc.*, **6**, 351 (1973).
95 K. Okamoto, A. Yano, S. Kusabayashi, and H. Mikawa, *Bull. Chem. Soc. Japan*, **47**, 749 (1974).
96 M. Yokoyama, Y. Endo, and H. Mikawa, *Chem. Phys. Lett.*, **34**, 597 (1975).
97 W. Klöpffer and H. Bauser, *Z. Phys. Chem. NF*, **101**, 25 (1976).
98 A. Itaya, K.-I. Okamoto, and S. Kusabayashi, *Bull. Chem. Soc. Japan*, **50**, 22 (1977).
99 G. Chryssomallis and H. G. Drickamer, *J. Chem. Phys.*, **71**, 4817 (1979).
100 W. Klöpffer, *Chem. Phys. Lett.*, **4**, 193 (1969).
101 W. Klöpffer and D. Fischer, *J. Polym. Sci. Polym. Symp.*, **No. 40**, 43 (1973).
102 K. Schmieder and K. Wolf, *Kolloid-Z.*, **134**, 149 (1953).
103 G. I. Lashkov, *Khim. Vysok. Energii*, **4**, 405 (1970).
104 A. Itaya, K.-I. Okamoto, and S. Kusabayashi, *Bull. Chem. Soc. Japan*, **49**, 2037 (1976).
105 A. Itaya, K.-I. Okamoto, and S. Kusabayashi, *Bull. Chem. Soc. Japan*, **49**, 2082 (1976).
106 Okamoto, K.-I., M. Yamada, A. Itaya, T. Kimura, and S. Kusabayashi, *Macromolecules*, **9**, 645 (1976).
107 A. J. Roberts, D. Phillips, F.A.M. Abdul-Rasoul, and A. Ledwith, in press.
108 A. J. Roberts, D. V. O'Connor, and D. Phillips, *Ann. N.Y. Acad. Sci.*, Vol. 366, 109 (1981).
109 G. Rippen and W. Klöpffer, *Ber. Bunsenges. Chem.*, **83**, 437 (1979).
110 G. Rippen, G. Kaufmann, and W. Klöpffer, *Chem. Phys.*, **52**, 165 (1980).
111 W. Klöpffer, *Chem. Phys.*, **57**, 75 (1981).
112 W. Hofberger and H. Bässler, *Phys. Status Solidi (B)*, **69**, 725 (1975).
113 R. D. Burkhart and R. G. Avilés, *Macromolecules*, **12**, 1073 (1979).
114 R. D. Burkhart and R. G. Avilés, *Macromolecules*, **12**, 1078 (1979).
115 M. Yokoyama, T. Tamamura, T. Nakano, and H. Mikawa, *J. Chem. Phys.*, **65**, 272 (1976).
116 M. Yokoyama, M. Funaki, and H. Mikawa, *J. Chem. Soc. Chem. Comm.*, 372 (1974).
117 G. E. Johnson, *J. Chem. Phys.*, **62**, 4697 (1975).
118 M. Keyanpour-Rad, A. Ledwith, and G. E. Johnson, *Macromolecules*, **13**, 222 (1980).

119 A. Itaya, K.-I. Okamoto, and S. Kusabayashi, *Bull. Chem. Soc. Japan*, **52,** 3737 (1979).
120 A. Itaya, K.-I. Okamoto, and S. Kusabayashi, *Chem. Lett.*, 131 (1978).
121 G. E. Johnson, *Macromolecules*, **13,** 145 (1980).
122 R. B. Fox, T. R. Price, R. F. Cozzens, and J. R. McDonald, *J. Chem. Phys.*, **57,** 534 (1972).
123 R. B. Fox, T. R. Price, R. F. Cozzens, and W. H. Echols, *Macromolecules*, **7,** 937 (1974).
124 N. Kim and S. E. Webber, *Macromolecules*, **13,** 1233 (1980).
125 E. A. Chandross and C. J. Dempster, *J. Am. Chem. Soc.*, **92,** 705 (1970).
126 R. B. Fox, T. R. Price, R. F. Cozzens, and J. R. McDonald, *J. Chem. Phys.*, **57,** 2284 (1972).
127 R. A. Beardslee and H. W. Offen, *J. Chem. Phys.*, **55,** 3516 (1971).
128 C. W. Frank and M. A. Gashgari, *Macromolecules*, **12,** 163 (1979).
129 M. T. Vala, Jr., J. Haebig, and S. A. Rice, *J. Chem. Phys.*, **43,** 886 (1965).
130 R. F. Cozzens and R. B. Fox, *J. Chem. Phys.*, **50,** 1532 (1969).
131 C. David, W. Demarteau, and G. Geuskens, *Eur. Polym. J.*, **6,** 1397 (1970).
132 F. Hirayama, L. J. Basile, and C. Kikuchi, *Mol. Cryst.*, **4,** 83 (1968).
133 C. David, M. Lempereur, and G. Geuskens, *Eur. Polym. J.*, **8,** 417 (1972).
134 C. David, M. Piens, and G. Geuskens, *Eur. Polym. J.*, **8,** 1019 (1972).
135 M. G. Schorr and F. L. Torney, *Phys. Rev.*, **80,** 474 (1950).
136 W. S. Koski, *Phys. Rev.*, **82,** 230 (1951).
137 L. Pichat, P. Pesteil, and J. Clément, *J. Chim. Phys.*, **50,** 26 (1953).
138 R. K. Swank and W. L. Buch, *Phys. Rev.*, **91,** 927 (1953).
139 L. Pichat and Y. Koechlin, *J. Chim. Phys.*, **48,** 225 (1951).
140 F. Hirayama, *J. Chem. Phys.*, **42,** 3163 (1965).
141 L. J. Basile, *J. Chem. Phys.*, **36,** 2204 (1962).
142 F. Heisel and G. Laustriat, *J. Chim. Phys.*, **66,** 1895 (1969).
143 P. F. Jones, S. Siegel, and R. S. Nesbitt, presented at the IUPAC Symposium in Leuven, 1972.
144 G. Geuskens and C. David, in *Degradation and Stabilisation of Polymers*, G. Geuskens, Ed., Applied Sciences, London, 1975, p. 113.
145 C. David, M. Piens and G. Geuskens, *Eur. Polym. J.*, **9,** 533 (1973).
146 L. J. Basile, *Trans. Faraday Soc.*, **60,** 1702 (1964).
147 C. David, N. Putman-De Lavareille, and G. Geuskens, *Eur. Polym. J.*, **10,** 617 (1974).
148 W. Klöpffer, *Eur. Polym. J.*, **11,** 203 (1975).
149 M. Nowakowska, J. Najbar, and B. Waligóra, *Eur. Polym. J.*, **12,** 387 (1976).
150 N. J. Turro and H.-C. Steinmetzer, *J. Am. Chem. Soc.*, **96,** 4677 (1974).
151 J. E. Guillet, C. E. Hoyle, and J. R. MacCallum, *Chem. Phys. Lett.*, **54,** 337 (1978).
152 R. C. Powell, *J. Chem. Phys.*, **55,** 1871 (1971).

REFERENCES

153 K. Jkezaki, *Jap. J. Appl. Phys.*, **13**, 437 (1974).
154 A. Hallam and J. B. Birks, *J. Phys. B, Atom. Mol. Phys.*, **11**, 3273 (1978).
155 C. David, M. Lempereur, and G. Geuskens, *Eur. Polym. J.*, **10**, 1181 (1974).
156 N. J. Turro, *Modern Molecular Photochemistry*, Benjamin/Cummings, Menlo Park, CA, 1978.
157 M. Heskins and J. E. Guillet, *Macromolecules*, **1**, 97 and 165 (1968).
158 W. Klöpffer, *Kunststoffe*, **60**, 385 (1970).
159 P. Hrdlovič, I. Lukáč, and Z. Maňásek, *Chem. Zwesti*, **26**, 433 (1972).
160 C. David, W. Demarteau, and G. Geuskens, *Eur. Polym. J.*, **6**, 1405 (1970).
161 E. Dan and J. E. Guillet, *Macromolecules*, **6**, 230 (1973).
162 I. Lukáč and P. Hrdlovič, *Eur. Polym. J.*, **14**, 339 (1978).
163 P. Hrdlovič, J. Trekoval, and I. Lukáč, *Eur. Polym. J.*, **15**, 229 (1979).
164 C. David, W. Demarteau, and G. Geuskens, *Eur. Polym. J.*, **6**, 537 (1970).
165 J. Simpson and H. Offen, *J. Chem. Phys.*, **55**, 4832 (1971).
166 R. Voltz, G. Laustriat, and A. Coche, *J. Chim. Phys.*, **63**, 1253 (1966).
167 R. M. Hochstrasser, *J. Chem. Phys.*, **40**, 1038 (1964).
168 T. F. Hunter, R. D. Mc Alpine, and R. M. Hochstrasser, *J. Chem. Phys.*, **50**, 1140 (1969).
169 C. David, V. Naegelen, W. Piret, and G. Geuskens, *Eur. Polym. J.*, **11**, 569 (1975).
170 M. Heskins and J. E. Guillet, *Macromolecules*, **3**, 224 (1970).
171 C. David, N. Putman, M. Lempereur, and G. Geuskens, *Eur. Polym. J.*, **8**, 409 (1972).
172 G. I. Lashkov and V. L. Ermolaev, *Opt. Spectrosc.* (Engl. Transl.), **22**, 462 (1967).
173 G. I. Lashkov and V. L. Ermolaev, *Theoret. Elsperim. Khim.*, **4**, 918 (1968) (trans- lated by K. N. Trirogoff, Aerospace Library Service, El Segundo).
174 G. A. Mokeeva and B. Ya. Sveshnikov, *Opt. Spectrosc.* (Engl. Transl.), **10**, 41 (1961).
175 P.-S.R. Cheung, C. W. Roberts, and K. B. Wagener, *J. Appl. Polym. Sci.*, **24**, 1809 (1979).
176 S. Irie, M. Irie, Y. Yamamoto, and K. Hayashi, *Macromolecules*, **8**, 424 (1975).
177 G. I. Lashkov, L. S. Shatseva, and E. N. Bodunov, *Opt. Spectrosc.*, **34**, 57 (1973).
178 W. Köpffer, in *Proceedings of EUCHEM Conference on Photopolymerization and Photocrosslinking of Organic Coatings*, Södergarn, Stockholm, August 25–27, 1980. to be published
179 W. Klöppfer, *Lenzinger Ber.*, **44**, 28 (1978).
180 *Chlorophyll Organization and Energy Transfer*, Ciba Foundation Symp. 61, Excerpta Medica, Amsterdam, 1979.

Chapter 6

PHOTOELECTRONIC PROPERTIES OF PHOTOCONDUCTING POLYMERS

J. Mort
Xerox Corporation
Webster, New York

G. Pfister
Cerberus Corporation
Männedorf, Switzerland

6.1	Introduction	215
6.2	Experimental Techniques	217
6.3	Charge Transport	220
	6.3.1 Introduction, 220	
	6.3.2 Experimental Results, 221	
	6.3.3 Transport Process, 227	
	6.3.4 Effects of Disorder and Morphology, 233	
	6.3.5 On Dispersive Transport and Concluding Remarks, 242	
6.4	Photogeneration and Photosensitization	246
	6.4.1 Photogeneration, 246	
	6.4.2 Photosensitization, 251	
6.5	Chemical Control of Conductivity	256
6.6	Summary	262
	References	263

6.1 INTRODUCTION

This chapter discusses recent progress in the understanding of the photoelectronic properties—charge generation and charge transport—in disordered organic solids. In particular, the focus is on poly(N-vinylcarbazole) (PVCA) and its complexes with trinitrofluorenone and on molecularly doped organic polymers. The molecularly doped systems are important in two key respects. The first is the opportunity they offer to study the characteristic photoelectronic properties of disordered organic solids and the role which molecular geometry and chemical structure play in determining these properties. They however also constitute a very simple and yet elegant way of studying the photoelectronic properties of disordered solids in general. Thus this novel class of materials is ideally suited for the study of the detailed microscopic processes involved in the transfer of an electronic charge between localized states.

Transfer of charge between localized states is believed to play a dominant role in charge transport and photogeneration in many organic and inorganic disordered solid materials. Until recently, however, the main experimental efforts were devoted to inorganic noncrystalline semiconductors, in particular, amorphous chalcogenides and tetrahedrally bonded solids where the localized states arise from atomic bond disorder. These studies led to significant progress in our understanding of the many novel aspects of charge transport and generation in disordered materials. For a review of the state of the art in the physics of amorphous semiconductors, the reader is referred to the biennial Conferences on Amorphous and Liquid Semiconductors (1).

In more recent years, this rapidly growing discipline has also been extended into the realm of disordered organic solids. It already seems apparent that amorphous organic solids exhibit rather fundamental differences from the chalcogenides and tetrahedrally bonded solids. Such differences stem from the fact that organic crystalline solids are typically characterized by very weak van der Waals-type bonding between the molecules which are the constituent building blocks. The weak intermolecular interactions lead to very narrow energy bands resulting in low mobilities for which band theories of transport are either questionable or invalid. Almost intuitively, therefore, one expects the noncrystalline organic state to be dominated by the molecules themselves and dynamic charge transfer in both photogeneration and transport to involve charge exchange between neighboring molecules.

It is here that the concept of molecular doping of organic polymers proves to be extremely powerful. This stems from the fact that it allows one to choose, at will, the density and kind of localized states since they are associated with the dopant molecules. Early work in this field were photoelectronic studies in solutions of the leuco base of malachite green in polystyrene (2), trinitrofluorenone in polyesters (3), and triphenylamine in hydrocarbon liquids (4). As will be described in more detail, a typical example of a molecularly doped polymer is the molecular dispersion of N-isopropylcarbazole (NIPCA) in a polycarbonate polymer (5). The samples are prepared by casting thin films of a solution of dopant molecule and host polymer. In the absence of NIPCA, no charge transport through the solid film can be resolved on a typical experimental time scale. On introduction of NIPCA, molecularly dispersed at concentrations in the range of $\sim 10^{20}$ molecules/cm^3, hole transport and generation can readily be observed, and their dependence upon NIPCA concentration is evidence that the underlying mechanism involves the transfer of electronic charge between localized states associated with the dispersed molecule. Hence by using this concept of doping, it has been possible to explore on a molecular level the steps involved in the transfer of charge associated with transport, generation, and photosensitization.

Polymers such as PVCA or a molecularly doped polymer have strong intrinsic optical absorption in the ultraviolet, and since practical usage employs visible light, the polymer photosensitivity must be extended into the visible. This can be accomplished by forming a charge-transfer complex with absorption in the visible, by dye sensitization with an appropriately absorbing dye, or by use of a thin contiguous sensitizing layer such as amorphous selenium (a-Se). Thus phenomena of considerable practical as well as scientific importance are those of photogeneration and photosensitization. Such processes are also basic in natural processes such as photosynthesis.

This chapter discusses central ideas and materials that give a perspective on the current understanding of transport, photogeneration, and photosensitization processes in photoconducting insulating polymer systems. It concludes with a discussion of very recent work on the fabrication of semiconductor systems produced by chemical treatment of such materials.

6.2 EXPERIMENTAL TECHNIQUES

The quantities of interest in a study of electronic properties of solids are the carrier mobilities (μ is the carrier velocity per unit applied electric field), majority carrier sign, carrier deep trapping lifetime, efficiency of photogeneration (Φ is the number of free carriers produced per absorbed photon), surface and bulk recombination lifetime and finally bulk conductivity.

Unlike in crystalline semiconductors, in disordered organic materials the properties of charge generation and transport are generally strong functions of the experimental variables such as applied electric field, temperature, and in some cases sample geometry. These functional dependences need to be carefully examined to gain insight into the physical processes underlying the electrical properties of these solids and to make meaningful predictions and interpretations of their performance in device structures. A well-known example is the application of organic photoconductors in xerography. Device modeling clearly demonstrates that the experimentally observed tailing of the time dependence of the surface potential decay curve may result either from a field-dependent photogeneration efficiency (emission-limited discharge) or a field-dependent carrier mobility (space-charge-limited discharge). For the scientists interested in fundamental questions, these dependences indicate the important role of localized states in the electrical properties of polymeric photoconductors.

A number of experimental techniques are available for the study of electrical conduction processes in solids, and the choice of a particular method depends mainly on sample shape, bulk conductivity, carrier lifetimes, and mobility values. In addition to the conventional dc dark and photoconductivity measurements, the techniques of time of flight and the xerographic discharge have proved to be most successful for the low-mobility, insulating thin films ($\mu \ll 0.1$ cm^2/V-s, $\sigma \ll 10^{-11}$ Ω^{-1} cm^{-1}) discussed in this article.

These techniques offer several advantages over the conventional bulk conductivity measurement ($\sigma = ne\mu$). They allow a separate and direct measurement of μ and n, do not require the often difficult-to-realize "ohmic contacts," and can readily be adapted for measurement of (a) carrier lifetimes with respect to deep traps or recombination, and (b) effects of nonuniform applied electric fields, for instance, due to trapped space charges.

The time-of-flight method was pioneered by Kepler (6) in his study of electronic transport in anthracene single crystals and by Spear (7) in his investigation of transport in amorphous selenium. The xerographic discharge technique was perfected in parallel with the development of xerography (8). Its main advantage with respect to time of flight is that one surface of the sample film does not have to be electroded. Hence this technique is particularly useful when films tend to destructively break down in the typically strong electric fields necessary in transport and generation experiments ($E \sim 10\text{--}100$ V/μm).

The time-of-flight experiment is performed by injecting a thin carrier sheet from the surface of the sample film and measuring the time t_T (transit time) it takes the sheet to drift across the sample under the influence of an applied dc electric field. The polarity of the applied field determines the sign of the carriers which drift across the sample. Hence, if the electrode where charge injection occurs is at a positive potential with respect to the back electrode, the drift of hole carriers is observed. In photoconductive materials, the sheet of charge is most conveniently produced by a flash of light which is absorbed within a small fraction of the sample length L. A typical experiment setup is shown in Fig. 6.1a. In order to time resolve the drifting charge sheet as a current pulse in the external circuit, the conditions $RC \ll t_T$ and $\tau \gg t_T$ have to be met, where R is the value of the load resistor, C is the capacitance of the sample, and $\tau = \rho\epsilon\epsilon_0$, the dielectric relaxation time (ρ is the bulk resistivity and ϵ is the permittivity). Both these conditions are easily satisfied for the samples under discussion. Typical values for the pertinent times are $t_T \sim 0.1\text{--}100$ ms, $RC < 0.1$ ms, and $\tau > 1$ s. If we assume, as is the case for all experiments described in this chapter, that the number of injected carriers is much less than the charge stored on the electrodes of the sample (CV_0), the electric field in the sample to a good approximation remains constant throughout the experiment. Hence in the external

(a)

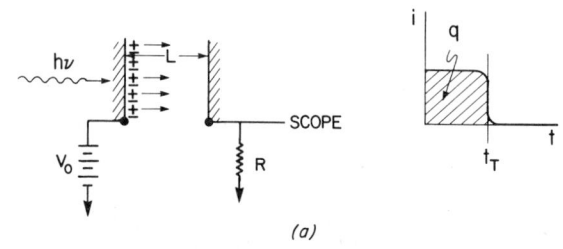

Figure 6.1a **Schematic representation of time-of-flight technique. The voltage across the sample is constant, V_0, and the drift of a carrier sheet across the sample is time resolved.**

EXPERIMENTAL TECHNIQUES

circuit, the drifting charge is manifested as a constant current $i = q\mu_d E/L$, which abruptly drops to zero at the time $t_T = L/(\mu_d E)$ at which the motion of the charge is stopped at the back electrode (Fig. 6.1a). In these equations, q is the total charge injected into the sample by the incident light flash F (absorbed photons per second), t_T is defined as the carrier transit time, and μ_d is the carrier drift mobility. Hence a simple analysis of the current pulse yields the carrier drift mobility μ_d and the generation efficiency $\Phi = q/eF$.

When interpreting drift mobility experiments, one has to recognize that, by virtue of its definition, μ_d is a measure of transport of carriers interacting with traps and therefore may be substantially smaller than the microscopic mobility μ_0 which characterized the transport of free carrier between trapping events.

Deviations from the ideal rectangular current pulse are frequently observed in disordered solids. These may arise from statistical fluctuations of the microscopic processes a carrier experiences as it drifts across the solid. It will be shown that the resulting shape of the current pulse reveals novel and important information on the statistics of charge transport which characterizes the disordered state.

The conventional time-of-flight technique requires electroded samples, and this often limits the upper range of applied fields due to destructive electrical breakdown. By using xerographic discharge techniques, these problems can be circumvented if the samples can be prepared in the form of thin films cast or evaporated on conducting substrates which prevent charge injection. The free sample surface is charged with a corona to a surface potential V_0, and the change in the surface potential $V(t)$ following a flash of strongly absorbed light is observed (Fig. 6.1b). For sufficiently low light intensities, the change in the sur-

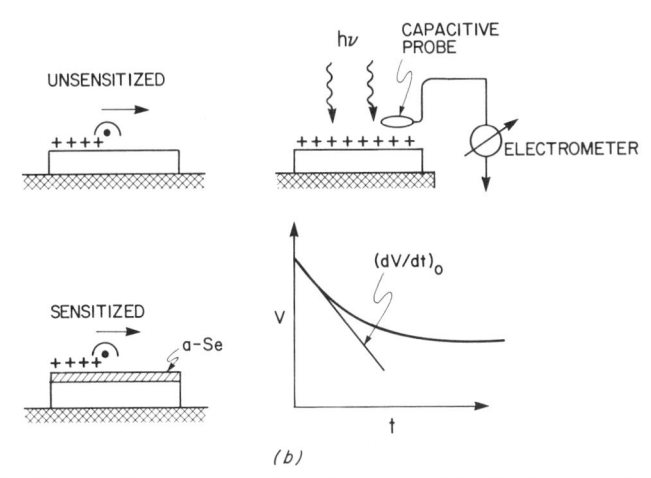

Figure 6.1b Schematic representation of xerographic discharge technique. The free unelectroded surface of the sample is charged with a corotron to V_0, and the decay of the surface potential upon exposure to light is observed.

face potential is small compared to V_0, in which case the discharge rate at $t = 0$ is given by $(dV/dt)_0 = \Phi eF/C$, where the symbols are the same as used before. Hence the initial dV/dt provides a direct measurement of the quantum efficiency. To obtain the drift mobility from a xerographic discharge measurement, the system has to be driven into space charge limitation, that is, the number of injected carriers has to be approximately equal to CV_0. For organic materials with typically low quantum efficiencies ($\Phi < 10^{-1}$), it is not easy to obtain space charge limitation since the required light intensities $F = CV/\Phi e$ often lead to photodecomposition of the organic solid. Space-limited discharge can be obtained by sensitizing the films with, for instance, a thin evaporated layer of a-Se (Fig. 6.1b) which has close to unity quantum efficiency for hole production. In many cases, holes generated by light absorbed in a-Se can efficiently be injected into an organic solid, for instance, poly(N-vinylcarbazole), such that the discharge of the multilayer structure proceeds under space charge-limited conditions (9).

Deep trapping of carriers and nonuniform fields lead to distortions of the current shape in time-of-flight experiments and make the interpretation of initial dV/dt less straightforward. These effects are not dealt with further, and the reader is referred to the extensive literature available on the xerographic discharge technique (8, 10).

6.3 CHARGE TRANSPORT

6.3.1 Introduction

Extensive time-of-flight and xerographic discharge measurements were performed on polymers based on carbazole such as poly(N-vinylcarbazole) (PVCA) (11, 12), their charge-transfer complexes, for instance, with trinitrofluorenone (TNF) (13), and molecularly doped polymers such as polycarbonate (Lexan) and triphenylamine (TPA) (14, 15) or N-isopropylcarbazole (NIPCA) (5). In all these systems, the same basic transport properties prevail. The charge carriers propagate through the solid by a hopping mechanism that proceeds among the chromophores in the intrinsically semiconducting polymeric solids and among the dopant molecules in the molecularly doped polymers. Pioneering work in intrinsically conducting polymers was performed by Gill (13) who investigated in detail the transport properties of PVCA and PVCA:TNF. These studies led to the concept of molecularly doping. By introducing carbazole side groups in the form of N-isopropylcarbazole into polycarbonate host polymers, Mort et al. (5) demonstrated that the vinyl backbone in the PVCA polymer does not contribute to charge transport but merely provides the mechanical stability of the films. With regard to the intrinsically conducting polymers, the molecularly doped polymers are much more flexible in material design since the concentration and kind of dopant molecule can be selected, which allows one to optimize certain transport properties. To some extent the mechanical properties of the host polymer

CHARGE TRANSPORT

can also be chosen. This flexibility proves to be an essential experimental leverage for unraveling the details of charge transport and generation in these systems.

It is not the purpose of this chapter to provide a detailed catalogue of the vast literature describing electrical properties of organic polymeric solids but rather to choose a few key experiments on model systems in order to illustrate the basic conduction phenomena that characterize the disordered organic state.

6.3.2 Experimental Results

The principle features of hopping transport in disordered solids are illustrated using as an example a polycarbonate (Lexan) polymer host matrix which contains a dispersion of triphenylamine (TPA) molecules. The preparation of these samples in the form of thin films is typical for organic polymers. The ingredients in our example, bisphenol A polycarbonate and triphenylamine, $(C_6H_5)_3N$, are dissolved in a common solvent, for instance, 1,2-dichloroethane. The well-stirred mixture is then cast with a doctor blade onto ball-grained clean aluminum substrates or onto glass slides coated with conducting electrodes. The films are then dried under vacuum at elevated temperatures to drive off the solvent molecules. The sample films are typically 10–20 μm thick, and the doping concentrations, expressed as weight ratio x of dopant molecule to polymer host, are in the range $x = 0.1$–0.5. For electrical measurements the films are overcoated with (semitransparent) electrodes 0.3 cm^2 in area.

Figure 6.2 shows absorption spectra of Lexan, TPA, and N-isopropylcarbazole (NIPCA) in methylene chloride solvent. For the doped systems, the absorp-

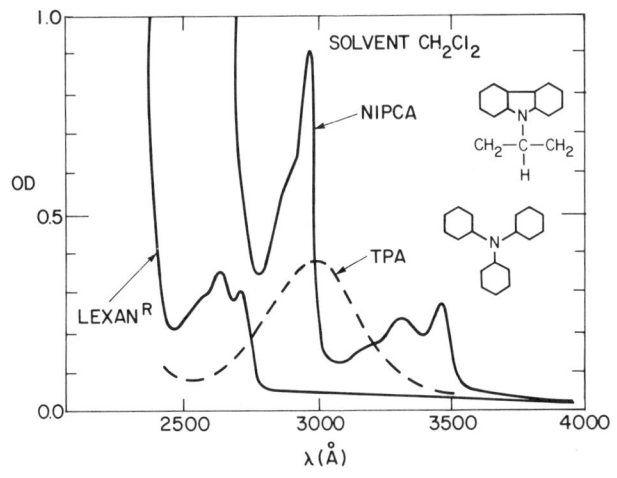

Figure 6.2 Optical absorption of polycarbonate (Lexan), triphenylamine (TPA), and N-isopropylcarbazole (NIPCA) in methylene chloride. For doped systems, the absorptions are additive, and no complex formation is noted.

tion is additive and no charge-transfer complex formation is observed. This is an important result for time-of-flight experiments, since it allows the selective excitation of the dopant molecule by the charge-generating light flash. In our experiments, we used a light pulse at 3361 Å wavelength derived from a, N_2 laser. For all concentrations studied, light at this wavelength is absorbed within a small fraction of the film thickness (i.e., close to the sample surface).

Only hole transport can be observed in TPA–Lexan. Figure 6.3 shows the hole drift mobility as a function of temperature in the form of an Arrhenius plot, $\log \mu$ versus $1/T$. The data were recorded for an applied field of 70 V/μm and various TPA concentrations. Three main conclusions can be drawn from Fig. 6.3: (1) Within the limits of the experimental temperature range, the hole mobility is activated with a single well-defined energy $\Delta(eV)$. (2) The activation en-

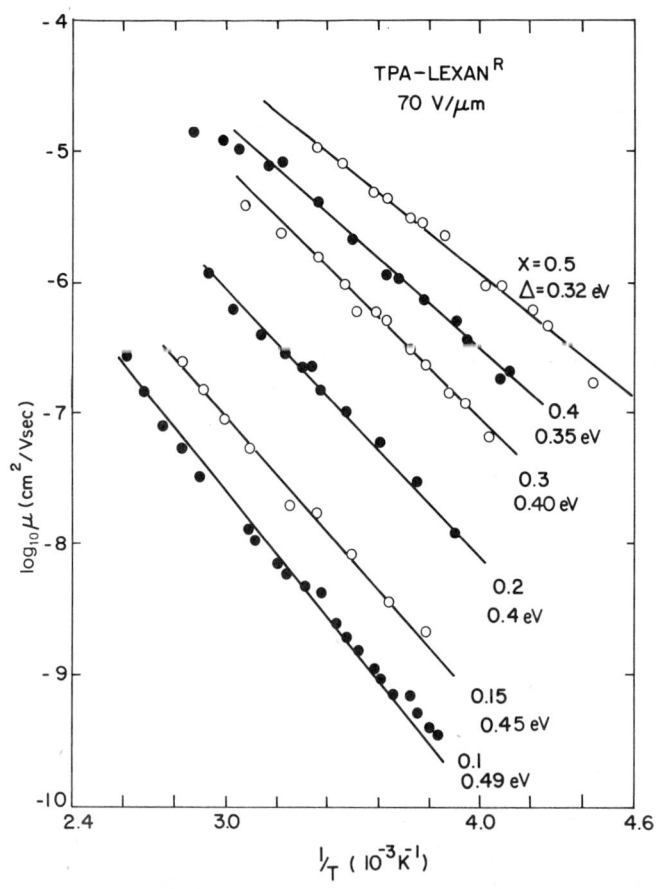

Figure 6.3 Hole drift mobility in TPA–Lexan as function of temperature. Data were recorded at a 70 V/μm applied field for various TPA concentrations X. (From Pfister [14].)

ergy increases with decreasing TPA concentration; compare, for instance, $\Delta = 0.32\ eV$, $x = 0.5$ and $\Delta = 0.49$ eV, $x = 0.1$. (3) For constant temperature, the mobility rapidly drops as the TPA concentration is reduced; for instance, at $T = 296$ K, the mobility decreases from $\sim 10^{-5}$ cm^2/V-s for $x = 0.5$ to $\sim 3 \times 10^{-9}$ cm^2/V-s for $x = 0.1$.

Thermally activated hole mobilities were also deduced from the space-charge-limited xerographic discharge. However in the reported concentration range, 0.67 to 0.25, no concentration dependence of the activation energy is observed (15). It is not clear at present whether this discrepancy reflects a fundamental difference of the results obtained from the two widely different measurement techniques and associated experimental conditions or whether the TPA concentration range used in the xerographic discharge measurement is not wide enough to clearly reveal the concentration dependence of the activation energy. A variation of the activation energy with concentration has also been observed for NIPCA–Lexan (16) but, interestingly, not for the charge transfer complex PVCA:TNF where the TNF concentration was changed (13).

The strong dependence of the hole mobility on TPA concentration indicates that charge transport occurs by a process in which the charge carriers hop among localized states associated with the dopant molecules. A very idealized description—assuming, for instance, the spatial extent of the localized state to be very much smaller than dopant molecule separation—predicts for the mobility the approximate (overlap) expression $\mu \propto \rho^2 \exp(-\gamma\rho) \exp(-\Delta/kT)$, where ρ is the average separation of the dopant molecules and γ is a measure of the exponential decay of the wave function outside the associated localized state. For a uniform dispersion, one may estimate ρ from the dopant molecule concentration n using $\rho \sim n^{-1/3}$.

Figure 6.4 shows the exponential dependence of the hole mobility on the separation between dopant molecules for a number of organic systems including the PVCA:TNF charge transfer complex. In the latter, hole transport occurs by hopping through the uncomplexed carbazole side groups while electrons hop through sites associated with the TNF acceptor. Adding TNF to PVCA removes uncomplexed carbazole units, which results in a rapidly decreasing hole mobility paralleled by a smaller increase of the mobility for electrons. These results are shown in Fig. 6.5. From the slope of the straight lines in the semilog plot (Fig. 6.4), one obtains the "localization" parameter γ that characterizes the decay of the wave function outside the "point-like" molecule. Using the conventional definition for the localization radius for hopping among point sizes $R_0 = 2/\gamma$, one obtains from Fig. 6.4 the values $R_0 \sim 1.1$, 1.54, and 1.0 Å for PVCA:TNF. NIPCA–Lexan, and TPA–Lexan, respectively. For a triphenylamine derivative (TPM) in polycarbonate, $R_0 \sim 1.16$ Å has been reported (17).

The drift mobility and its activation energy are field-dependent. Figure 6.6 shows log μ versus log E for TPA–Lexan doped at the level $x = 0.5$. The solid lines are theoretical fits and are discussed in Section 6.3.4. In the field range 20–80 V/μm, one finds that the activation energy decreases with applied field

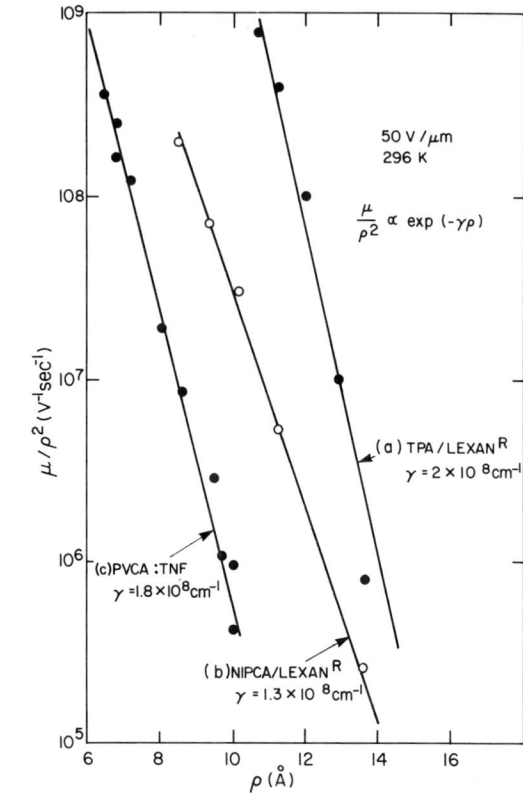

Figure 6.4 Concentration dependence of hole drift mobility in the molecularly doped polymers NIPCA–Lexan and TPA–Lexan and the charge transfer complex PVCA:TNF. (From Pfister [14](a). Mort et al. [5](b) and Gill [13](c).

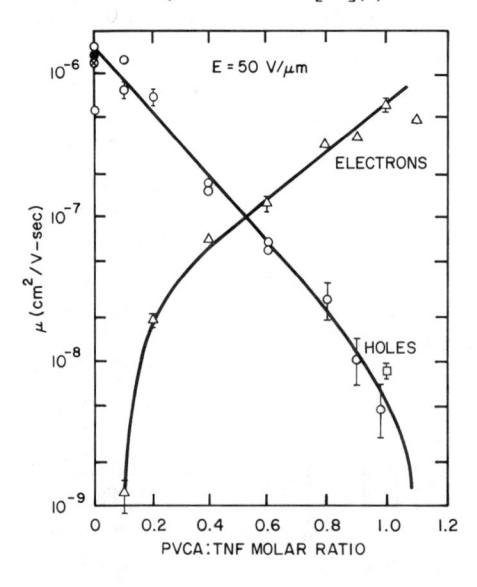

Figure 6.5 Variation of hole (μ_p) and electron (μ_n) drift mobility with TNF-PVCA molar ratio. (From Gill [13] and references therein.)

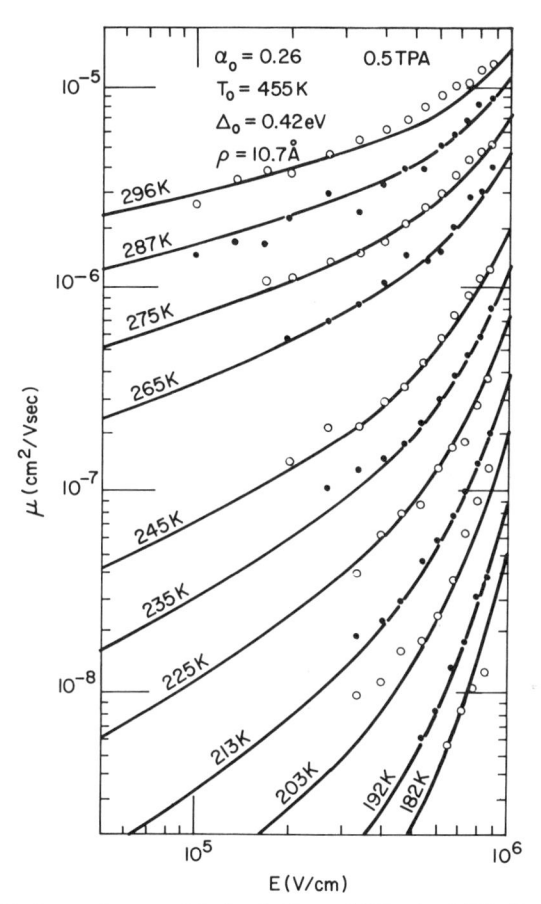

Figure 6.6 Field dependence of hole drift mobility in TPA–Lexan at various temperatures. Solid lines are theoretical fits of Eq. 6.3 using the listed parameters. (From Pfister [14].)

approximately as $\Delta(E) \sim \Delta_0 - \beta E$ (Fig. 6.7). The value β increases slightly with increasing hopping distance ρ, but for fixed concentration it does not appear to significantly depend upon the kind of dopant molecule. This is of interest since the field dependence of the drift mobility at room temperature has been observed to vary from $\mu \sim$ const to $\mu \propto E^3$ for various doped polymer systems. On the other hand, the zero field intercept Δ_0 can significantly change with concentration and polymer system. A typical range is $\Delta_0 \sim 0.3$–0.7 eV.

For applied fields smaller than ~ 10–20 V/μm, the activation energy for many systems increases faster than predicted from the linear relationship (see PVCA in Fig. 6.7). It is often found that a plot of $\Delta(E)$ versus $E^{1/2}$ yields a straight line over a wider field range than the proposed linear relationship (Fig. 6.7) (14). A discussion based on this finding is given in Section 6.3.4.

The range of Δ_0 from ~ 0.3 to 0.7 eV observed for various dopings, namely,

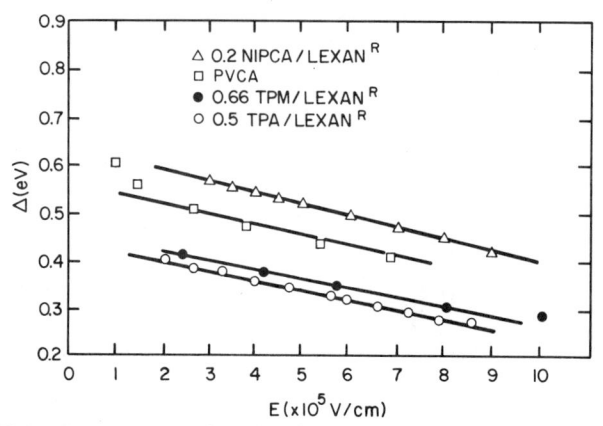

Figure 6.7 Field dependence of activation energy for a number of disordered organic solids. (From Pfister [14] and Limburg et al. [17].)

the parallel shift of the straight lines in Fig. 6.7, would amount to a variation of the drift mobility by a factor of $\sim\exp(0.4\,\text{eV}/kT) \sim 10^7$ at room temperature. This variation is much more than actually observed and is addressed in the following.

For all systems reported so far, the field dependence of the mobility is much weaker than would be expected on the basis of the Arrhenius temperature term $\exp[-\Delta(E)/kT]$ calculated for the measured values $\Delta(E)$. This deviation is observed at all temperatures. Hence the strong exponential field dependence has to be offset by a field dependence which to a first approximation is temperature independent. This can be done by introducing in the Arrhenius term an effective temperature, T_{eff}, rather than the laboratory temperature T, where $T_{\text{eff}} > T$. The early studies of Gill (13) on PVCA:TNF and later extensive measurements on TPA–Lexan (14) suggest the approximate phenomenologic relationship $1/T_{\text{eff}} = (1/T) - (1/T_0)$, where T_0 is an experimental temperature which appears to characterize the transport system.

Accordingly, in a plot of log μ versus $1/T$, the straight lines drawn through the experimental points measured at various fields intersect approximately at the temperature T_0. An example is shown in Fig. 6.8 for hole transport in PVCA–TNF (13). Hence the smaller the difference between T and T_0, the larger the effective temperature T_{eff} and the weaker the field dependence $\mu(E)$. It then follows that although the field dependence of the activation energy $\Delta(E)$ is not much dependent upon doping level and doping molecule (Fig. 6.7), the field dependence of $\mu(E)$ can vary significantly due to variations of T_{eff}. This is experimentally observed. For instance, $\mu(E)$ varies more strongly for PVCA than for TPA–Lexan and, indeed, $T_0(\text{PVCA}) > T_0(\text{TPA–Lexan})$. These arguments suggest that molecules giving rise to weakly field-dependent mobilities can be characterized by low T_0 (or high T_{eff}). Hence T_0 is an important parameter in the study of transport in organic polymer systems.

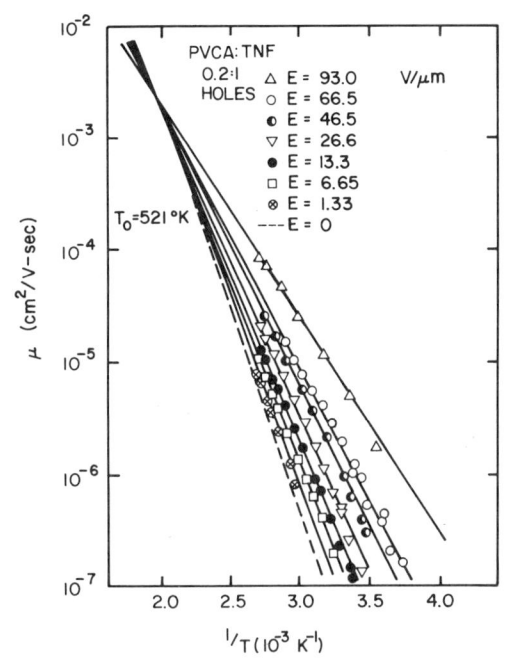

Figure 6.8 Temperature dependence of hole drift mobility with applied field as parameter in a 0.2:1 TNF:PVCA film. (From Gill [13].)

6.3.3 Transport Process

The strong dependence of the hole drift mobility upon intersite distance ρ suggests that the carriers propagate by a hopping motion. This mechanism can be visualized as either an oxidation–reduction or a donor–acceptor process. Take hole transport, for example, and assume that as a result of the photogeneration process some dopant molecules are positively charged (radical cations). Under the influence of the applied electric field, neutral molecules will repetitively transfer electrons to their neighboring cations (Fig. 6.9). The net result of this process is the motion of a positive charge across the bulk of the sample film. Note that this is strictly an electronic and not an ionic transport process since no mass displacement is involved. Hence for hole transport to occur, one expects that the dopant molecule is donor-like in its neutral state. On the other hand, for electron transport where electrons hop from the radical anions to their neighboring neutral molecules, the dopant molecules are acceptor-like. Indeed, for the donor-like molecules NIPCA and TPA dispersed in Lexan, only hole transport is observed (5, 14, 15), while for the acceptor molecule TNF dispersed in polyester, transport is carried by electrons (3). Thus the chemical nature of the dopant molecule, preserved in the disordered molecular solid, largely determines the sign of the mobile carriers. Since these molecules are

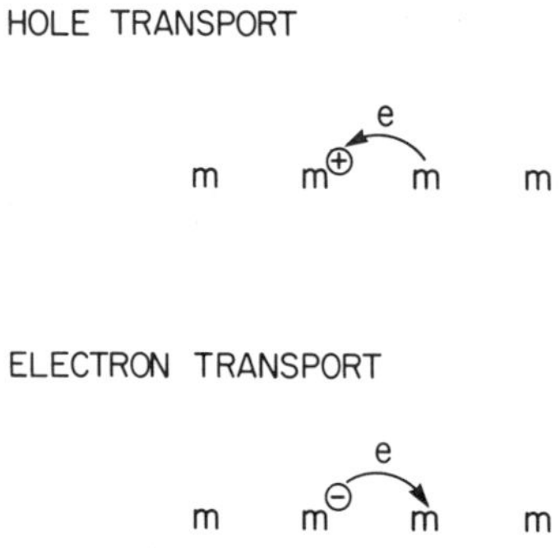

HOLE TRANSPORT

m   m⊕   m   m
         ‾e‾

ELECTRON TRANSPORT

m   m⊖   m   m
         ‾e‾

Figure 6.9 Charge transport in doped polymer systems visualized as a donor–acceptor or an oxidation–reduction process.

typically either donor or acceptor-like, their associated charge-transport capabilities are typically unipolar. In this respect the donor–acceptor complex PVCA:TNF is of interest, being ambipolar with electron transport via the TNF acceptor states and hole transport via the PVCA chromophore donor states (3). Note that the transport properties of organic *disordered* solids are completely different from those of the molecular *crystals*. Here the chemical nature of the constituent molecules is lost and transport is typically ambipolar, with both carriers having similar and largely temperature-independent mobilities (18) which are of the order of ~ 1 cm^2/V-s.

In the ideal hopping process, one thinks of a charge hopping among point-like localized sites whose geometric dimensions are much smaller than their average separation. These conditions are not met in the doped polymers or intrinsically semiconducting polymers. The localized sites are dopant molecules or side groups attached to a polymer backbone, and the spatial dimensions of these groups are typically in the range 5–10 Å. This is to be compared with the average separation of 10–20 Å calculated from the density of the molecules as they are introduced into the films. Due to the proximity and nonspherical shape of the transport-active molecules, one expects that the integral of the wave functions of a carrier localized on neighboring molecules, the overlap term, is very sensitive to the relative orientation of these molecules. This overlap term can vary by many orders of magnitude as a function of the relative orientation of two molecules for a fixed distance of the respective centers of mass. It is to be expected then that the thermal motions of molecules modify the overlap term, leading to stronger localization at lower temperatures. Indeed, Fig. 6.10 shows that the localization parameter γ for hole transport in TPA–Lexan increases from $\gamma \sim 2 \times 10^8$ cm^{-1} at 296 K to 2.8×10^8 cm^{-1} at 255 K (14).

The observation of a temperature-dependent overlap $\gamma(T)$ suggests an interesting microscopic transport step. Assume a carrier arrives at a molecule whose overlap with neighboring molecules is extremely poor due to mutual misalignment. Such a situation might be called a "conformational trap" (19, 20). Under these conditions, the hopping time to the next molecule might be longer than the time it takes to optimize the overlap by relative molecular motion of adjacent

CHARGE TRANSPORT   229

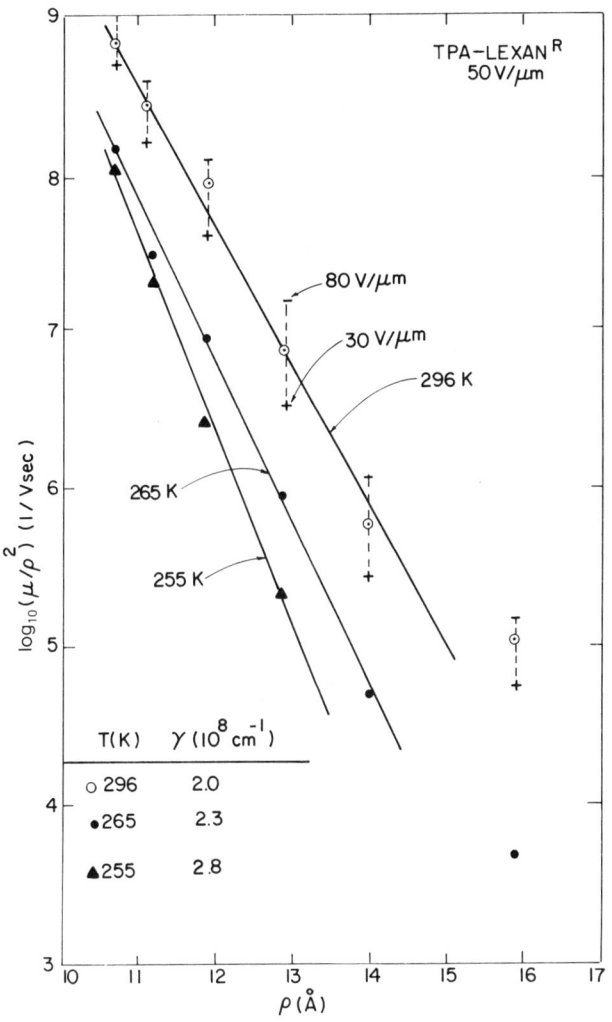

Figure 6.10 Concentration dependence of hole transport in TPA–Lexan at various temperatures. (From Pfister [14].)

molecules. The carrier would await such a coincidence and then rapidly complete the hop. Indeed, the typically large transport activation energies observed in disordered organic solids suggest that contributions from molecular motions may add to the energies of electronic polarization (polaron binding energy) and disorder. A further contribution to the activation energy might result from trapping, an effect which is described in the following.

One expects that in hole transporting systems only those neutral impurities act as hole traps which have an ionization potential smaller than that of the transport-active molecule. Furthermore, impurities with larger ionization energies will in general not be recognizable in transport studies because of the additional activation energy necessary for holes to frequent these localized sites. Figure 6.11 shows an experimental verification of these predictions for the hole transporting system NIPCA–Lexan to which various amounts of TPA were added (21). TPA has a lower ionization energy than NIPCA. Referring to Fig. 6.11, two sets of samples were measured. In one set, the NIPCA concentration was zero and the TPA concentration was varied as described before. The strong concentration dependence of the hole velocity shown by the solid circles in Fig. 6.11 reflects the discussed hopping transport.

In the other series of samples, Lexan was doped with NIPCA molecules (with $n_{NIPCA} = 1 \times 10^{21}$ cm^{-3}) and the TPA concentration was varied. The n_{TPA} was kept below $\sim 3 \times 10^{20}$ cm^{-3} to ensure that the average intersite distance $\rho_{NIPCA} \sim (n_{NIPCA})^{-1/3}$ among the NIPCA molecules was not changed by the TPA molecules. The drift velocity results at 50 V/μm are shown in Fig. 6.11 as open

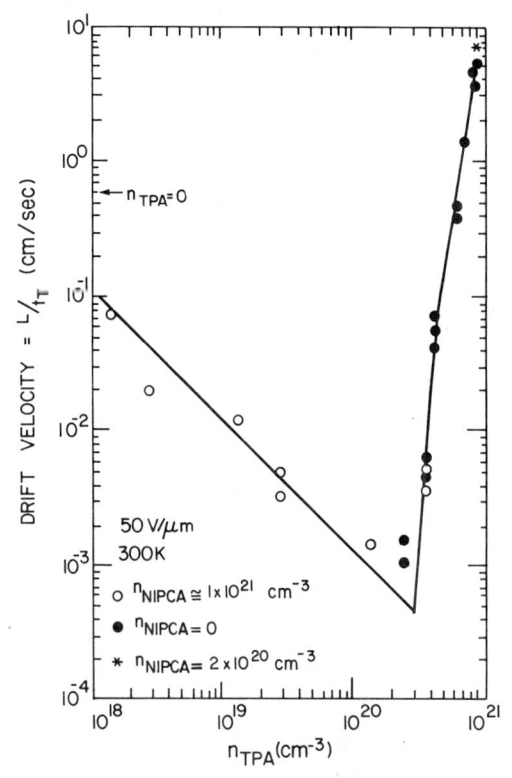

Figure 6.11 Hole velocity as a function of TPA concentration in Lexan (●) and in Lexan doped with fixed NIPCA concentration of 1×10^{21} cm^{-3} (O). The data point marked with * pertains to a sample containing $\sim 9 \times 10^{20}$ TPA/cm^3 and 2×10^{20} NIPCA/cm^3 to show that NIPCA does not act as a hole trap in the TPA–Lexan transport system. The applied field was 50 V/μm. (From Pfister et al. [21].)

circles. For $n_{TPA} = 0$ (arrow), transport occurs via hopping among NIPCA molecules. Note that for the same concentration of $\sim 1 \times 10^{21}$ cm^{-3}, transport via TPA molecules exceeds that via NICPA molecules by more than one order of magnitude.

The addition of TPA reduces the drift velocity from the value at $n_{TPA} = 0$ in a manner which, for $n_{TPA} < 10^{18}$ cm^{-3}, is approximately proportional to n_{TPA}^{-1}. With further increase in TPA concentration, the velocity goes through a minimum at $\sim 3 \times 10^{20}$ cm^{-3} and then increases to approach the values obtained with films of the first sample series which contained no NIPCA. Samples with $n_{TPA} > 4 \times 10^{20}$ cm^{-3} cannot be prepared without changing the average intersite distance among NIPCA molecules. Also for total concentrations in excess of 2×10^{21} cm^{-3}, crystallization effects become apparent.

It is possible to explain the general features of these observations in a relatively simple way. For $n_{TPA} = 0$, the hole transport occurs via hopping among the NIPCA molecules present in a fixed concentration (Fig. 6.12a). As TPA molecules are introduced at low concentrations, carriers occasionally become localized on a TPA molecule, which, because its ionization potential is lower than NIPCA, acts as a trap for holes. Since the overlap between TPA molecules is so small at these concentrations, further drift of the charge localized on TPA must await thermal excitation back to a NIPCA molecule (Fig. 6.12b). The data points for $n_{TPA} \sim 2 \times 10^{20}$ cm^{-3} pertain to this mechanism which has been termed "trap-controlled hopping" (20). At sufficiently high TPA concentrations, the overlap among the TPA molecules becomes sufficiently large that TPA–TPA hopping begins to compete with the hopping among NIPCA–NIPCA and TPA–NIPCA pairs observed at low TPA loading. This process causes the drift velocity to rise for $n_{TPA} > 2 \times 10^{20}$ cm^{-3}, and it appears from the data shown in Fig. 6.11 that for $n_{TPA} > 2 \times 10^{20}$ cm^{-3}, hopping among TPA completely dominates the charge transport (Fig. 6.12c). TPA at low concentrations inhibits hole transport through NIPCA because its ionization potential is lower. It follows then that, in the converse case, charge transport through TPA should not be influenced by NIPCA as long as the intersite distance $(n_{TPA})^{-1/3}$ remains constant. This is experimentally confirmed in Fig. 6.11 by the coincidence of the velocities measured for $n_{TPA} \sim 4 \times 10^{20}$ cm^{-3} for both sample series and by the point identified by an asterisk which pertains to a sample series with the loadings $n_{NIPCA} \sim 2 \times 10^{20}$ cm^{-3} and $n_{TPA} \sim 8.5 \times 10^{20}$ cm^{-3}.

The results on the carbonate polymer system containing the two active molecules NIPCA and TPA suggest that the influence of a foreign molecule on charge transport can be assessed from a knowledge of the respective ionization potential and electron affinities. Thus, for hole transporting systems, impurities may act as traps if their ionization energy is smaller than that of the main transporting molecule. Similar conclusions were reached from transport measurements in doped organic crystals (22).

Early drift mobility measurements by Gill (3, 13) on the charge-transfer complex PVCA:TNF showed that to a good approximation the field dependence

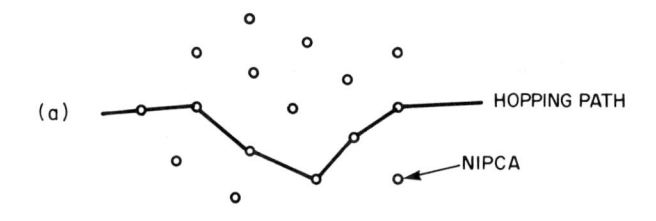

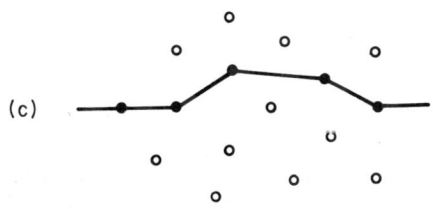

Figure 6.12 Schematic representation of the hole transport process observed for various TPA concentrations in the TPA–NIPCA–Lexan system. Compare Fig. 6.11. (a) Hopping transport through NIPCA in the absence of TPA. (b) Trap-limited hopping through NIPCA with TPA being the hole trap. (c). Hopping through TPA hole traps with NIPCA present, but is inconsequential for transport efficiency.

of the drift mobility can be described by the relationship $\exp(\beta E^{1/2}/kT_{\text{eff}})$, where T_{eff} is the effective temperature introduced in Section 6.3.3. Compare also Figs. 6.8 and 6.17. [It should be recognized, however, that over the experimental field range of drift mobility measurements, expressions of the form to be given by Eq. (6.3), Section 6.3.4, fit the data equally well.] The apparent square-root field dependence appearing in the exponent is typical for a Poole–Frenkel effect (23) where the motion of the carriers is controlled by their release from Coulomb traps, that is, centers which are charged when unoccupied. Indeed, the Poole–Frenkel coefficient β obtained for PVCA:TNF is within a factor of ~ 3 equal to the theoretical value for a reasonable static permittivity of $\epsilon \sim 3.5$. Despite this apparent agreement, the interpretation of the drift mobility in terms of Coulomb traps was not favored (3) since the origin of such positively and negatively charged traps (of equal density) was unclear, and furthermore the reproducibility of mobility measurements observed for samples prepared from various solvents and by different procedures seemed to question a dominant role of extrinsic impurities. The issue has been recently revisited by Hirsch (24), who showed that a consistent interpretation of Gill's drift mobility data can be obtained if one assumes the presence of the order of 10 ppm Coulomb traps. A numerical analysis in terms of this model yields for the trap-free mobility (the polaron hopping mobility) an estimate of the order of $\sim 10^{-3}$ cm^2/V-s which is activated with an energy of the order of ~ 0.1 eV. Hence, according to this interpretation, the activation energy for holes in PVCA, typically of the order of ~ 0.6 eV, is largely due to trapping at extrinsic Coulomb traps. It should be stressed, however, that the site energy of the negatively charged Coulomb traps may be well below the hole-transporting valence levels, and therefore drifting carriers will not hop onto these centers—a situation similar to the case of NIPCA centers in the TPA– Lexan hole transport system discussed above. However, while the neutral NIPCA molecules are invisible for holes in TPA–Lexan (Fig. 6.12), Coulomb centers, by virtue of their strong electric field, may significantly increase the capture cross section of transport-active traps located within the Coulomb sphere. These traps may indeed be of structural origin (for instance, excimer forming sites, discussed in Chapter 5, Sections 5.2.4 and 5.4). The value for the intrinsic hopping mobility derived by Hirsch (24) is remarkably close to the values estimated by Hughes (25) on the basis of x-ray-induced currents in PVCA during the first 20 ns after excitation, by Reimer and Bässler (26) from an analysis of the photoinduced current peak in PVCA, and by Szymanski and Labes (27) from drift mobility measurements in what might have been (?) trap-free PVCA. At this time, the relative importance of extrinsic Coulomb traps and intrinsic traps associated with sample morphology is not clear. This issue, which is of central importance for understanding transport process in disordered organic solids in general and in the model polymer PVCA in particular, is again raised in the following section.

6.3.4 Effects of Disorder and Morphology

In a disordered molecular solid, the mutual orientation and intermolecular distances of the constituent molecules (dopant molecules as in TPA–Lexan, pendant groups as in PVCA) exhibit a distribution that generates so-called diagonal and off-diagonal disorder (see also Chapter 2, Section 2.3). The former refers to the fluctuation of site energies $\Delta \epsilon$, the latter to that of the inter- and intramolecular transition matrix elements $\Delta \Gamma$. If $\Delta \epsilon \gg \Delta \Gamma$ the carriers become localized on individual sites; hence the conventional band picture breaks down and transport of carriers has to be treated as a sequence of hops in an assembly of localized states with fluctuating site energies and transition matrix elements.

For molecular (van der Waals) solids, the intra- and intermolecular transition elements are typically of the order 0.01 and 0.1 eV, respectively (hence the narrow bands of the molecular crystal!). Estimates of the magnitude of diagonal disorder vary from ~0.1 to ~0.5 eV (28–30). The former value is based on a calculation in which a Gaussian distribution of site energies results from fluctuations of the polarization energy due to fluctuations of intermolecular separations. These are assumed to be of the order of $\Delta\rho/\rho \sim 0.01$. The latter value stems from a theoretical analysis of the anomalously broad photoemission lines in pendant groups polymers. In either case, $\Delta\epsilon/\Delta\Gamma > 1$, meaning that charge carriers will be localized on dopant molecules such as in TPA–Lexan or pendant

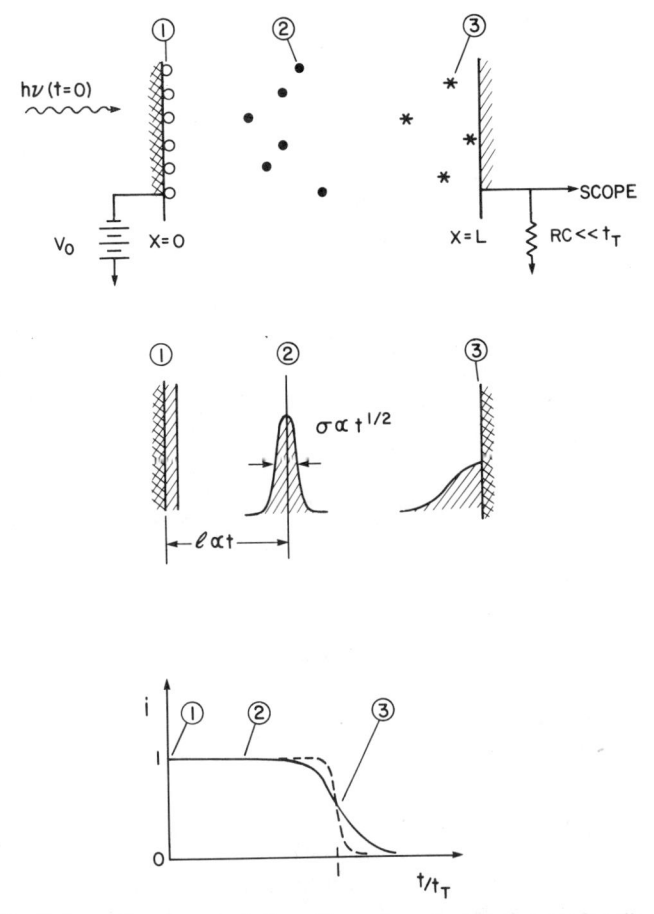

Figure 6.13 Schematic representation of carrier propagation under Gaussian conditions. Top: Position of representative carriers in the sample bulk at $t = 0$ (O), $t < t_T$ (●), and $t \sim t_T$ (*). Middle: Charge distribution in sample bulk at $t = 0$, $t < t_T$, and $t \sim t_T$. Bottom: Current pulse in external circuit induced by charge displacement. Units normalized to t_T and $i_T = i(t_T)$. Dashed line represents transient current for lower applied bias field (longer transit time).

groups such as in PVCA discussed in the previous section. The magnitude of $\Delta \epsilon$, however, will play a key role in the interpretation of the statistics of carrier transport in the disordered molecular solid, as is shown in the following paragraph.

As a result of diagonal and off-diagonal disorder, carriers propagating through the solid experience a distribution of hopping times. Consequently, the carrier packet which in a time-of-flight experiment is injected as a thin sheet at the surface of the sample (Section 6.3.1) will broaden as it penetrates the bulk. In the conventional Gaussian case, the distribution of the hopping times cuts off at a time much shorter than the experimental observation time. In the time-of-flight experiment, the experimental time is determined by the transit time t_T. As shown in Fig. 6.13, the mean position of the carrier packet increases in time, $\langle l \rangle \propto t$, whereas the width of the Gaussian spread increases only in proportion to the square root of time, $\sigma \propto t^{1/2}$. The spreading of the carrier packet does not manifest itself as long as $t < t_T$, since the entire charge q injected at $t = 0$ drifts with the average velocity $v_d = \mu_d E$, where μ_d is the drift mobility. In the Gaussian regime, a drop of the current for $t < t_T$ would indicate a loss of carriers in deep traps where they are immobilized for a time $\gtrsim t_T$. When the leading edge of the carrier packet reaches the substrate electrode, the transient current begins to drop and exhibits a tail whose width equals the width of the carrier packet at the transit time $\sigma \sim t_T^{1/2}$. Therefore in normalized time units t/t_T, the width of the current tail has the time dependence $t_T^{-1/2}$ (Fig. 6.13).

For many disordered solids, this behavior, generally predicted on the basis of Gaussian statistics, is not observed. The current pulse does not show the rectangular shape in Fig. 6.13 but is characterized by a rather featureless decay. In fact, a transition from the well-defined current transient to a featureless trace is often observed as a function of decreasing temperature. As an example, we show in Fig. 6.14 the hole current traces for PVCA. Note that at 420 K the pulse shape is nearly rectangular with well-defined t_T, whereas below room temperature the transit time cannot be extracted from the current pulse without resorting to the special methods to be described below. The featureless current pulse indicates a wide spread of the hole packet as it arrives at the substrate electrode. In fact, it is so wide that it no longer exhibits the Gaussian spread as shown in Fig. 6.13 but is asymmetrically skewed, with a leading edge penetrating into the bulk and a sharp cutoff on the backside of the packet (Fig. 6.15). Such current traces have become known as being indicative of "dispersive transport."

Dispersive transport behavior was first noted in inorganic amorphous solids, notably the chalcogenide glass a-As_2Se_3. The dependence of the transit time t_T for holes on sample geometry and applied electric field as well as the shape of the transient current indicated that transport in this material cannot be described in terms of conventional theories which rely on Gaussian statistics and which ascribe the featureless current traces to excessive bulk trapping or delayed release from surface traps. The first to recognize these important new features were Scher and Montroll [31], who provided a general theoretical framework

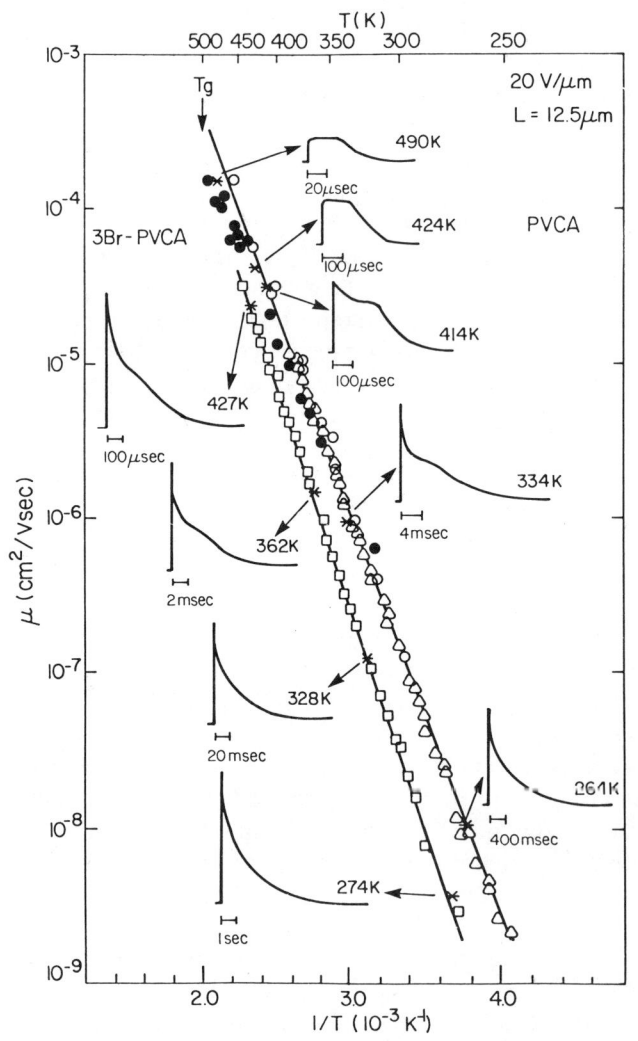

Figure 6.14 Hole mobility in PVCA and 3Br-PVCA as function of temperature. Several representative hole current signals are shown. $E = 20$ V/μm. (From Pfister and Griffiths [12].)

which allows the analysis of dispersive transient current and furthermore predicts interesting relationships between pulse shape and thickness and field dependence of the transit time. The basic argument is that trivial fluctuations of positional disorder (hopping distance, activation energy) can introduce a large spread of individual hopping times since these fluctuations are exponentially amplified [$(\exp(-\gamma\rho)$, $\exp(-\Delta/kT)$]. The resulting spread of the hopping times can be so broad that it extends into the time range of the experiment, that is, the

CHARGE TRANSPORT 237

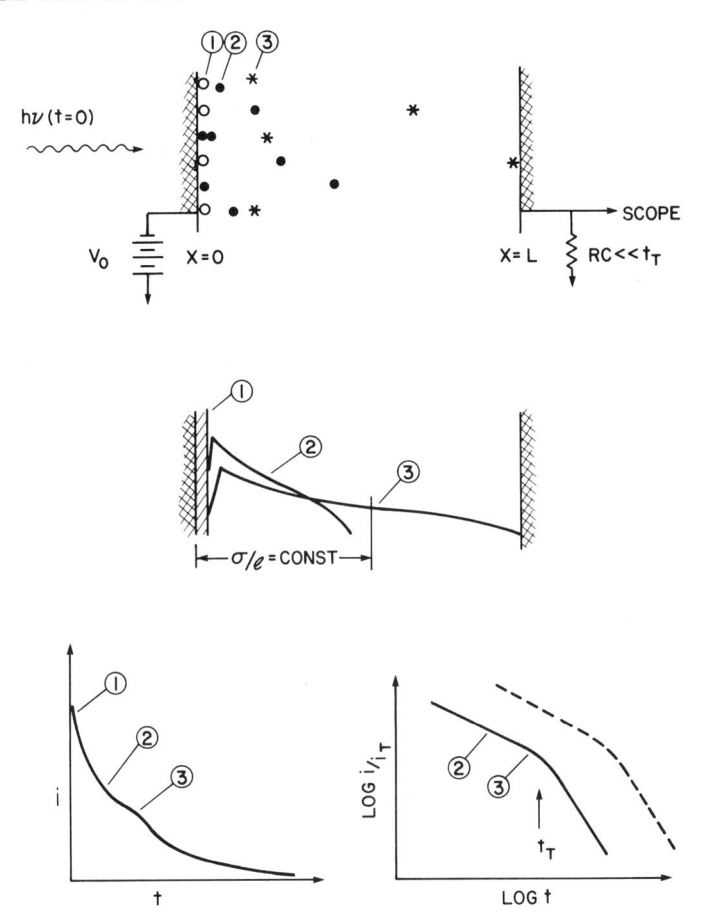

Figure 6.15 Schematic representation of carrier propagation under ideal non-Gaussian conditions. Top: Position of representative carriers in the sample bulk at $t = 0$ (O), $t < t_T$ (●), and $t \sim t_T$ (*). Middle: Charge distribution in sample bulk at $t = 0$, $t < t_T$, and $t \sim t_T$. Bottom: Current pulse in external circuit induced by charge displacement in linear units (left) and logarithmic units (right). Dashed line represents transient current for lower applied bias field (longer transit time). (From Pfister and Scher [37].)

transit time t_T is a time within the distribution. It is clear that under these circumstances Gaussian statistics no longer apply.

The mathematical analysis by Scher and Montroll uses the formalism of continuous-time random walk (CTRW) in which a carrier hops among a random set of localized sites. Non-Gaussian dispersive transport, however, is not restricted to hopping motion. Noolandi (32) and Schmidlin (33) have analytically shown that conventional multiple trapping transport, in which carriers move in extended states and interact with a distribution of traps, formally reproduces the results of CTRW, the major difference being the analytical expression for the parameter characterizing the dispersion. Since in the present case we deal with hopping systems, we adopt the Scher–Montroll picture to illustrate the essential features of dispersive non-Gaussian transport.

The CTRW treated by Scher and Montroll simulates the hopping of a carrier in a random distribution of localized sites by a carrier hopping on a regular lattice with hopping events governed by a hopping time distribution $\psi(t)$. The distribution function $\psi(t)$ determines the statistics of the carrier transport. If $\psi(t) \sim \exp(-t/\tau)$, transport can be described by conventional Gaussian statistics for $t > \tau$. However, if $\psi(t) \sim t^{-(1+\alpha)}$, where α is a parameter describing the dispersion, non-Gaussian transport results as there is no characteristic time in $\psi(t)$. The algebraic power law has been treated by Scher and Montroll. Other hopping distribution functions are possible but the power law is the only one which allows an exact analytical solution of the CTRW problem, and, in retrospect, the law has been justified by numerous experiments on different systems. In the derivation of current shape and transit time for the non-Gaussian case, Scher and Montroll assumed $\psi(t)$ to be temperature independent, that is, the dispersion is due to static off-diagonal disorder only. This assumption has often been challenged, and it now appears that, in accord with $\Delta\epsilon/\Delta\Gamma > 1$, the dispersion due to diagonal disorder dominates, hence α is temperature dependent. Exact analytical treatment of a carrier hopping in a random array of localized sites characterized by a distribution of site energies and transfer integrals has not been published. However, very recently Monte Carlo computer studies by Silver et al. (34, 35), and Marshall (36) have shed a lot of insight into the more general properties of dispersive transport.

Before addressing these newer results in the next section, the findings of Scher and Montroll for a carrier hopping in a three-dimensional random array of isoenergetic sites will be discussed, as they contain the essential features of the more general case. The time dependence of the transient current and transit time are given by

$$i(t) \sim \begin{cases} t^{-(1-\alpha)} & t < t_T \\ t^{-(1+\alpha)} & t > t_T \end{cases} \qquad (6.1)$$

and

$$t_T \sim \left[\frac{L}{l(E)}\right]^{1/\alpha} \exp\left(\frac{\Delta_0}{kT}\right) \qquad (6.2)$$

where $0 < \alpha < 1$. Hence the parameter α relates the shape of the current pulse with the field E and the thickness L dependence of the transit time. Note that the transit time exhibits a superlinear thickness dependence, $L^{1/\alpha}$, which clearly contradicts Gaussian statistics (where $t_T \propto L$). The more disordered the system, the smaller the α-value and from Eqs. (6.1) and (6.2), the more dispersive the cur-

CHARGE TRANSPORT

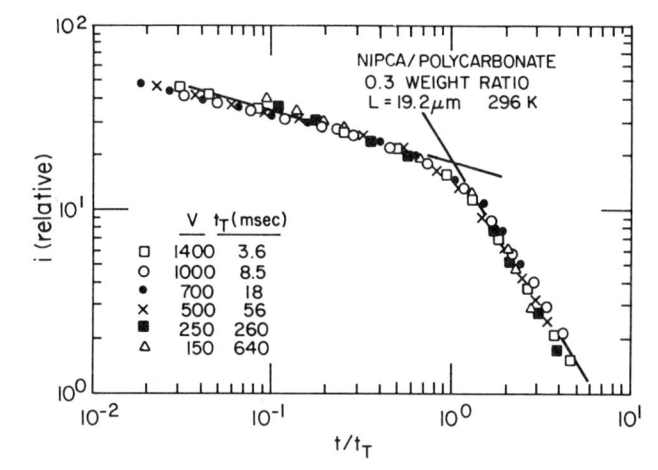

Figure 6.16 Master plot of transient hole traces in NIPCA–Lexan at room temperature in units log i vs. log t/t_T. The traces were normalized to t_T and shifted along the log i axis for best superposition. (From Mort et al. [5].)

rent shape and the stronger the thickness (and field) dependence; $l(E)$ in Eq. (6.2) is the mean displacement in field direction per hop.

Figure 6.16 shows an experimental verification of Eq. (6.1) for NIPCA-doped Lexan. The "masterplot" was produced by shifting individual current traces along the axes of the log i versus log t plot. Similar master curves were produced for a wide range of disordered materials including PVCA (31) and a-As$_2$Se$_3$ (37).

A most important indication of dispersive non-Gaussian transport is the superlinear thickness dependence of the transit time, $L^{1/\alpha}$. For the chalcogenide glasses a-As$_2$Se$_3$ (37) and a-Se (38), this relationship could be verified for evaporated samples extending over the range of ~2 to >100 μm. Similar experiments on organic films either did not show the thickness dependence (PVCA, Ref. 39), or the uncertainty of the experimental results did not allow one to conclusively attribute the observed $t_T(L)$ relationship to non-Gaussian statistics, that is, Eq. (6.2) (TPA–Lexan, Ref. 14). Until a superlinear thickness dependence of the transit time can be shown to be true for a disordered organic solid of the kind discussed here, doubt will remain whether the observed transit time dispersion originates from non-Gaussian statistics as described by Scher and Montroll. This very important issue should be further addressed in future experiments, notably on doped polymer systems.

A calculation of the mean displacement $l(E)$ requires knowledge of the microscopic transport step. Assuming hopping among point sites, the phenomenologic relationship (14, 34)

$$\mu \sim \frac{1}{E} \sinh \left[\frac{e\rho E}{2kT}\right]^{1/\alpha} \exp\left(\frac{-\Delta_0}{kT}\right) \qquad (6.3)$$

has been derived, where ρ is the hopping distance introduced earlier. By comparison with the experimental results which can be represented by introducing an effective temperature (Section 6.2.2), the empirical relationship

$$\alpha = \frac{\alpha_0}{1 - (T/T_0)} \qquad (6.4)$$

was assumed to hold over the limited temperature range accessible in the measurements; T_0 is the characteristic temperature introduced in Section 6.3.3. Hence T_0 relates the temperature dependence of the dispersion $i(t)$ with the field dependence of the transit time. Using Eqs. (6.3) and (6.4), the solid lines in Fig. 6.6 were calculated for the parameters Δ_0, T_0, and α_0 listed in the figure. Note that ρ was determined from the concentration of the dopant molecule TPA and that the field dependence at the various temperatures is determined by T_0 and α_0 only.

A temperature dependence of the dispersion parameter is indicative of a spread of the activation energy of the hopping carriers, that is, diagonal disorder. For transport in extended states with trapping in a trap distribution with a density of states $N(\epsilon)$ decaying exponentially from the band edge,

$$\alpha = \frac{T}{T_0} \qquad (6.5)$$

has been derived, where kT_0 is the energy below the transport channel where the trap density has decayed to one-third (36, 40). For carriers hopping in a Gaussian energy distribution of width σ, Monte Carlo simulations yield (34)

$$\alpha^{-1} = \left(\frac{T_0}{T}\right)^2 + 1 \qquad (6.6)$$

where $T_0 = \sigma/4k$ relates to the width of the Gaussian energy distribution. In all these models, the temperature dependence of the dispersion is associated with a spread in activation energies of the hopping carrier. An interesting different interpretation of T_0 has been given by Hirsch (24), who relates T_0 to a temperature dependence of the dielectic permittivity observed below the glass transition in the PVCA:TNF charge-transfer complex.

With disorder increasing, one expects from Eqs. (6.1), (6.2), and (6.3) the shape of the transient current pulse to become more dispersive (more featureless) and the field dependence of the mobility to become steeper. Furthermore, increased disorder is expected to lead to a larger activation energy. Observations consistent with these expectations are indicated when comparing the transport properties in the polymer series PVCA, 2Br-PVCA, and 3Br-PVCA and the doped polymer NIPCA–Lexan (42). Pertinent results from these measurements are now used to illustrate the effects of morphology on transient transport and to further substantiate the general correctness of the dispersive transport theory.

Although the polymer backbone in carbazole polymers is not directly involved in the transport process, it may play an indirect role by influencing the polymer morphology. Thus in amorphous PVCA, considerable intrachain ordering of the carbazole rings into a pseudo 3/1 helix and interchain parallelizing are observed (43). This morphology results in two different characteristic carbazole hopping site distances, that is, the intrachain intersite distance (4.1 Å) and the interchain intersite distance (9.3 Å) (43). In this picture, it is the interchain distance which controls the motion of carriers, and not the previously assumed overall carbazole intersite distance of 6.5 Å (2). Indeed, the hole mobility in PVCA is equal to that of Lexan doped with the PVCA monomer NIPCA at a concentration for which the average NIPCA intersite distance is 9 Å. (The activation energy for both systems is approximately equal and the dispersion of the monomer in Lexan is assumed to be random.)

The influence of hopping site ordering on charge-carrier mobility has been studied by thermally crystallizing the PVCA and by placing substituents on the carbazole ring system which modify the intrachain crystallization and therefore also the interchain ordering. Bromine has been substituted in both the 2- and 3-positions of the carbazole ring system. These polymers were obtained by free radical polymerization of the respective bromo-N-ethylcarbazoles. In the 2-position, bromine has only a minor influence on intra- and interchain ordering. The x-ray scattering pattern and the basic structure of 2Br-PVCA is similar to that of PVCA, and both polymers are amorphous as prepared but can be crystallized by heating above T_g. Substitution of bromine in the 3-position, however, results in a very different polymer structure. The interchain and intrachain ordering is no longer present and the polymer cannot be crystallized. Although the degree of ordering present in the two polymers is very different, the measured densities (1.498 g/cm^3) indicate little difference in the average packing in the amorphous phase.

Figure 6.17 shows a plot of the room temperature hole mobility as a function of applied field in PVCA, paracrystalline PVCA, 2Br-PVCA, and 3Br-PVCA. It is evident that as the degree of order decreases, the mobility decreases also and, simultaneously, the field dependence increases. Turning now back to Fig. 6.14, we note additional evidence for the influence of disorder on transport. In both PVCA and 3Br-PVCA, the mobility is thermally activated; however, the activation energy in the more disordered polymer is larger. Furthermore, in PVCA a clear transition from dispersive to nondispersive transport is evident at 424 K. In the case of 3Br-PVCA, however, the dispersion is much less temperature sensitive and persists up to the highest feasible measurement temperature. These correlated observations are theoretically expected, Eqs. (6.1), (6.2), and (6.3), and, together with similar observations in the inorganic model systems a-Se (38) and a-As$_2$Se$_3$ (37), furnish an internally consistent interpretation of transport in the disordered solid state.

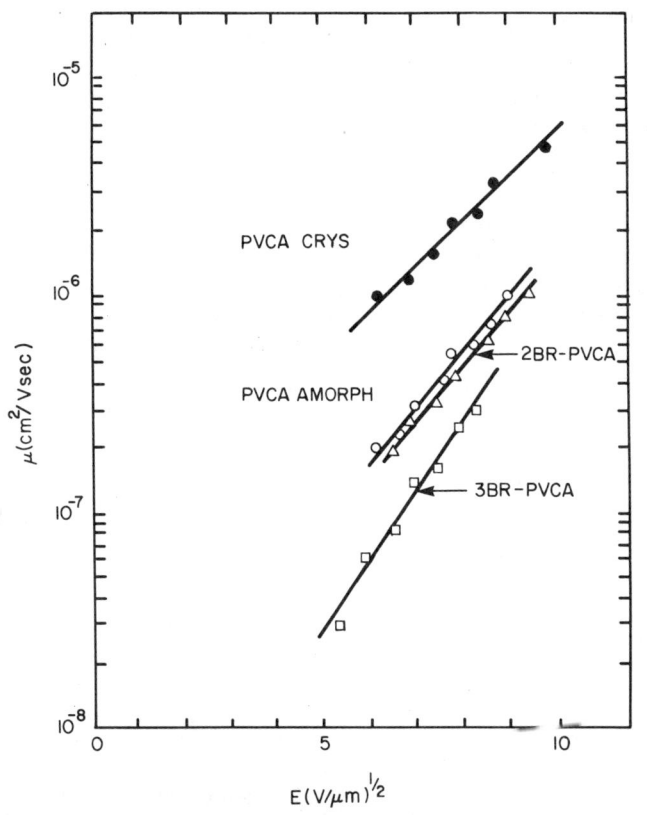

Figure 6.17 Charge carrier (hole) mobility in vinylcarbazole polymers. (From Griffiths et al. [42].)

6.3.5 On Dispersive Transport and Concluding Remarks

Since Scher and Montroll's pioneering article on dispersive transit times, there has been an ongoing discussion about the microscopic physical mechanisms underlying such behavior. The Scher–Montroll report treated hopping between isoenergetic spatially random sites yielding the expressions for current and transit time given in Eqs. 6.1 and 6.2. The dispersion parameter α is temperature independent, and the sum of the pre- and posttransit time power exponents of the transient current, $-(1 - \alpha) - (1 + \alpha) = -2$, is independent of materials and experimental parameters. The theoretical findings were in excellent agreement with hole transit-time studies in amorphous $As_2 Se_3$ (37) and the charge transfer complex PVCA:TNF (31) which lead to the interpretation that in both materials, charge transport occurs by a hopping process. For the organic solids, concentration studies provide independent information for this transport mechanism (2, 12); but for the chalcogenide glasses, this interpretation is still the subject of debate since it has not been established by direct measurement. (Note the elegant way by which trans- port processes among localized states can be established in the molecularly doped polymer systems described in this chapter.)

It was soon recognized that the Scher–Montroll approach should not be restricted by hopping among isoenergetic spatially random sites but should be equally applicable to hopping among localized sites distributed in energy and extended-state transport impeded by a distribution of traps (37). Indeed, the formal equivalence of the CTRW treatment and conventional multiple trapping theory (master transport equation, extended-state transport, traps distributed in energy) was derived by Noolandi (32) and Schmidlin (33). More recently, a theoretical analysis by Tiedje and Rose (41) for band transport in the presence of a trap distribution decaying exponentially in energy into the band gap essentially reproduced Scher–Montroll results but with the temperature-dependent dispersion parameter given by Eq. 6.5. This latter analysis was successfully applied to dispersive carrier transport in amorphous silicon hydride a-Si:H. Also, some very interesting recent measurements by Orenstein and Kastner (44) on hole photocurrent decay in surface cells of amorphous $As_2 Se_3$ brought to light a temperature dependence of the dispersion parameter similar to that found in a-Si:H and stronger than previously assumed (37), thus allowing for a reinterpretation of hole motion in terms of band transport in this binary chalcogenide model compound.

While there is general agreement that dispersive (non-Gaussian) transport can be observed in systems where carriers move in extended states and interact with a distribution of traps, the existence of such behavior in pure hopping systems has been vividly debated. Pollak (45) argued that for the isoenergetic randomly spaced sites treated by Scher and Montroll, non-Gaussian transport does not occur because a site from which a carrier has a low probability to hop to a neighboring site, due to its spatial separation, has a correspondingly low probability for being entered. (Note the conceptual difference between hopping motion and multiple trapping where the capture cross section of chemically or structurally similar trapping sites does not depend upon the trapping depth, i.e., the trap release rate. Furthermore, in multiple trapping transport, a trapping/release event is not associated with a spatial displacement of the carrier.) Pros and cons for the existence of non-Gaussian behavior in hopping systems were recently discussed in a number of excellent articles, notably by Marshall et al. (36), Silver et al. (34, 35), and Bässler (28), which are all based on computer studies using Monte Carlo techniques to simulate hopping transport on spatially random sites with or without a distribution of site energies.

For isoenergetic randomly spaced sites, Marshall (36) found no evidence for non-Gaussian dispersion on the time scale of a transit time; that is, the transient current rapidly ($<0.1 t_T$) approached a time-dependent value for $t < t_T$. However, he observed a non-Gaussian current tail following $t > t_T$ which was much longer than expected for a Gaussian spread carrier packet, thus indicating some kind of dispersion. Using isoenergetic variable-range percolation in which hopping sites of different concentration are randomly placed on a cubic lattice,

Silver and Datta (35) found non-Gaussian dispersion when the time for carriers to hop onto a "giant" cluster of "denser" localized sites was much longer than the time for a carrier to transit the giant cluster.

The existence of non-Gaussian dispersion for hopping in a spatially random isoenergetic system is mostly of theoretical interest, as it becomes increasingly clear that in real systems the dispersion narrows as the temperature is increased. Indeed, in the chalcogenide glass selenium, a transition from non-Gaussian to Gaussian hole transport behavior at about 180 K is observed (38). A similar narrowing of the transit time dispersion is noted for PVCA, and, to a lesser degree, for 3Br-PVCA, shown in Fig. 6.14.

While in glassy selenium the transit time exhibits a superlinear thickness dependence at low temperatures which approaches a linear dependence above about 180 K, thus supporting the interpretation of a transition from non-Gaussian to Gaussian statistics, similar evidence has not yet been found for the organic model system PVCA (see Section 6.3.4). Furthermore, the analysis of the current shape for PVCA and 3Br-PVCA reveal deviations from the power laws [Eq. (6.1)] and, as shown in Fig. 6.18, exhibit different pre- and posttransit time dispersion parameters (12). Hence $(1 - \alpha_i) - (1 + \alpha_f) \neq -2$. Against this experimental background, further theoretical studies on the possibility of non-Gaussian transport in hopping systems as forecast by Scher and Montroll become of central importance.

Again computer simulations have brought much insight to this problem, and again, as for the case of isoenergetic hopping, non-Gaussian dispersion is possible for certain experimental conditions and assumptions pertaining to the energy distribution. For hopping energies distributed randomly in an energy interval, dispersive transport is not predicted on the time scale of the transit time experiment (36). However, a significant increase of the dispersion is predicted when the dimensionality of the hopping process is restricted, for instance, to two- or one-dimensional hopping. Such an effect, it has been suggested, could possibly explain the current shapes in polymers such as PVCA where polymer chains may introduce dimensional restrictions in the hopping process. (36).

An extensive computer analysis of carrier (and exciton) hopping was carried out by Silver et al. (34) for a system with a Gaussian energy distribution of Gaussian width σ. Based on arguments presented in Section 6.3.4, these authors consider a width of ~0.1 eV to be typical for a disordered van der Waals solid. Silver et al. find the following results pertinent in the present context: (a) Non-Gaussian dispersion is observed over a considerable range in time. However, for $\sigma \sim 0.1$ eV, the time frame for that dispersion to occur is much shorter ($t < 10^{-8}$ s) than the experimental transit time at which Gaussian behavior prevails. (b) In the dispersive time range power dependences of the form of Eq. (6.1) can be observed; however, the power laws of pre- and posttransit time current shapes do not add up to -2, that is, $\alpha_i \neq \alpha_f$. (c) The initial dispersion parameters α_i can be expressed by Eq. (6.6). (d) At long times the mobility follows a $\ln \mu \sim (\sigma/T)^2$ rather than an Arrhenius-type temperature relationship. Over the

CHARGE TRANSPORT 245

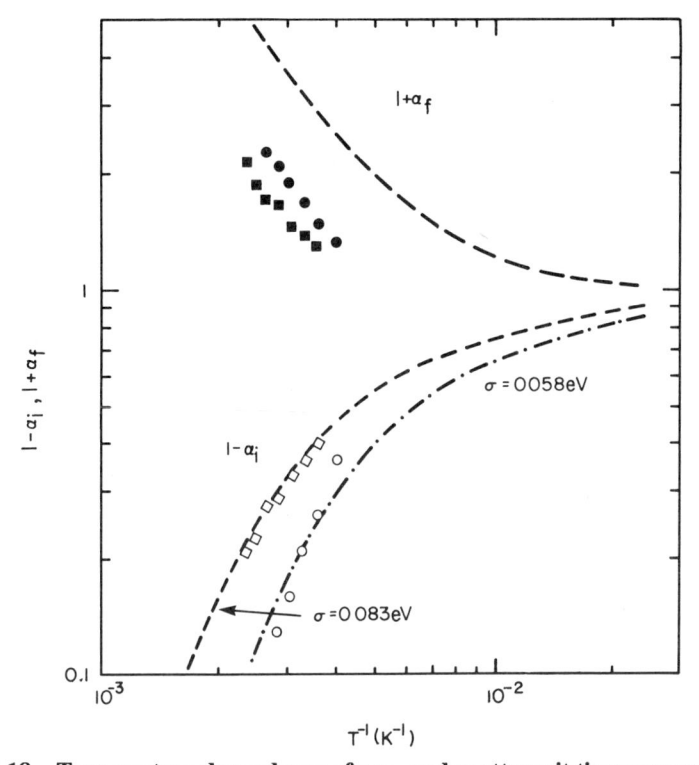

Figure 6.18 Temperature dependence of pre- and posttransit time current shapes approximated by $t^{-(1-\alpha_i)}$ and $t^{-(1+\alpha_f)}$, respectively, for hole transport in PVCA (circles) and 3Br-PVCA (squares). The lines are predicted from Monte Carlo computer simulations for carriers hopping in an energy distribution of Gaussian width σ. (From Silver et al. [34]; data from Pfister and Griffiths [12].)

temperature range of the typical transit time experiment, an activation energy can however be obtained from the data. (e) A field dependence of the transit time is predicted, resulting from a field dependence of the dispersion parameter, which over the experimental field range approximates the observed Poole–Frenkel exponential square-root field dependence. Hence in this model, no Coulomb traps are required to produce the observed field dependence. (Note that the interpretation given by Eq. (6.3) does not require the presence of such traps either.)

Applying these findings to PVCA and 3Br-PVCA, one finds that the simulation approximates the dispersion parameters reasonably well, in particular those determined from the pretransit time pulse shape (Fig. 6.18). However, the activation energy of about 0.14 eV which results from the Gaussian distribution of width about 0.1 eV is much smaller than experimentally observed (Fig. 6.14), and the time range where non-Gaussian transport prevails is much too short. In fact, the computer simulation predicts conventional Gaussian behavior for $t \sim t_T$. A Gaussian width of ~0.23 eV would be required in order to obtain an apparent activation energy of about 0.6 eV for holes in PVCA. In terms of the arguments put forth by Silver et al., so broad an energy distribution is unlikely to be present in the molecular solid. Also for $\sigma \sim 0.23$ eV, Eq. (6.6) yields $\alpha \sim 0.01$, which predicts far more dispersive behavior than is actually observed (Fig. 6.14). On the other hand, a broader energy distribution would be more reasonable in terms of the theoretical framework by Duke et al. (28, 29) (see Chapter 2, Section 2.3).

The computer results suggest that transport in these carbazole polymers is controlled by traps leading to a larger activation energy and shifting non-Gaussian behavior to longer times. Thus in PVCA, a trap-limited hopping process of the type discussed for the NIPCA–TPA–Lexan in Section 6.3.3, Fig. 6.11, could prevail. The traps may possibly derive from structural defects, such as the excimer-forming sites (EFS) described in Chapter 5. This interpretation in terms of non-Gaussian transport would again require a superlinear thickness dependence of the transit time as given by Eq. (6.2). An excellent review by Bässler (28) on these computer simulation results, applied to both carrier and exciton transport in a random set of localized sites, will appear shortly.

All theoretical treatment and computer studies approximate the hopping system by a random array of point sites. This assumption appears particularly severe for the system discussed here, since the transport active molecules, TPA in Lexan or the pendant carbazole side groups of the PVCA systems are large molecules with dimensions of the order of the intersite distance. The resulting steric hindrances ("conformational traps," see Section 6.3.4) are likely to play an important role in transport activation energy and dispersion.

Dispersive non Gaussian transport has become an issue of central importance and interest in the study of transient phenomena such as carrier transport and exciton motion in disordered solids. Experimental evidence for non-Gaussian properties exists, notably for the inorganic compounds a-Se, a-Si:H, and a-As_2Se_3. It may be some time before unequivocal experimental evidence for dispersive transport in a hopping system, the transport mechanism originally proposed in disordered molecular solids and which gave rise to the development of so many exciting new ideas, is found.

6.4 PHOTOGENERATION AND PHOTOSENSITIZATION

6.4.1 Photogeneration of Carriers

The photogeneration of free charge carriers in many molecular solids, including crystalline systems, is electric-field and temperature dependent (46). Thus these dependences are not uniquely characteristic of disorder since they are seen in crystals. The primary characteristic of the systems in which these phenomena are observed is that they exhibit relatively short thermalization lengths associated with low mobilities.

The fate of an excited molecule can be manifold, as is discussed in detail in Chapter 5. The excitation can either relax internally within the molecule via non-radiative or radiative (fluorescence, phosphorescence) processes or it can "thermalize" to form a bound electron–hole pair where the electron and hole are localized on the same or different molecules. All these steps might be preceded by exciton migration which can proceed over many hundreds of angstroms in these systems. The bound electron–hole pair itself can recombine due to mutual Coulomb attraction, or it can be separated, particularly in the presence of an externally applied electric field, into a free electron and a hole.

A commonly used theory to explain this phenomenon is one originally developed by Onsager (47) to describe the probability of dissociation of a thermalized ion pair in an isotropic medium. Assuming that the photogeneration of free carriers involves the dissociation of a bound electron–hole pair, the quantum efficiency $\Phi(E, T)$ can be expressed as

$$\Phi(E, T) = \Phi_0 \int p(r, \theta, E, T)\, g(r, \theta)\, d^3 r \qquad (6.7)$$

where Φ_0 is the yield of thermalized bound pairs and is assumed independent of field; $p(r, \theta, E, T)$ is the probability of dissociation of the thermalized pairs as a function of initial separation, field, and temperature; and $g(r, \theta)$ represents the initial pair distribution. It is commonly assumed that $g(r, \theta)$ is isotropic and that all pairs have the same thermalization distance r_0.

With these assumptions, the integration of Eq. (6.7) yields (48)

$$\Phi(r_0, E) = \Phi_0 \left[1 - \left(\frac{kT}{eEr_0}\right) \sum_{g=0}^{\infty} I_g\left(\frac{e^2}{\epsilon_0 kTr_0}\right) I_g\left(\frac{eEr_0}{kT}\right) \right] \qquad (6.8)$$

where $I_g(x) = I_{g-1}(x) - [\exp(-x) x^{g/g'}]$ and $I_0(x) = 1 - \exp(-x)$. A widely used technique for testing the applicability of the Onsager theory to a particular system involves the analysis at low fields of the field dependence of the photogeneration efficiency (46). Equation (6.8) predicts that the field dependence in the low-field regime is linear, with a slope-to-intercept ratio of $e^3/2\epsilon k^2 T^2$. Since this slope-to-intercept ratio is independent of the particular distribution of the initial thermalized pair distribution and contains no adjustable parameters, it constitutes a critical test of theory. Figure 6.19 shows the relative quantum efficiency Φ/Φ_0 as a function of electric field for different values r_0 at room temperature predicted by the Onsager theory using $\epsilon = 3$.

The photogeneration efficiency is usually measured by employing the time-of-flight or xerographic discharge techniques described in Section 6.2 (see also Ref. 49). Several studies have been made of the photogeneration process in poly(N-vinylcarbazole), PVCA. One of the earliest quantitative studies was due to Regensburger (50), who found that the xerographic gain (proportional to photogeneration efficiency) replicated the spectral dependence of the absorption spectrum

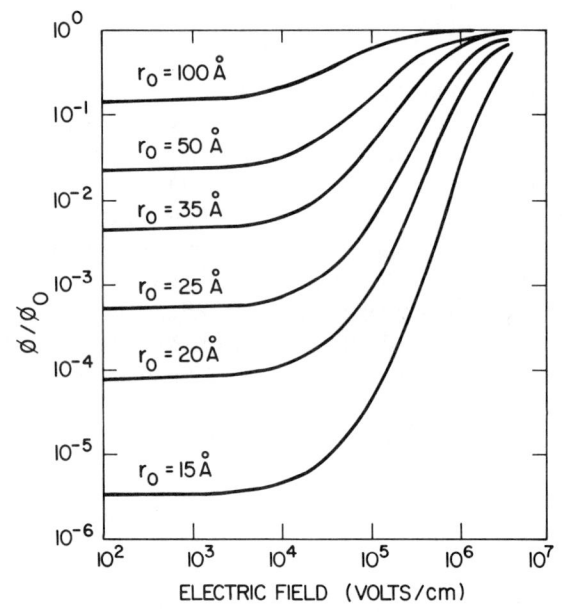

Figure 6.19 Escape probability predicted by Onsager theory for isotropic initial distributions of electron–hole pairs. $T = 296$ K, $\epsilon = 3$.

of PVCA, even though in the experiments all the light was strongly absorbed by the samples. This correspondence between absorption and photoconductivity action spectra, which is often observed in organic solids, is strongly indicative that the photogeneration mechanism involves exciton diffusion to the illuminated surface where the dissociation into free carriers occurs at surface impurity or defect states.

Pfister and Williams (51) expanded on this work by looking at the effects on photogeneration efficiency of surface history, photodecomposition, and the controlled addition of acidic impurities. They found that the symbiotic relationship between the xerographic gain and the absorption disappeared for highly purified PVCA films that have been kept in the dark and exposed only to very low light levels during measurements. The gain now showed a stepwise increase toward higher excitation energies, with the steps occurring at the transitions of excitation into higher-lying singlet states ($S_0 \rightarrow S_1$, $S_0 \rightarrow S_2$, $S_0 \rightarrow S_3$). These results were reproduced by Borsenberger and Ateya (52) and are shown in Fig. 6.20. For undoped, purified PVCA, a value of $\Phi_0 \sim 0.14$ was found. Thermalization distances r_0 ranged from 22.5 to 30 Å, depending on whether the excitation was into the first singlet state (photon energy 4.2 eV) or third singlet state (photon energy 4.77 eV). In the case of the PVCA film doped with TCAA, $\Phi_0 \sim 0.14$, and r_0 for excitation into the first singlet state was 30 Å. This increased thermalization length in the acid-doped films indicates that the charge separation in the charge-transfer state m^* is enhanced by acids, perhaps by stabilization of the

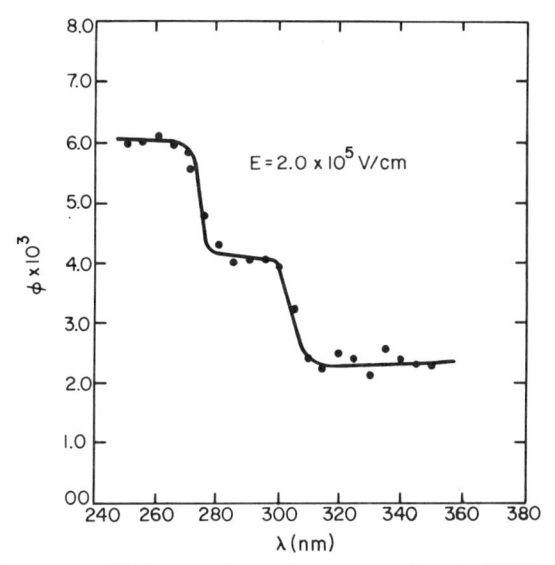

Figure 6.20 Quantum efficiency of carrier generation in PVCA vs. excitation wavelength λ. Note the steps when excitation energy allows transitions in the second and third excited singlet states. (From Borsenberger and Ateya [52].)

charge-transfer state by the polarizable acid molecule or from its acceptor properties.

The formation of the charge-transfer complex of PVCA and TNF in which PVCA is a donor and TNF acts as an acceptor according to

$$D + A \rightleftharpoons D^+ + A^-$$

is well known. In this particular charge-transfer complex, there is little active character unless an excited state of the complex is achieved by light absorption in the charge-transfer absorption band. For PVCA:TNF, this charge-transfer band extends through the visible part of the spectrum and has led to practical application of films of this complex in electrophotography. Similar studies on this system led to values of $\Phi_0 = 0.23$ and $r_0 = 25$ Å. These are not very different from the values for undoped PVCA, although Φ_0 for both is surprisingly larger than for crystalline anthracene, where $r_0 \sim 35$ Å and Φ_0 is only 2×10^{-2}. Apparently the much higher value for the photogeneration efficiency in films of PVCA or PVCA:TNF is due to the much higher probability for autoionization of the optically excited states.

More detailed studies of a similar system, TPA–Lexan, including a description of the field dependence of the photogeneration efficiency in terms of the Onsager model, have been made by Borsenberger et al. (53). Figure 6.21 shows a fit of the Onsager theory to experimental data for a fixed triphenylamine concentration. The parameters required to produce the fit are $\Phi_0 \sim 0.26$ and $r_0 = 26$ Å.

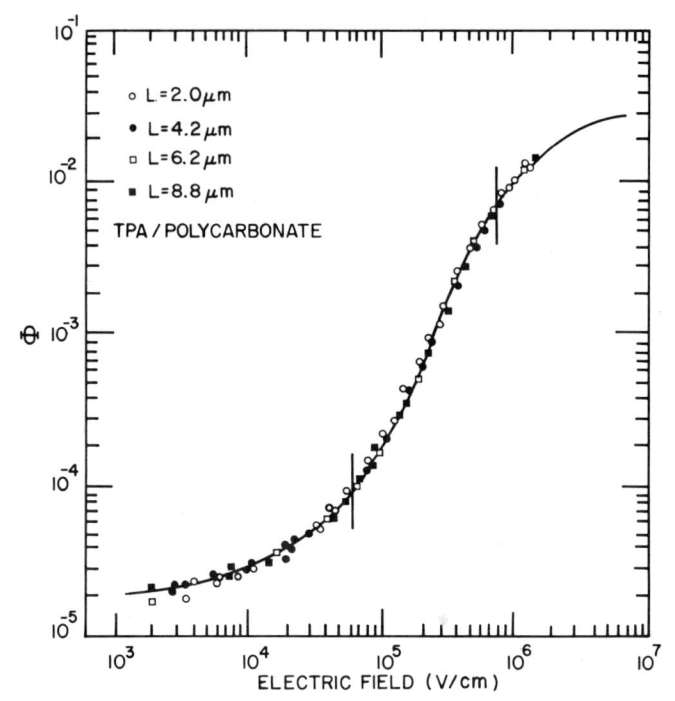

Figure 6.21 Field dependence of the photogeneration efficiency Φ. The temperature was 23°C and the triphenylamine concentration 40%. The solid line was calculated from the Onsager theory with a thermalization distance of 26 Å and primary quantum yield of 0.026. (From Borsenberger et al. [53].)

By means of the ability to control the dopant molecule concentration, it was possible to measure the concentration dependence of Φ_0 and r_0. Borsenberger et al. (53) speculated that the observed increase in Φ_0 with increase in the concentration of TPA is associated with the existence of preferred sites, for example, at the surface, for pair formation. The transfer of energy to these sites could be increasingly more probable as the TPA concentration increases. The thermalization process is believed to occur by the loss of excess kinetic energy by the photoexcited pair by emission of phonons.

Mort and Pfister (54) have recently pointed out that although the Onsager formalism appears to give an excellent fit to the experimental data, considerable uncertainty remains about the physical description of the photogeneration process and its relationship to bulk or macroscopic transport parameters, particularly in disordered organic solids. Evidently, these critical photoelectronic phenomena are characteristically associated with charge transport. Indeed, the concentration dependence of the photogeneration efficiency which is indicative of nearest-neighbor interactions may disguise the fact that this ultimate interaction leading to pair formation may be preceded, for example, by energy transfer over several molecular sites.

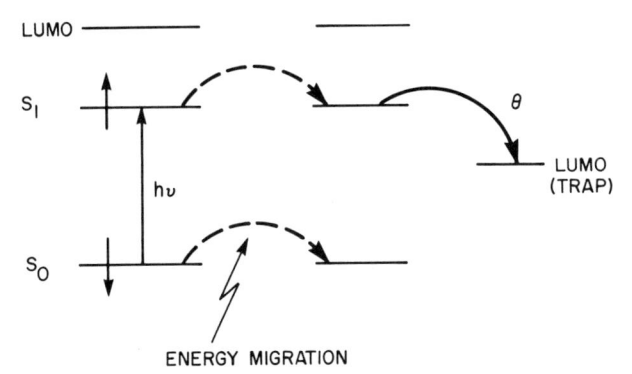

Figure 6.22 Photogeneration process in disordered organic solids. (From Mort and Pfister [54].)

Energy transfer almost certainly is involved if the lowest singlet absorption for which carrier production is observed lies below the lowest unoccupied molecule's orbital (LUMO). This situation, in fact, prevails in most organic systems. For instance, in PVCA, the LUMO level is estimated to be approximately 0.75 eV above S_1. Since any excess carrier placed onto a molecule has to go into the LUMO, it is unlikely that the photoexcited carrier is thermally excited from S_1 to LUMO such that an electron from a neighboring molecule can fill the hole left in the ground-state orbital (Fig. 6.22). Rather, it is expected that the excited state migrates via energy transfer among various sites until it finds a trap—structural or extrinsic such as oxygen (see also Chapter 5)—that has a LUMO below the optically excited level (Fig. 6.22). Now the electron can be transferred from S_1 to the trap LUMO, which amounts to the reaction $m^* \to M^+$ and trap $\to$ trap$^-$. Hence a bound electron–hole pair has been formed, with the hole residing on a chromophore of the polymer (or dopant molecule in molecularly dispersed systems) and an electron residing on a trap. As discussed, the dissociation of this bound pair can now be described in terms of the Onsager formulation.

6.4.2 Photosensitization

Since polymers such as PVCA are typically only photosensitive to ultraviolet radiation, there has been a natural interest in extending the spectral sensitivity into the more useful visible regime (55). Consequently, an enormous literature exists dealing with this subject, and yet despite this, major questions remain unanswered about the details of this scientifically and technologically important phenomenon.

Dye sensitization of polymers involves a molecular solution of the sensitizer throughout the polymer film. In dye sensitization, therefore, the photosensitization involves electronic charge transfer between a photoexcited dye molecule and a polymer chromophore.

Several investigators have reported that the sensitization in the absorption region of the dye passes through a maximum with increasing dye concentration, and different models were proposed to account for this behavior. Meier et al. (56) have suggested that the formation of dye dimers and aggregates is responsible for the decrease in sensitivity observed with methylene blue-sensitized PVCA. Sato and Ikeda (57) have studied PVCA sensitized with a pyrylium dye and proposed that the sensitivity decrease with increasing dye concentration is related to a decrease in hole range.

The dual role of the sensitizing agent as both carrier generator and desensitizer was also observed by Okamoto et al. (58) when sensitizing PVCA with malachite green dye. The loss of hole carriers with increasing dye concentration was attributed to recombination and trapping in the bulk.

Although the mechanism of dye sensitization is presumably electron transfer between the photoexcited dye molecule and a nearby donor (or acceptor) molecule, the criteria for selection of efficient sensitizing dyes remain a significant question. Apart from the internal relaxation processes within the sensitizing molecule which compete with the photoelectron transfer, relative redox potentials of the involved molecules which reflect their relative electronic energy levels are vitally important. Ikeda et al. (59) studied the sensitization efficiency of four classes of dyes with PVCA. Included were triarylmethane, pyrylium, monomethine cyanine, and trimethine cyanine dyes. Quantum efficiencies were highest for the pyrylium dyes, reaching 0.5 at 10^6 V/cm in one case. The conclusion was that the magnitude of the sensitization effect depends on the electron affinity of the dyes, although no quantitative information was presented.

Molecularly doped polymer systems offer a significant advantage over other polymeric systems for studying the fundamentals of dye sensitization, and recent work has been carried out in this area. One can control the absolute and relative concentrations of both the sensitizing agent and transport molecule in the polymer matrix. Sensitization studies utilizing the hole transporting material triphenylamine (TPA) dispersed in a Lexan polycarbonate matrix (discussed in Section 6.3) have recently been described. The transport matrix was doped with either a commercial dye, rhodamine B, or thiapyrylium dye (60).

Carrier generation efficiencies were determined from xerographic measurements for films containing various concentrations of dyes by the techniques described earlier. In addition, the trapping effects of each dye system were determined using the same dye-sensitized films on which a thin amorphous selenium layer was deposited onto the dyed transport layer. Carriers were then generated in the selenium layer by excitation with light strongly absorbed by selenium (4300 Å). The carriers (holes) were injected into the unexcited dye-transport layer and the residual potential after xerographic discharge was measured. With this technique, the trapping effects of each dye in its own ground state could be observed independently of the photogeneration process.

Two distinct phenomena were identified and studied. The first was the dual role of a sensitizer molecule as a sensitizer and a potential trap for mobile carriers. The second was the photosensitization step itself. In the latter experiments, the dye concentration was kept fixed and at a sufficiently low value to avoid significant perturbation of the measurement by the trapping function of the dye. The effect of dye molecule–transport molecule separation was then studied by varying the transport molecule concentration (dye–dye molecule distance $\gg$ transport–transport molecule distance). An advantage of this approach is that in the absorption maximum of the dye where the photosensitization was monitored, there is no change in absorption from sample to sample. A complication is that the potential variation of the photosensitization process as the dye–transport molecule distance is increased is invariably accompanied by an increase in transport–transport molecule separation and its accompanying decrease in carrier transport. Meaningful measurements of the concentration dependence of the photosensitization process therefore required that the transport effect be inconsequential in the measurements. This was accomplished by ensuring that the photosensitization efficiency η, the number of free carriers produced per photon absorbed by the dye molecules, was measured for all transport molecule concentrations at sufficiently low light intensities that the discharge was emission limited and not space-charge limited (transport limitation) (9, 49). The time resolution used in these experiments did not allow a measure of the electron transfer rate itself. This is in contrast to mobility or transport measurements utilizing the time-of-flight technique. In this case, the actual distribution of carrier velocities, proportional to the transfer probabilities between neutral molecules and ground states of molecule ions, is directly resolved (5).

The dual role of the dye molecule as sensitizer and trap is exemplified in Fig. 6.23, where the results of two experiments as a function of rhodamine B concentration (TPA concentration being fixed) are shown. The curve marked (a) was achieved by measurements at 5500 Å on the structure indicated. It represents the inverse optimum exposure (ordinate) required to achieve maximum contrast potential for unity input density for films with equal thickness. These were doped with a fixed TPA and variable rhodamine B concentration. As can be clearly seen, the inverse optimum exposure or sensitivity first increases and then goes through a maximum at a dye concentration of ~0.0025 g rhodamine B per 0.5 g TPA. Curve (b) shows data measured on the same samples as in (a), except that they were overcoated with 1 μm of amorphous Se. The films were step-function illuminated with 4300 Å light (essentially total absorption in the a-Se layer), and the residual potential, which is inversely proportional to the carrier range, is seen to increase dramatically as the dye concentration increases. This directly reveals the effect of the unexcited dye molecule on the hole range. In curve (b), the reciprocal of the residual potential is plotted versus dye concentration over the concentration range at which the dye sensitization goes through a maximum. Thus rhodamine B serves dual and conflicting roles. Initially, the introduction of the dye leads to a photosensitization due to the increased absorption. Ultimately, however, sufficient dye concentration is achieved to decrease the hole range, and this produces a decreasing sensitivity. The sum of these two competing processes accounts for the observed maximum versus concentration in sensitization.

Unlike the case of rhodamine B, it was found that even for relatively large

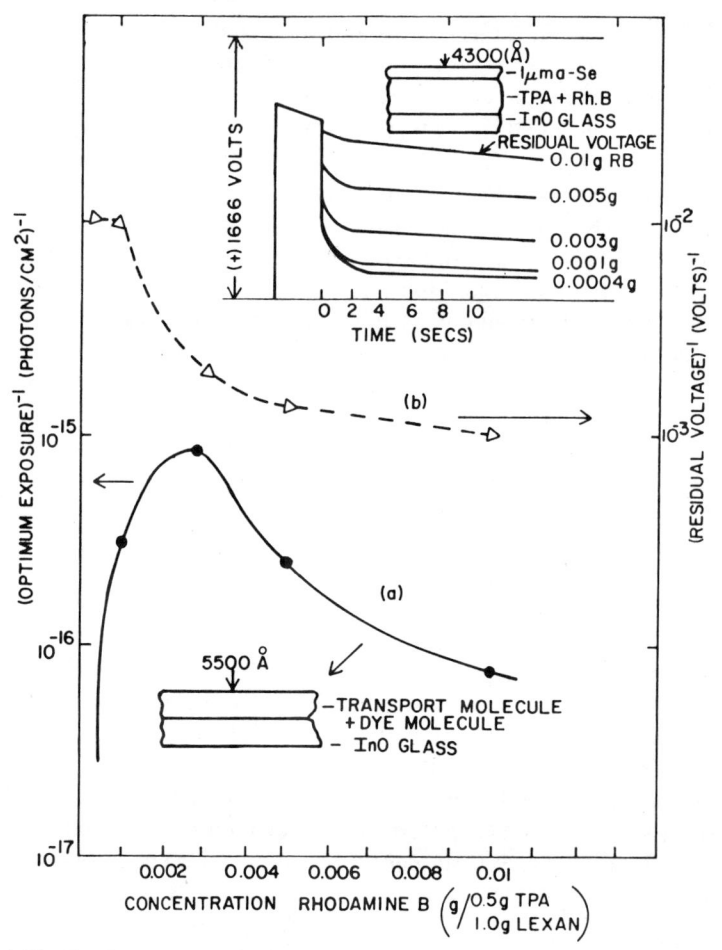

Figure 6.23 Reciprocal exposure (left ordinate) and reciprocal residual voltage (right ordinate) vs. concentration of rhodamine B in a fixed TPA concentration-doped Lexan. Upper insert shows discharge curves for the different rhodamine B concentrations. Illumination occurs for $t > 0$, (in a step-function mode). (From Grammatica and Mort [60].)

thiapyrylium dye concentration, there was no significant increase in the residual potential. It was therefore concluded that this dye is not a very effective trap for holes relative to TPA, and the results were consistent with the thiapyrylium dye acting as a shallow trap which through a trap-controlled hopping process reduces average carrier velocities. In comparison, the rhodamine B acts as a deep trap because of an effectively infinite release time. The difference between the two dyes presumably resides in the effective trap depth they present to the holes hopping between TPA molecules.

The study of the dependence of the photosensitization on dye molecule–

PHOTOGENERATION AND PHOTOSENSITIZATION

transport molecule separation was made on the thiapyrylium-doped films. For these experiments, the thiapyrylium dye concentration was maintained constant for all samples and at sufficiently low level to ensure a molecular solution. Since the earlier measurements showed that this dye did not significantly trap holes in TPA-doped Lexan, there was no problem in this regard.

The effective efficiency of photosensitization of Lexan films doped with varying concentrations of the transport molecule TPA has been measured. As discussed earlier, the TPA concentration is always much greater than the fixed dye concentration. Varying the TPA concentration is the means of varying the dye molecule–transport molecule separation.

Figure 6.24 shows data for films of fixed thiapyrylium dye concentration but varying transport molecule concentration. This figure shows the strong dependence of the photosensitization process on the dye molecule–transport molecule

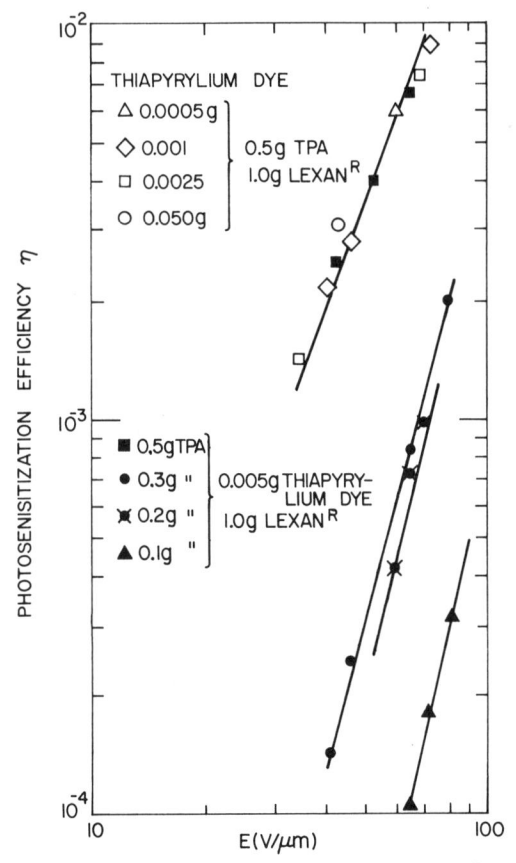

Figure 6.24 Photosensitization efficiency η vs. electric field for a Lexan film containing a fixed thiapyrylium dye concentration and a variable TPA transport molecule concentration. (From Grammatica and Mort [60].)

concentration in two ways: At a given field, there is a strong decrease in the absolute photogeneration efficiency which is paralleled by a discernible increase in the field dependence of the photosensitization step. The restricted field ranges of the data are dictated by the ability to do the measurements.

A plot of the relative photosensitization efficiency/ρ^2 (ρ is the average dye molecule–transport molecule separation) versus ρ on a semilogarithmic plot is linear, indicating that the photosensitization process is proportional to ρ^2 exp$(-\gamma^*\rho)$, where $R_0^* = 2/\gamma^*$, is a localization radius and therefore controlled by wave function overlap between the photoexcited dye molecule and the unexcited transport molecule. A value of ~3 Å is found, which is essentially the same as for the photogeneration process where only TPA molecules are present and photogeneration involves excitation of TPA molecules in the ultraviolet.

Note the similar concentration dependence of transport (Fig. 6.10) and photosensitization in these molecularly doped systems which exemplifies the dominant role of charge exchange between localized states. It is a reasonable finding that the localization parameter characterizing ground-state charge exchange γ is found to be smaller than γ^*, the localization parameters for charge exchange in the generation step which involves excited-state molecules.

With increasing dye molecule–transport molecule separation, the photosensitization step exhibits a stronger field dependence (Fig. 6.24). This could be interpreted by a field-induced change of the localization parameter. Specifically, at increasingly high fields, the charge becomes more delocalized and leads to an increase in charge transfer and hence photosensitization.

6.5 CHEMICAL CONTROL OF CONDUCTIVITY

In earlier sections, the photoelectronic properties of photoconducting insulating polymers either of the intrinsic type such as poly(N-vinylcarbazole) or molecularly doped matrices were described. The photogeneration, photosensitization, and transport properties were studied by employing transient photoconductivity. This was induced by the transient photooxidation of the system and the time resolution or integration of charge displacement. By analogy, it should therefore be possible in these molecular systems to produce controlled steady-state dark conductivities by the use of variable-equilibrium chemical oxidation and/or reduction of the system. This approach and idea has been clearly recognized in the past and has been studied in a number of molecular solids, particularly crystals such as the studies of the violanthrene–I_2 system (61, 62). There have also been some studies on reasonably characterized polymeric systems such as PVCA–I_2 (63) and other systems of varying degrees of characterization and tractability (64). Lupinski (65) made one of the first reports of variable controlled conductivity in a relatively stable system, namely, TCNQ polymer salts. The polymer cations were derived from poly(4-vinylpryridine), atactic and isotatic poly(2-vinylpyridine), poly(N-vinylimidazole), and a copolymer of 4-vinylpyridine and styrene. The materials produced were soluble in organic solvents, and in the course of removing residual neutral TCNQ (TCNQ^0), Lupinski et al. (65) noted a progressive increase in resistivity. Subsequent back doping with controlled concentrations of TCNQ^0 led to the results shown in Fig. 6.25. Thus the polymer cation can be viewed as an electron donor to the TNCQ^0, and the rapid rise in conductivity with increasing amounts of TCNQ^0 suggests that the conduction process involves electron hopping between the TCNQ^0 molecules with the possible involvement of ionized cation states of the polymer. The observed conductivity maximum is reminiscent of that seen in chemically oxidized molecularly doped polymers which has been recently discussed (66).

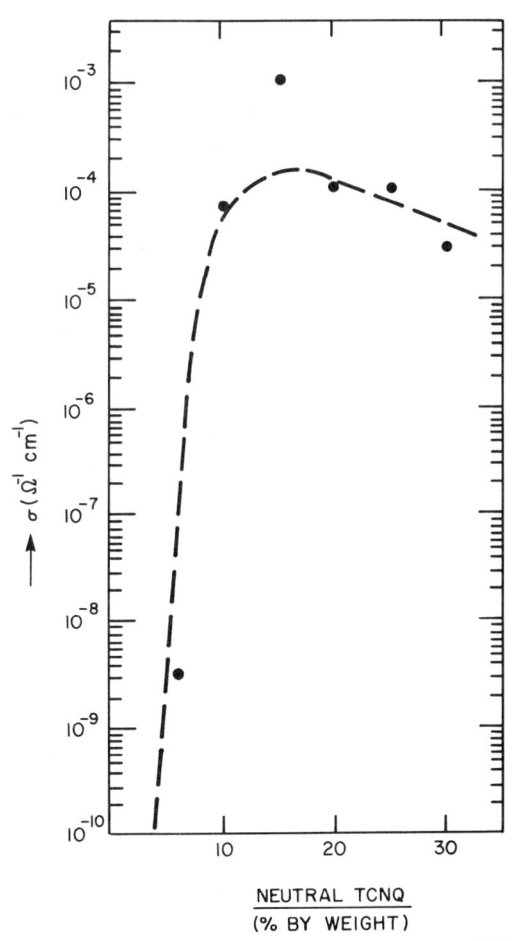

Figure 6.25 Conductivity as function of added neutral TCNQ–(styrene-2-vinyl-1-butylpyridinium–TCNQ$^-$) copolymer with 80% pyridine rings quarternized. Sample with 0% TCNQ is not shown; it has $\sigma \sim 10^{-10}\ \Omega^{-1}\ cm^{-1}$. (From Lupinski et al. [65].)

These systems are particularly powerful from the point of view of elucidating underlying mechanisms and are discussed in more detail for that reason. The advantage these systems offer is that both neutral and charged species (either cation or anion) can be varied independently. As in the case of the conduction studies in the insulating materials, already discussed, the polymer matrix is to first order electrically inert. Figure 6.26 shows results of studies on the system tri-p-tolyl-amine (TTA) doped into Lexan polycarbonate (66). By oxidation with antimony pentachloride ($SbCl_5$), it is possible to form the salt $TTA^+SbCl_5^-$. This salt exhibits remarkable solubility in Lexan such that it is possible to disperse on a molecular basis concentrations $> 10^{20}$ cm^{-3} and yet obtain homogeneous single-phase films. The curves in Fig. 6.26 represent three different total concentrations of molecules. It is possible to controllably alter the concentrations of TTA^+ ions which are paramagnetic and thus can be measured by ESR spin-counting techniques. As can be seen, taking, for example, the lower curve, at TTA^+ of $\sim 8 \times 10^{19}$ cm^{-3} and a total $TTA^+ + TTA^0$ concentration of 3.8×10^{20} cm^{-3} the conductivity is already significantly larger than for a system containing only nominally pure TTA^0 [$\sigma < 10^{-14}$ (Ω^{-1} cm^{-1}]. As the TTA^+ concentrations are progressively increased, the conductivity rises but ultimately goes through a maximum, after which the conductivity drops rapidly. The other curves demonstrate the same dependencies on TTA^+ concentrations but exhibit overall higher conductivities. These effects can be understood in the context of the concentration dependence of mobility discussed in Section 6.3 for a molecularly doped polymer.

Thus, for a given spin (carrier) concentration, the conductivity increases from

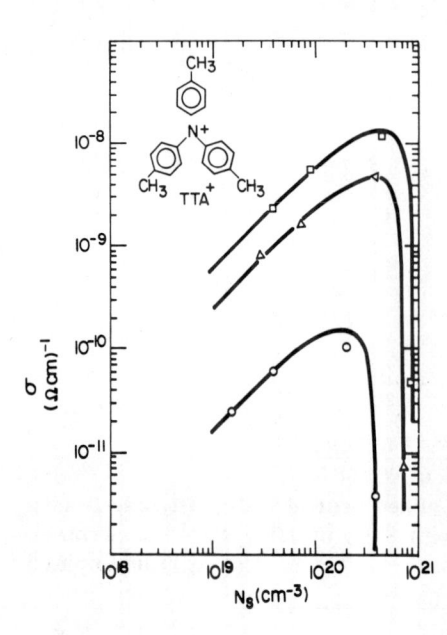

Figure 6.26 Plot of dc conductivity of TTA–Lexan films vs. spin density for the different concentrations of total molecules (□), 8.6×10^{20} cm^{-3}, (△) 7.2×10^{20} cm^{-3}, (○), 3.9×10^{20} cm^{-3}. (From Mort et al. [66].)

the lower to the upper curve because of decreased TTA^0–TTA^+ separation and hence an increased probability for hopping transport. This steady-state variable dark conductivity is the equilibrium analogy to the transient conductivity induced by flash photooxidation. On the basis of the known carrier concentrations deduced from the ESR measurements and the mobility measured by time of flight in an insulating films of TTA, the conductivities observed are lower by four orders of magnitude than expected. This casts doubt on the assumption that all the TTA^+ species contribute to the conductivity. This is an unanswered question at this time.

This type of system has also allowed the first observation of magnetic superexchange in a molecular solid (67). Superexchange magnetic interaction, where the pairwise coupling of the spins of two magnetic species is mediated by the involvement of a singlet atom, is an important phenomenon observed in inorganic ionic solids (68). Until recently, no unambiguous observation of superexchange had been made in disordered molecular ionic solids. In well-studied systems, such as MnO, MnF_2, the sensitivity of the superexchange to the overlap and charge-transfer properties of the wave functions of the constituent species is a prime question (69).

The work on molecularly doped polymers in which monomeric molecules are doped into a host matrix to form homogeneous solid molecular solutions of extremely high dopant concentrations ($\sim 10^{20}$ cm^{-3}) indicates that these are interesting systems in which to seek superexchange phenomena. As already discussed, transient electronic transport, photogeneration, and photosensitization involve, at a microscopic level, the electronic transfer of charge between a transiently oxidized molecule and a neighboring neutral molecule (54). By fractional oxidation of the dopant molecules, it is possible, as described above, to produce amorphous films with variable and controlled concentrations of both neutral molecules and their ions. In the case where the molecular ion is stable and open-shell in character, it is thus possible to simultaneously vary and study the steady-state electrical and magnetic properties of the film since the ions, which are paramagnetic, are also integrally involved in the electronic conduction process.

The potential advantages of a suitably chosen system are (a) the ease with which homogeneous, solid films can be prepared, (b) its good stability at very high concentrations, (c) its uncomplicated and symmetrical ESR line shapes, and (d) the large magnitude of the superexchange interactions and the wide concentration range, which cannot be matched in inorganic materials, over which it can be observed. The simultaneous occurrence of these four features in a single system is very unusual and offers a unique opportunity to study superexchange interactions. A major drawback is that because of the complexity of a disordered system such as this, any meaningful quantitative analysis seems intractable. However, as in the study of electronic transport processes, one can capitalize on the ability to controllably vary the nature and concentration of independent dopant species to study magnetic processes controlled by wave function overlap.

The results reported by Troup et al. (67) were for the TTA–Lexan system described earlier. The counterion $SbCl_5^-$, in addition to the cation TTA^+, should be paramagnetic, but Yoshioka et al. (70) have shown, using NMR techniques, that the chloroantimonate must have a diamagnetic formula, probably corresponding to units such as $SbCl_4^-$ and $SbCl_6^-$.

Figure 6.27 shows plots of the linewidth versus spin concentration for a variety of samples. The curve denoted by the open circles was obtained by dissolving only $TTA^+SbCl_5^-$ powder into Lexan in known concentrations and constitutes a lower limit to the exchange process since all electrostatic interactions are direct exchange processes between the TTA^+ species. It was assumed that

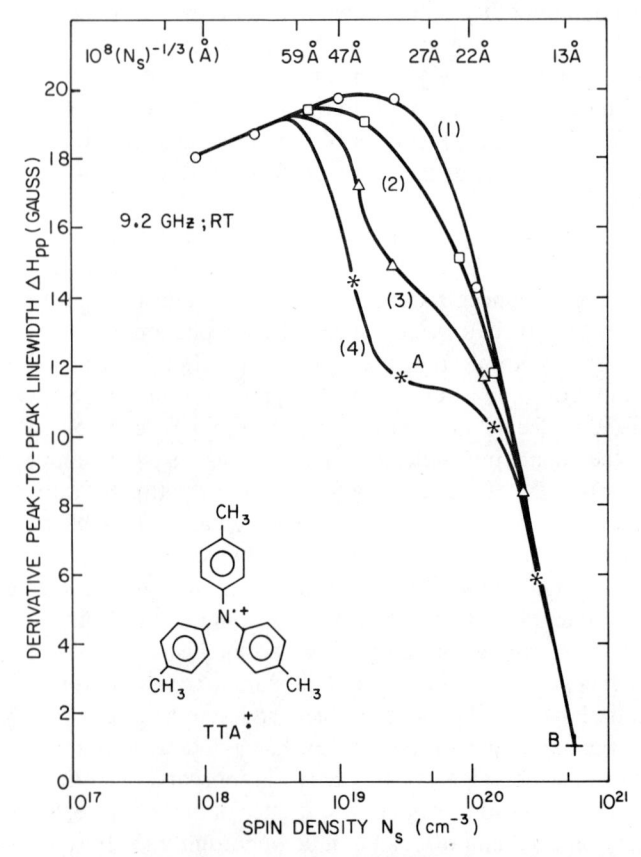

Figure 6.27 Derivative peak-to-peak linewidth (gauss) for four series of films of TTA–$TTA^+SbCl_5^-$ polycarbonate vs. spin density N_S (cm^{-3}) at room temperature. Curve (O) corresponds to pure $TTA^+SbCl_5^-$ dissolved in Lexan. For curves (□), (△), and (A), the total concentrations of neutral plus oxidized molecules N are respectively 3.86×10^{20}, 7.19×10^{20}, and 8.58×10^{20} cm^{-3}. Point B is for pure $TTA^+SbCl_5^-$ powder. Film designated A was used for the linewidth temperature dependence. (From Troup et al. [67].)

the polycarbonate matrix played no role magnetically. Initially, it can be seen that the linewidth increases up to a spin concentration of 3×10^{19} cm^{-3} due to dipole–dipole interactions which are expected to broaden the line by 2–3 G. This is followed by a rapid decrease in the linewidth as the direct exchange between the randomly distributed TTA$^+$ species becomes important. As a consistency check, the linewidth versus concentration curve extrapolates to the value for pure TTA$^+$ SbCl$_5^-$ powder (point B). The other curves correspond to films with different but constant total numbers of molecules (N) but where a systematic and controlled one-for-one replacement of neutral TTA by TTA$^+$SbCl$_5^-$ molecules occurs. Clearly, at high spin concentrations all the curves coalesce since direct exchange ultimately always dominates. At low spin concentrations the system approximates to an ensemble of well-separated paramagnets with very weak electrostatic interactions, and the curve again follows a common line.

In the intermediate range of spin concentrations, 5×10^{18} cm$^{-3} < N_s < 2 \times 10^{20}$ cm^{-3}, the paramagnets are still sufficiently separated that direct exchange interactions are weak. There is, however, a significant probability that it is now possible for a neutral TTA molecule to lie between two spin sites and lead to an enhanced magnetic coupling of the otherwise interacting spins. The number of spin pairs so coupled by a neutral TTA molecule will increase rapidly as the concentration of neutral TTA molecules is increased for a given spin density. Such configurations change the higher-order moments of the absorption envelope and can thus lead to the observed family of narrowed lines.

Motional effects such as electronic hopping were considered as a source of the line narrowing since dc electrical conduction in a molecularly doped polymer is known unambiguously, as discussed in Section 6.3.3, to involve electron hopping between neutral molecules and their cations (5). A feature of this electronic conduction process critical to the intrepretation is that the mobility is strongly temperature dependent [μ = const. exp $(-\Delta/kT)$; $\Delta = 0.4$ eV, $\sigma_{RT} < 10^{-8}$ Ω^{-1} cm^{-1}]. Thus if the line narrowing observed in Fig. 6.27 were due to diffusive hopping of spins (carriers), then at lower temperatures this process would freeze out, with a consequent rebroadening of the resonance line.

The temperature dependence of the linewidth shown in Fig. 6.28 demonstrates that varying the temperature from 300 to 60 K only results in a small increase of ~1.5 gauss in linewidth in the most strongly exchange-narrowed case which is observed in the film denoted by the asterisks in Fig. 6.27. The rapid line broadening below 60 K is probably due to freezing out of phonon modes, resulting in longer relaxation times. By contrast, the electrical conductivity decreases by many orders of magnitude and the carriers are effectively immobile even at 150 K. It was therefore concluded by Troup et al. (67) that motional narrowing associated with electronic hopping (the sole operative transport mechanism) was not a significant source of the observed line narrowing. Thus, the line narrowing in this disordered molecular solid was identified with superexchange interactions involving the TTA$^+$ spins mediated by intervening neutral TTA molecules.

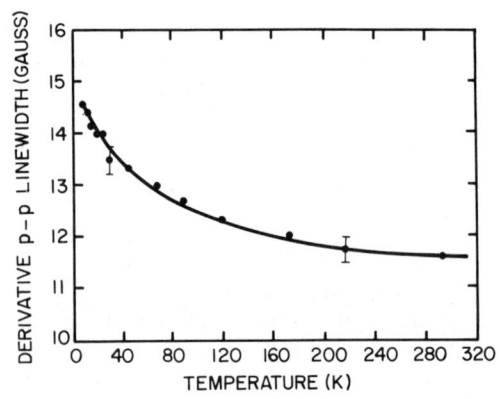

Figure 6.28 Derivative peak-to-peak linewidth (gauss) of film denoted (A) in Fig. 6.27 on an expanded scale vs. temperature (K). (From Troup et al. [67].)

The ability to observe and identify this phenomenon again highlights the remarkable power of the technique of molecularly doped polymers. By analogy with the case of electronic conduction, this technique allows the extremely simple fabrication of magnetic systems of varying composition and concentration and thus provides a direct method for studying nearest-neighbor interactions of which magnetic superexchange is one example.

6.6 SUMMARY

As has been intimated throughout this chapter, a prime motivation in the study of the fundamental properties of photoconductive polymers has been the potential and demonstrated value of polymeric-based photoreceptors in the electrophotographic industry. Polymers that are electrically insulating in the dark but can transport nonequilibrium carriers produced by light are well suited to function as photoreceptor elements (71). Polymers such as PVCA or molecularly doped polymers have strong intrinsic optical absorption only in the ultraviolet; and since visible light is employed in practice, the polymer photosensitivity must be extended into the visible. This can be done by (a) formation of a charge-transfer complex with absorption in the visible, (b) dye sensitization with an appropriately absorbing dye, or (c) use of a thin, contiguous sensitizing layer such as amorphous selenium. The ability to make large-area flexible polymer films at relatively low cost by solution coating accounts for the application of these materials in electrophotography. Although in pendant-group polymers or polymers doped with aromatic molecules, charge transport is a hopping process with inherently low mobilities, less than 10^{-3} cm^2/V-s, this is not a limitation for electrophotographic usage. In this process, the most important parameter, within limits, is how far the carriers move before they are immobilized rather than how fast they travel (49). For most process speeds, such mobilities are adequate provided the photogenerated carriers can transverse the total device thickness, that is, $\mu E \tau >$ sample thickness, where μ is the mobility, E is the electric field, and τ is the lifetime of carriers with respect to deep traps. Remarkably, such polymers do exhibit large carrier ranges because of their very large deep-trapping lifetimes.

For systems where the photogeneration occurs by photoexcitation within a polymer, such as the charge-transfer complexes or dye-sensitized systems, the photogeneration efficiency can be controlled by a geminate recombination mechanism. In such a process, the initial photoexcitation leads to electron–hole pairs, which diffuse within their mutual Coulomb well. There is a finite probability, which decreases with applied field, that the pair will recombine on the initial excitation site rather than thermally dissociate into a free electron and a hole (53). Carrier recombination in molecular systems can lead to quantum efficiencies for photogeneration that are substantially less than unity and are strongly field dependent. This may result in a photosensitivity limitation, depending on the particular system, and can be overcome to some degree by increasing the light exposure in a machine.

For other electronic applications, the magnitudes of the carrier mobilities are of paramount importance, since they determine the frequency response of devices or, through the related diffusion lengths, determine ultimate collection efficiencies in devices such as photovoltaic cells. Pendant-group polymers and disordered molecularly doped systems are not likely to find applications in these areas because of their low mobilities and the probable importance of geminate recombination processes.

REFERENCES

1 Proceedings of the 8th International Conference on Amorphous and Liquid Semiconductors, Cambridge, Mass, 1979, W. Paul and M. Kastner, Eds. North Holland, New York, 1980.
2 W. Mehl and N. E. Wolff, *J. Phys. Chem. Solids*, **25**, 1221 (1964).
3 W. D. Gill, in *Proceedings of 5th International Conference on Amorphous and Liquid Semiconductors*, Garmisch-Partenkirchen, 1973. J. Stuke and W. Brenig, Eds., Taylor and Francis, London 1974, p. 901.
4 E. Pitts, G. C. Terry, and F. W. Willets, *Trans, Faraday Soc.*, **62**(10), 2851 (1966); **62**(10), 2858 (1966).
5 J. Mort, G. Pfister, and S. Grammatica, *Solid State Comm.*, **18**, 693 (1976).
6 R. G. Kepler, *Phys. Rev.*, **119**, 1226 (1960).
7 W. E. Spear, *Proc. Phys. Soc. Lond.*, **Sect. B70**, 669 (1975); *ibid.*, **76**, 826 (1960).
8 See, for instance, *IEEE Transactions on Electron Devices*, special issue on electrophotography, ED-19 (1972).

9   J. Mort, I. Chen, R. L. Emerald, and J. H. Sharp, *J. Appl. Phys.*, **43**, 2285 (1972).
10 See for instance, I. Chen and J. Mort, *J. Appl. Phys.*, **43**, 1164 (1972); I. Chen, *ibid.*, **43**, 1137 (1972); I. Chen, R. L. Emerald, and J. Mort, *ibid.*, **44**, 3490 (1973).
11 J. Mort, *Phys. Rev.*, **5**, 3329 (1972).
12 G. Pfister and C. H. Griffiths, *Phys. Rev. Lett.*, **40**, 659 (1978).
13 W. D. Gill, *J. Appl. Phys.*, **43**, 5033 (1973).
14 G. Pfister, *Phys. Rev. B*, **16**, 3676 (1977).
15 P. M. Borsenberger, W. Mey, and A. Chowdry, *J. Appl. Phys.*, **49**, 273 (1978).
16 G. Pfister, unpublished results.
17 W. Limburg, D. M. Pai, J. M. Person, D. Renfer, M. Stolka, R. Turner, and J. Yanus, *Org. Coatings Plast. Chem.*, **38**, 534 (1978).
18 L. B. Schein, *Phys. Rev. B*, **15**, 1024 (1977).
19 J. Slowik, *Bull. Am. Phys. Soc.*, **21**, 314 (1976).
20 F. Gutmann, *J. Appl. Phys.*, **8**, 1417 (1969).
21 G. Pfister, J. Mort, and S. Grammatica, *Phys. Rev. Lett.*, **37**, 1360 (1976).
22 K. M. Probst and N. Karl, *Phys. Status Solidi (a)*, **27**, 499 (1975).
23 J. Frenkel, *Phys. Rev.*, **54**, 647 (1938).
24 J. Hirsch, *J. Phys. C, Solid State Phys.*, **12**, 321 (1979).
25 R. C. Hughes, *J. Chem. Phys.*, **58**, 2212 (1973).
26 B. Reimer and H. Bässler, *Phys. Status Solidi (a)*, **51**, 445 (1979).
27 A. Szymanski and M. Labes, *J. Chem. Phys.*, **50**, 3568 (1969).
28 H. Bässler, *Phys. Status Solidi (b)*, **107**, 9 (1981).
29 C. B. Duke, W. R. Salaneck, T. J. Fabish, J. J. Ritsko, H. R. Thomas, and A. Patton, *Phys. Rev. B*, **18**, 5717 (1978).
30 C. B. Duke, *Mol. Cryst. Liq. Cryst.*, **50**, 63 (1979).
31 H. Scher and E. W. Montroll, *Phys. Rev. B*, **12**, 2455 (1975).
32 J. Noolandi, *Solid State Comm.*, **24**, 477 (1977); *Phys. Rev. B*, **16**, 4466, 4474 (1977).
33 F. W. Schmidlin, *Solid State Comm.*, **22**, 451 (1977); *Phys. Rev. B*, **16**, 2362 (1977).
34 M. Silver, G. Schönherr, and H. Bässler, *Phil. Mag.*, **43**, 943 (1981); see also Ref. 27.
35 M. Silver and T. Datta, *J. Non-Cryst. Solids*, **35, 36**, 111 (1980); M. Schaffman and M. Silver, *Phys. Rev. B*, **19**, 116 (1979); .
36 J. M. Marshall, *Phil. Mag.*, **38**, 335 (1978); *ibid.*, **43**, 401 (1980); A. C. Sharp, *J. Non-Cryst. Solids*, **35, 36**, 99 (1980).
37 G. Pfister and H. Scher, *Phys. Rev. B*, **15**, 2026 (1977); *Adv. Phys.*, **27**, 747 (1978).
38 G. Pfister, *Phys. Rev. Lett.*, **36**, 271 (1976).
39 A. R. Tahmasbi and J. Hirsch, *Solid State Comm.*, **34**, 75 (1980).
40 G. Pfister, *Phils. Mag.*, **36**, 1147 (1977).
41 T. Tiedje, and A. Rose, *Solid State Comm.*, **37**, 49 (1981).
42 C. H. Griffiths, G. Pfister, and A. Ledwith (to appear).

REFERENCES

43 C. H. Griffiths, *J. Polym. Sci. Polym. Lett. Ed.*, **16**, 271 (1978).
44 J. Orenstein and M. A. Kastner, *Solid State Comm.*, **40**, 85 (1981).
45 M. Pollak, *Phil. Mag.*, **36**, 1157 (1977).
46 R. H. Batt, C. L. Braun, and J. F. Hornig, *J. Chem. Phys.*, **49**, 1967 (1968).
47 L. Onsager, *Phys. Rev.*, **54**, 554 (1938).
48 A. Mozumder, *J. Chem. Phys.*, **60**, 4300 (1974).
49 J. Mort and I. Chen, *Appl. Solid State Sci.*, **5**, 69 (1975).
50 P. J. Regensburger, *Photochem. Photobiol.*, **8**, 429 (1968).
51 G. Pfister and D. Williams, *J. Chem. Phys.*, **61**, 2416 (1974).
52 P. M. Borsenberger and A. I. Ateya, *J. Appl. Phys.*, **49**, 4035 (1978).
53 P. M. Borsenberger, L. E. Contois, and D. C. Hoesterey, *J. Chem. Phys.*, **68**, 637 (1978).
54 J. Mort and G. Pfister, *Polym.-Plast. Technol. Eng.*, **12**, 89 (1979).
55 K. Meier, *Spectral Sensitization*, Focal Press, New York, 1968.
56 H. Meier, W. Albrecht, and A. Tschirwitz, *Current Problems in Electrophotography*, N. F. Berg and K. Hauffe, Eds., Walter deGrugter, New York, 1972, p. 163.
57 H. Sato and M. Ikeda, *J. Appl. Phys.*, **43**, 4108 (1972).
58 K. Okamoto, A. Itaya, and S. Kusobayashi, *J. Polym.*, **7**, 622 (1975).
59 M. Ikeda, H. Sato, K. Morimoto, and Y. Murakami, *Photogr. Sci. Eng.*, **19**, 60 (1975).
60 S. Grammatica and J. Mort, *J. Chem. Phys.*, **67**, 5628 (1977).
61 W. A. Barlow, G. R. Davies, E. P. Goodings, R. L. Hand, G. Owen, and M. Rhodes, *Mol. Cryst. Liq. Cryst.*, **33**, 123 (1976).
62 T. Uchida and M. Akamatu, *Bull. Chem. Soc. Jap.*, **35**, 981 (1962).
63 A. Rembaum and J. Moacanin, *Polymeric Semiconductors*, Jet Propulsion Laboratory, Pasadena, California, 1964.
64 E. P. Goodings, *Chem. Soc. Rev.*, **5**, 95 (1976).
65 J. M. Lupinski, K. D. Kopple, and J. J. Hertz, *J. Polym. Sci. C*, **16**, 1561 (1967).
66 J. Mort, S. Grammatica, D. J. Sandman, and A. Troup, *J. Electronic Materials*, **9**, 411 (1980).
67 A. Troup, J. Mort, S. Grammatica, and D. J. Sandman, *Solid State Comm.*, **33**, 91 (1980).
68 C. Herring, *Magnetism,* ed G. T. Rado and K. Suhl, Eds., Academic, New York, 1966, Vol. 11B, p. 173.
69 K. N. Shrivastava and V. Jaccarino, *Phys. Rev. B,* **13**, 299 (1976).
70 T. Yasmoko, K. Wanatuba, and M. Ohya–Nishiguchi, *Bull. Chem. Soc.* Japan, **48**, 2533 (1975).
71 J. Weigl, *Angew. Chem. Int. Ed. Engl.*, **16**, 374 (1977).

Chapter 7

CONDUCTING POLYMERS: POLYACETYLENE

D. Baeriswyl, G. Harbeke, H. Kiess, and W. Meyer
RCA Laboratories Ltd.
Zurich, Switzerland

7.1	Introduction	268
7.2	Preparation and Characterization of Polyacetylene	270
	7.2.1 Acetylene Polymerization, 270	
	7.2.2 Cis and Trans Isomers in Polyacetylene, 271	
	7.2.3 Morphology, 273	
	7.2.4 Structure of Polyacetylene, 274	
7.3	Theoretical Approach	276
	7.3.1 Molecular-Orbital and Crystal-Orbital Theories, 277	
	7.3.2 Simple (Hückel-Type) Model Hamiltonian for trans-Polyacetylene, 279	
	7.3.3 Bond Alternation in Even Chains, 280	
	7.3.4 Bond Alternation Defects in Odd Chains, 285	
	7.3.5 Continuum Description and the Soliton Concept, 287	
	7.3.6 The Soliton Lattice and the Commensurate–Incommensurate Transition, 289	
	7.3.7 Correlation Effects, 291	
	7.3.8 Predictions for Polyacetylene, 292	
	7.3.9 Recent Developments, 294	
7.4	Pristine and Lightly Doped Polyacetylene: A One-Dimensional Semiconductor	295
	7.4.1 Magnetic Properties, 295	
	7.4.2 Optical Properties, 299	
	7.4.3 Electrical Transport Properties, 303	
7.5	The Semiconductor–Metal Transition and Heavily Doped Polyacetylene	305
	7.5.1 Dopants and Doping Procedures, 305	
	7.5.2 The Nature of the Highly Conducting State and the Semiconductor–Metal Transition, 308	
7.6	Other Conducting Polymers	313
	7.6.1 Undoped Polymers, 313	
	7.6.2 Doped Aromatic Polymers, 314	
7.7	Concluding Remarks	316
	Acknowledgment	317
	References	317

7.1 INTRODUCTION

For a long time, organic materials were found to be electrically insulating or at most semiconducting, the reason being that these systems usually consist of well-separated molecules. Only in the extreme case of graphite where an individual "molecule" consists of a whole sheet of carbon atoms was the conductivity found to be nearly metallic, at least parallel to the layers. The comparison of large classes of organic materials in the early sixties allowed a systematic study of charge transport. Semiconductive behavior not only showed up in the value of the room-temperature conductivity and its exponential temperature dependence, but also in photoconductive and optical absorption experiments—reviews of early work have been published by Inokuchi and Akamatu (1), Berlin (2), and Pohl (3).

Within a given class of materials, the conductivity usually was enhanced if the size of the molecules was increased, but this was not sufficient to guarantee a reasonable conductivity. In addition, unsaturated bonds leading to delocalization of the valence electrons (the π-electrons) over the molecules were necessary; and, in order to achieve a high mobility, the π-orbitals were required to overlap appreciably with those of neighboring C atoms (4). Furthermore, adding strong donors (simple metals) or strong acceptors (halogens) was found to increase the conductivity by several orders of magnitude (1).

Little's proposal in 1964 to synthesize a room-temperature superconductor consisting of a conducting backbone and large polarizable side groups (5) stimulated a great deal of activity leading to the discovery of new organic compounds with unusually high electric conductivity. Some of them consist of segregated stacks of planar donor and acceptor molecules. They are highly conducting if the distance between successive molecules on a stack is small. A representative example is TTF–TCNQ, which shows metallic behavior parallel and nonmetallic behavior perpendicular to the stacks. (A collection of recent review articles is contained in Ref. 6.)

In contrast to these mixed-valence compounds where the mobility arising from the *intermolecular* overlap of π-orbitals is small, a high mobility is anticipated for π-electrons moving along a polymer chain since it is due to *intramolecular* coupling of orbitals. Apart from the inorganic polymeric conductor $(SN)_x$, which is unique due to its superconducting transition around 1 K (7), the most widely studied material is polyacetylene, $(CH)_x$. In an ideal infinite chain of $(CH)_x$, one would expect the π-electrons to form a half-filled band leading to metallic behavior. However, as was found a long time ago in optical studies on finite chains, the frequency of maximum absorption, after first decreasing with increasing chain length (as expected), tends to saturate at a finite value of about 2 eV. This suggests that a system of infinitely long $(CH)_x$ chains will be a semiconductor with an electronic gap of about 2 eV. Kuhn (8) has advanced the idea of bond alternation as the origin of this gap. This hypothesis was subsequently worked out in more detail by several authors within molecular orbital

INTRODUCTION

theory. (See Ref. 9 for an excellent review of the theory). But later, the hypothesis was questioned by Ovchinnikov and co-workers (10), who proposed instead the electronic correlation as the main origin for the gap.

Experimentally, a breakthrough was achieved by Ito, Shirakawa, and Ikeda in 1974 (11) when they succeeded in obtaining flexible silvery films of $(CH)_x$, showing semiconducting behavior. Three years later, Chiang et al. (12) were able to show that, by doping the material, the conductivity could be increased from the semiconducting to the metallic regime. The range of conductivities obtained upon doping is compared with conductivities of other materials in Fig. 7.1. Since then, a huge number of papers has appeared dealing with the various physical, chemical, and structural properties of polyacetylene and giving a fair insight into the whole topic of conductivity in polymers. Consequently, many intricate questions arose such as the nature of spinless charge carriers which led to new concepts and speculations about the mechanism of charge transport in

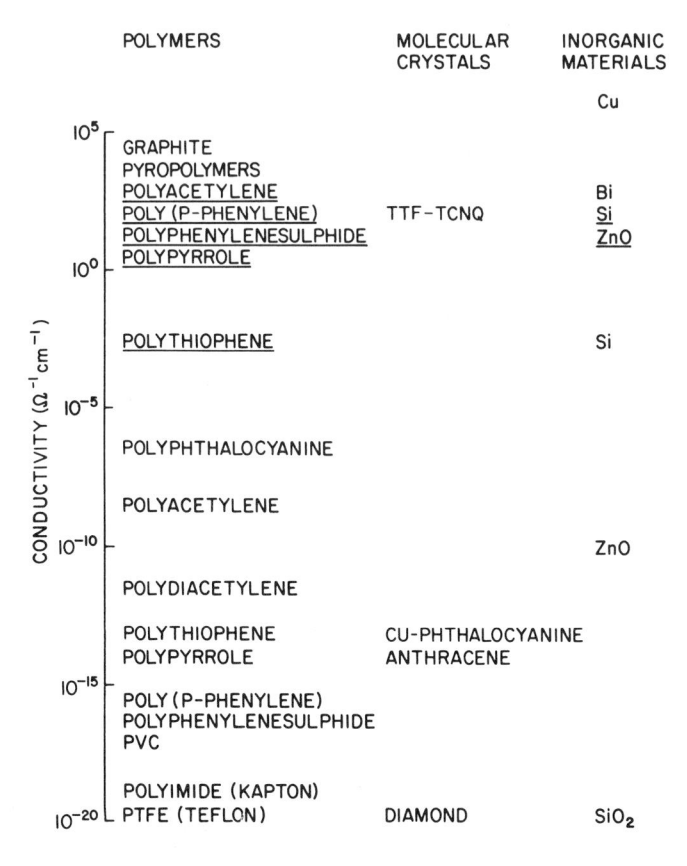

CONDUCTIVITY ($\Omega^{-1} cm^{-1}$)	POLYMERS	MOLECULAR CRYSTALS	INORGANIC MATERIALS
			Cu
10^5	GRAPHITE PYROPOLYMERS POLYACETYLENE POLY (P-PHENYLENE) POLYPHENYLENESULPHIDE POLYPYRROLE	TTF-TCNQ	Bi Si ZnO
10^0			
10^{-5}	POLYTHIOPHENE POLYPHTHALOCYANINE POLYACETYLENE		Si
10^{-10}	POLYDIACETYLENE		ZnO
10^{-15}	POLYTHIOPHENE POLYPYRROLE POLY (P-PHENYLENE) POLYPHENYLENESULPHIDE PVC	CU-PHTHALOCYANINE ANTHRACENE	
10^{-20}	POLYIMIDE (KAPTON) PTFE (TEFLON)	DIAMOND	SiO_2

Figure 7.1 Approximate conductivities of various organic and inorganic materials. Species that can be doped are listed in the undoped and heavily doped state (underlined).

organic systems. Due to the one-dimensionality of an ideal $(CH)_x$ chain, many effects such as nonlinearity, disorder, and fluctuations are expected to be strongly enhanced. To what extent the real system is one-dimensional is difficult to answer and will depend on experimental conditions and on the physical quantity to be studied. On the other hand, the experimental conditions are harder to define than, say, in crystalline silicon, and a parameter judged irrelevant for the characterization of these conditions may later turn out to be of primary importance. Such effects, which are unavoidable in this early stage of a new field of research, should always be borne in mind if contradictory results have to be discussed.

Due to the basic simplicity of $(CH)_x$, the large amount of publications dealing with this material, and the relatively dispersed and preliminary studies on the other related compounds, we have decided to review mainly the work on $(CH)_x$ as a model substance and to mention only briefly some results on other conducting polymers. The chapter is organized as follows. A brief characterization of the material is given in Section 7.2. The main theoretical models are reviewed in Section 7.3. In Sections 7.4 to 7.6, a critical discussion of experiments is given. We therefore concentrate mainly on those results which are commonly accepted by different groups working in the field and mention only briefly contradictory results obtained in the same type of experiment. In particular, Section 7.4 deals with pristine and lightly doped $(CH)_x$, Section 7.5 deals with heavily doped $(CH)_x$ and the nature of the semiconductor–metal transition, and Section 7.6 gives a short summary of work on other doped polymers. Finally, we conclude with discussion of future work to be done and on possible applications of conducting polymers.

7.2 PREPARATION AND CHARACTERIZATION OF POLYACETYLENE

7.2.1 Acetylene Polymerization

A wide variety of catalyst systems have been described for the polymerization of acetylenes (13–20). Besides the route via acetylene polymerization, these polymers also can be obtained by dehydrochlorination of poly(vinyl chloride) (PVC) (21) or derived from other polymers (22–24).

Already in 1958, Natta and co-workers (13) succeeded in the synthesis of partially crystalline polyacetylene using a soluble Ziegler-type catalyst. However, the material obtained was an unprocessable powder until Ito et al. described the synthesis of highly crystalline polyacetylene films (11) using a very similar catalyst but in a much higher concentration than Natta. The method of polyacetylene film synthesis, which provoked a revival of interest in polyacetylene, is briefly described below.

PREPARATION AND CHARACTERIZATION OF POLYACETYLENE

The catalyst solution consists of 0.25 mol/liter tetrabutoxytitanium in toluene, to which 1.00 mol/liter triethylaluminum is added. The mixture is stirred at room temperature for several minutes, then cooled to Dry-Ice temperature and degassed under reduced pressure for at least 30 min (11).

In a typical experiment, purified acetylene gas under inert conditions is blown over the quiet surface of the catalyst solution. The polyacetylene film immediately forms at the gas–liquid interface, and the film thickness can be controlled by the duration of the acetylene flow. The polymerization can be interrupted by evacuation of the system. The polyacetylene is purified by extensive washing with toluene and a methanol–hydrochloric acid mixture. To avoid oxidation, polyacetylene should be treated under protective gas only and stored under high vacuum. The procedure described gives a black, insoluble polymer film with a metallic luster, which does not melt until decomposition near 200°C.

7.2.2 Cis and Trans Isomers in Polyacetylene

Acetylene polymerized at temperatures below RT gives a cis-rich material; at temperatures higher than RT, trans-rich $(CH)_x$ is formed (11). DSC thermograms of cis-rich polyacetylene exhibit an exothermic peak around 140°C, which is assigned to cis → trans isomerization. Thus, *cis*-$(CH)_x$ can be converted into *trans*-$(CH)_x$ by heat treatment (25).

The cis and trans isomers can be distinguished and their content determined by characteristic absorption in the IR and Raman spectra (11, 26, 27). Theoretically, four different structures of $(CH)_x$ isomers may be possible, as indicated in Fig. 7.2. The trans-rich $(CH)_x$ consists mainly of the trans-transoidal (or "all-trans") isomer. The comparison of calculated vibrations for copolymers of C_2H_2 and C_2D_2 with the observed IR spectra lead to the result that *cis*-$(CH)_x$ consists of the cis-transoidal isomer; the trans-cisoidal form could be ruled out (26). The cis-cisoidal isomer for geometric reasons seems to be negligible.

It is generally assumed that bond alternation exists in polyacetylene, but up to now this has not been experimentally established. The number of reflections in electron- or x-ray diffraction experiments is still too small to determine the atomic positions in the $(CH)_x$ crystal lattice (see Sect. 7.2.4). The question of bond alternation also cannot be decided from IR (26) or Raman spectra (28), since the number of active modes is independent of the existence of bond alternation.

Within the bond alternation model, the two strongest Raman lines at wavenumbers of about 1500 and 1100 are commonly referred to as C=C and C—C stretching modes, respectively, although both eigenvectors contain considerable C—H in plane bending (29). The Raman frequency due to trans sequences was found to depend on the wavelength of the excitation (30). This dependence was used to estimate the number of double bonds $N_{C=C}$ from the $\nu_{C=C}$ versus $N_{C=C}$ relationship for polyene homologues of known chain length. It has been found

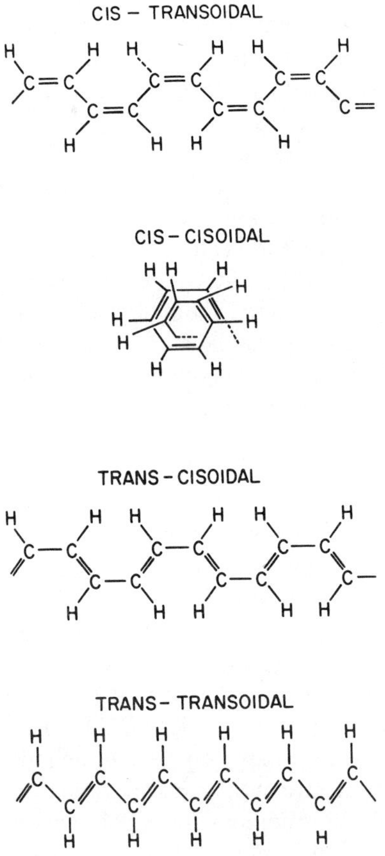

Figure 7.2 Theoretical isomers of polyacetylene.

that the chain length distribution peaks at about $N_{C=C} = 30$, while the width of distribution is rather broad (27). It should be noted, that no shift of vibrational frequencies has been observed in the lines characteristic of cis-$(CH)_x$. From the pronounced enhancement of the Raman lines with red excitation it was concluded that the cis-$(CH)_x$ consists mainly of long chains with $N_{C=C} > 50$ (31). On the other hand, the degree of polymerization was determined after conversion of $(CH)_x$ into soluble polymers by chlorination (32) and hydrogenation (33) and was found to range typically from 200 to 500, corresponding to $(CH)_{400}$ to $(CH)_{1000}$, respectively. These numbers are consistent with the data derived from the Raman experiments, since after chemical conversion the methods used for the determination of the degree of polymerization do not probe the effective conjugation length of conjugated double bonds in $(CH)_x$, but the total length of the C—C chain. The difference in the data derived from Raman experiments for cis- and trans-$(CH)_x$ for $N_{C=C}$ can be interpreted as the introduction of conjugation defects during cis–trans isomerization (31).

7.2.3 Morphology

Unoriented Polyacetylene

Polyacetylene films typically consist of a fleece of randomly oriented fibers with diameters of 200 Å and indefinite length (11), as shown in Fig. 7.3. The bulk density obtained from the dimensions and weight of $(CH)_x$ films range from 0.3 to 0.6 g/cm^3 compared with their flotation density of 1.15 g/cm^3, indicating that the polymer fibers fill only one-fourth to one-half of the film volume (34, 35). Depending on the catalyst concentration, the acetylene pressure, and the temperature during synthesis, the morphology of the $(CH)_x$ may deviate from that of the well-known films, as is indicated by the formation of gel- or foam-like $(CH)_x$ (35–37).

Oriented Polyacetylene

Partial orientation of the polyacetylene fibers can be achieved by stretch alignment of the films with a combined mechanical and thermal treatment (38). An electron micrograph of a stretch-aligned $(CH)_x$ film is shown in Fig. 7.4. Stretch-

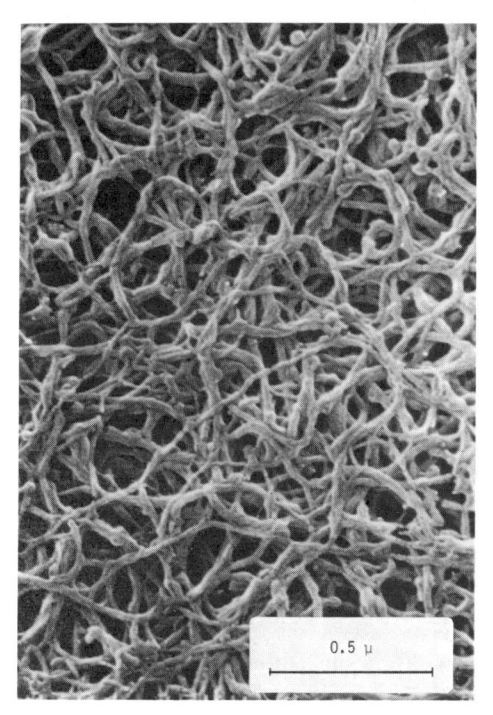

Figure 7.3 Electron micrograph of polyacetylene film. (From Shirakawa and Ikeda [35].)

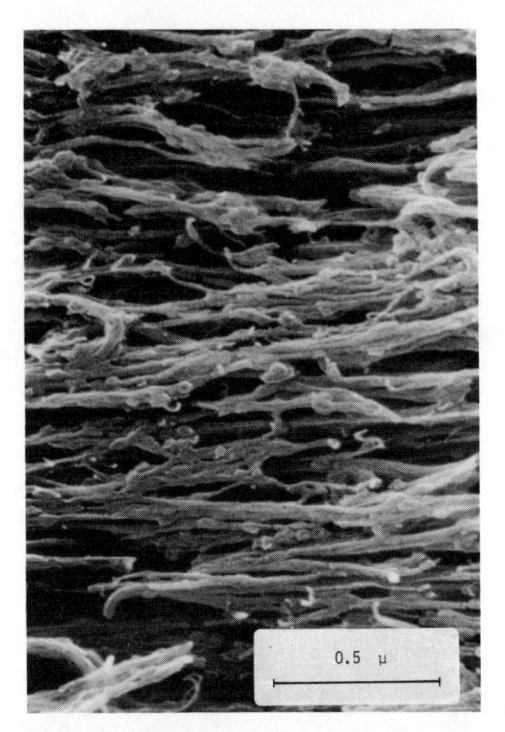

Figure 7.4 Electron micrograph of a highly stretch-aligned polyacetylene film. (From Shirakawa and Ikeda [35].)

oriented polyacetylene films exhibit considerable electrical (39) and optical (40) anisotropy. The tensile properties depend strongly on both the cis–trans content and the density of the films (35, 41).

The degree of orientation obtained in polyacetylene films by the stretching method is limited by the fracture of the material at elongation ratios of typically 3 (35). Another way to produce oriented $(CH)_x$ is the simultaneous polymerization and the crystallization under shear flow conditions (42). With this method, $(CH)_x$ films consisting of highly parallel aligned fibers can be obtained as evidenced by the scanning electron micrograph shown in Fig. 7.5.

7.2.4 Structure of $(CH)_x$

Polyacetylene synthesized with the procedures described above is a polycrystalline material with a high degree of crystallinity as indicated by strong x-ray reflections around $2\theta = 24°$ ($Cu^{K\alpha}$) (13, 11). For both the cis and the trans isomer, crystal structures have been proposed to be analogous to that of orthorhombic polyethylene on the basis of packing calculations and limited information from x-ray powder diagrams (43, 44).

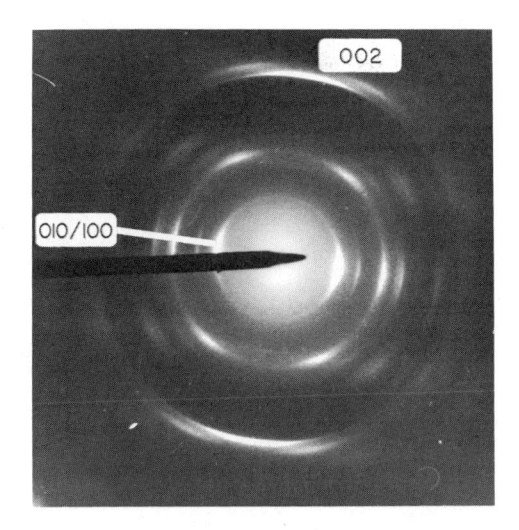

Figure 7.6 Electron diffraction pattern of shear flow-polymerized $(CH)_x$.

trans modification in shear flow polymerized $(CH)_x$ a new modification with an unexpectedly long chain repeat has been detected also (45).

A series of papers (48–51) deal with measurements of the specific heat and thermal conductivity of doped and undoped $(CH)_x$. The thermal conductivity of $(CH)_x$ is rather high at 300 K and, as a function of temperature, it shows a kink at 50 K which has been related to the morphology of polyacetylene (50).

7.3 THEORETICAL APPROACH

Given the geometric arrangement of polyacetylene, the question arises what the electronic structure will be, and in particular what the ground state and the low-lying excited states will look like. This problem has been discussed already in the early work of Lennard-Jones (52) and Coulson (53) but still is not

Table 7.1 Unit Cell Parameters of *cis*- and *trans*-Polyacetylene

Type	Reference	Crystal System	a (Å)	b (Å)	c (Å)[a]	Angle (deg.)
cis-PA	32	orthorhombic	7.74	4.32	4.47	—
trans-PA	45	monoclinic	3.73	3.73	2.44	$\gamma/98 \pm 2$
trans-PA	178	monoclinic	4.25	7.33	2.46	$\beta/91.4$
trans-PA	179	orthorhombic	7.32	4.24	2.46	—

[a] Chain axis.

PREPARATION AND CHARACTERIZATION OF POLYACETYLENE

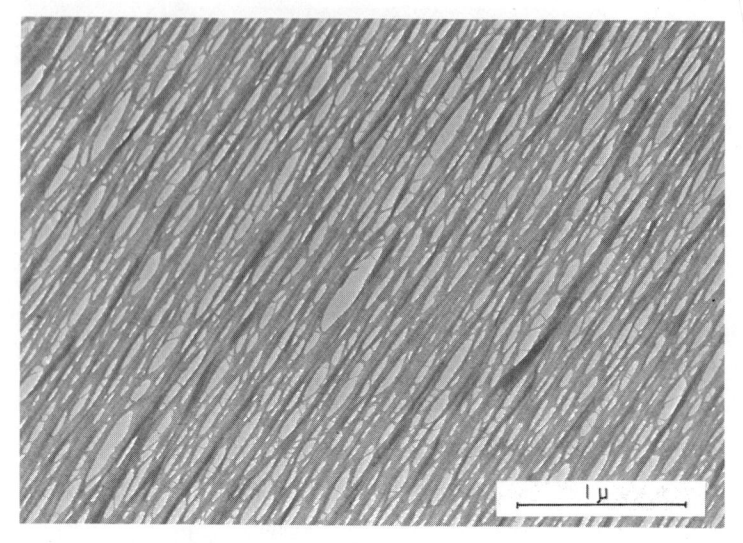

Figure 7.5 Electron micrograph of shear flow-polymerized polyacetylene.

New results from electron diffraction and x-ray experiments are in good agreement with the earlier data of Baughman and co-workers for *cis*-polyacetylene only, but they disagree with the space group assignment. The new data are obtained with a material synthesized with a catalyst described by Luttinger (16). This cis-$(CH)_x$ does not differ in its main features from the $(CH)_x$ obtained with the Ziegler–Natta catalyst but it is claimed to be more stable against thermal isomerization (32). Electron diffraction experiments show that the polymer chain axes in the lamellar subunits of such $(CH)_x$ are oriented normal to the lamellar surfaces (32).

Shear flow-polymerized $(CH)_x$ contains nearly single crystalline and well-oriented material which gives rise to sharp fiber patterns, as shown in Fig. 7.6. So far, only trans material has been examined. The fiber patterns give clear evidence that the chain axes in shear flow-polymerized $(CH)_x$ are oriented parallel to the fiber direction (45). Recent x-ray studies with stretch-aligned *trans*-polyacetylene and selected area electron diffraction experiments with unstretched samples also claim the polymer chains to be oriented parallel to the $(CH)_x$ fiber axis (46, 47).

Although the experimental data of different sources seem to merge, the crystal lattice parameters for *trans*-$(CH)_x$ are still controversial, since they are based on data derived from both diffraction experiments and packing calculations. Thus, at least three models have been proposed for usual *trans*-polyacetylene, the data of which, together with those of *cis*-$(CH)_x$, are compiled in Table 7.1. It should be considered, however, that different workers may deal with different crystal modifications of $(CH)_x$; for example, besides the usual completely solved. The reason for this difficulty lies in the various instabilities which can occur in a one-dimensional system of delocalized electrons, such as the Peierls instability induced by the electron–phonon coupling (54) or the localization due to correlation (55). Despite the fact that these instabilities strongly affect the electronic and optical properties, the energy differences involved are small and depend sensitively on the specific values of certain coupling constants and on approximations made during the calculations.

7.3.1 Molecular-Orbital and Crystal-Orbital Theories

In planar organic complexes such as graphite or the two isomers of polyacetylene, three of the four valence electrons of the carbon atom are in saturated bonds pointing to neighboring carbon or hydrogen atoms within the plane (σ-bonds). The remaining electron (the π-electron) has a p_z wave function perpendicular to the plane and overlapping with at least two neighboring π-orbitals. Therefore, in contrast to the σ-electrons, the π-electrons are not localized in directional bonds. This simple picture is the basis for the semi-empirical descriptions of unsaturated hydrocarbons. It has been initiated by Hückel (56) and is extensively used in the following (see, for example, Ref. 9).

Its basic assumptions are as follows:

1. Only the π-electrons are explicitly considered. The σ-electrons are contained in the parameters of the theory, for instance, the elastic constant is associated with σ-bond compression.
2. The π-electron states are calculated with respect to fixed nuclear configurations (Born–Oppenheimer approximation).
3. The complicated many-body Hamiltonian (including electron–ion and electron–electron interactions) is replaced by a sum of single-particle terms leading to an effective Hamiltonian H_{eff}. The electronic ground state is then a single Slater determinant of eigenfunctions of H_{eff}.
4. The eigenfunctions of H_{eff} are taken as linear combinations of atomic orbitals $|n\rangle = \varphi(\mathbf{r} - \mathbf{R}_n)$, where $\mathbf{R}_n$ is the position vector of the nth C atom. Furthermore, the overlap matrix elements are neglected, that is, $\langle n|m\rangle = 0$ if $n \neq m$, and it is assumed that $\langle n|H_{\text{eff}}|m\rangle = 0$ except for $n = m$ and for nearest neighbors (tight-binding approximation).

In the simple Hückle theory, no attempt is made to calculate the matrix elements. Instead, they are considered as free parameters to be adjusted to measured quantities. Various refinements have been proposed. In particular, assumption 3 has been improved by representing the ground state as a superposition of different Slater determinants leading to "configuration interaction" as described, for instance, by the Pariser–Parr–Pople model (57, 58).

Attempts to calculate *ab initio* the ground state and the band structure of polyacetylene have been made only recently, by taking all s- and p-electrons explicitly into account. The Bloch functions of the ideal infinite chain are taken as linear superpositions of atomic orbitals which are usually chosen in the form of Slater orbitals approximated by a few Gaussians (59). As in Hückel theory, the nuclei are assumed to be fixed in a given configuration, but now the matrix elements are explicitly calculated, at least within a certain interaction range (and put equal to zero outside this range). The calculation follows the Hartree–Fock scheme as developed for molecules by Roothaan (60) and extended to periodic structures by Del Re, Ladik, and Biczó (61). The first explicit numerical calculation using this "self-consistent field, linear combination of atomic orbitals, crystal orbital" (SCF LCAO CO) method for a *trans*-$(CH)_x$ chain has been carried out by André and Leroy (62). Some of the conclusions of their report, in particular about the stability of the regular structure with respect to bond alternation, have been questioned (63, 64), but the overall band structure for a given ionic configuration seems not to be very sensitive with respect to the degree of sophistication used during the computation. In the absence of bond alternation, the π-electron band, which overlaps only very weakly with a σ-band, has a bandwidth of about 20–25 eV and, as expected from Hückel theory, is half filled. However, in agreement with the Peierls "theorem," this configuration is unstable with respect to bond alternation which produces an electronic gap at the Fermi level of the order of 5–10 eV. The amount of bond alternation has been either taken from semiempirical work and x-ray data (62, 63, 65) or optimized by computing the ground-state energies for various bond lengths and angles (64).

As shown by Karpfen and Petkov, the optimal bond lengths, the π-bandwidth, and the electronic gap depend quite sensitively on the type of atomic wave functions chosen at the beginning (64). But in any case the bandwidths as computed *ab initio* are larger by about a factor of 2 than the characteristic values obtained from semiempirical considerations (9); and, in particular, the values of the electronic gap exceed largely the optical gap as determined experimentally on polyacetylene. It is tempting to associate this discrepancy with the limitations of any Hartree–Fock calculation since it is well known that correlations may strongly modify the value of the absorption edge (66). On the other hand, the Born–Oppenheimer approximation may also lead to an overestimation of the amount of bond alternation, in particular since the effective double-well potential in which the nuclei move appears to be quite shallow (64).

In contrast to the trans isomer, in *cis*-$(CH)_x$ the Fermi level is expected to lie always in the gap between π-bonding and antibonding states, also in the absence of bond alternation, simply because of the symmetry of the unit cell (67). Nevertheless, bond alternation is predicted to occur also in the cis isomer, and the cis-transoid is found to be more stable than the trans-cisoid form (65, 68). If the amount of bond alternation is optimized, the most stable isomer is found to be the all-trans form (68), in agreement with experiment.

The three-dimensional band structure corresponding to the proposed crystal

THEORETICAL APPROACH

structure for the cis-transoid $(CH)_x$ (43) has been calculated by Grant and Batra (67) within a tight-binding scheme. The interchain band dispersion near the bandgap is never larger than 0.3 eV, compared to conduction and valence bandwidths of nearly 6 eV parallel to the chain direction.

7.3.2 Simple (Hückel-Type) Model Hamiltonian for trans-$(CH)_x$

Despite the merits of the available first principle calculations and the need for more efforts in this direction (e.g., to clarify the role of correlation), the Hückel theory, due to its simplicity and flexibility, appears to yield not only a more manageable ordering principle for existing data, but also a more efficient framework for theoretical predictions. In the following, the assumptions of the Hückel theory presented in Section 7.3.1 are cast into the form of a simple model Hamiltonian as proposed by Su, Schrieffer, and Heeger (69).

Consider a *trans*-$(CH)_x$ chain as in Fig. 7.7. Taking only those ionic coordinates into account which are (linearly) involved in the bond length alternation, we are left with the displacements u_n of the CH groups parallel to the chain axis. The Hamiltonian then consists of two parts, the energy describing σ-bond compression,

$$\mathcal{H}_\sigma = \tfrac{1}{2} K_0 \sum_n (u_{n+1} - u_n)^2 \tag{7.1}$$

and the tight-binding Hamiltonian for the π-electrons,

$$\mathcal{H}_\pi = -\sum_{ns} t_{n,n+1} (c_{ns}^+ c_{n+1s} + c_{n+1s}^+ c_{ns}) \tag{7.2}$$

where we have introduced creation and annihilation operators c_{ns}^+ and c_{ns}, respectively, for π-electrons in the atomic orbital n and with spin projection $s = \pm 1$. Sometimes an additional term is added in order to stabilize the chain against overall contraction (70), but such a term is not relevant as long as one is not concerned with specific end chain effects (71). Equation (7.2) shows that the important electronic operator is the bond order

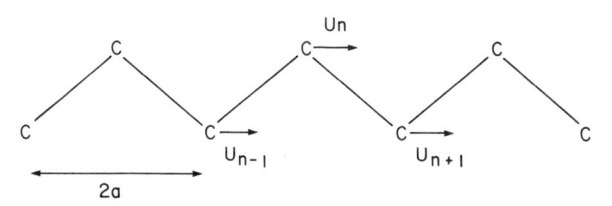

Figure 7.7 Relevant displacement coordinates of a *trans*-$(CH)_x$ chain.

$$P_{n,n+1} = \tfrac{1}{2} \sum_{s} (c_{ns}^+ c_{n+1s} + c_{n+1s}^+ c_{ns}) \tag{7.3}$$

which has a mean value of $2/\pi$ for an infinite undistorted chain. The "resonance integral" $t_{n,n+1}$ depends on the bond length and is expanded in powers of the displacement coordinates. Due to the finite value of the bond order, this expansion should be carried out up to the quadratic term which would be of the same order as the elastic energy, Eq. (7.1). For simplicity, we absorb this quadratic term within the force constant K_0 and use the expansion

$$t_{n,n+1} = t_0 - \alpha(u_{n+1} - u_n) \tag{7.4}$$

Equations (7.1), (7.2), and (7.4) define our model Hamiltonian which depends on three parameters: the screened force constant K_0, the constant t_0 related to the π-bandwidth, and the coupling constant α. It is tempting to try to fit these parameters on measured properties of small molecules such as ethane and benzene. But one has to keep in mind that, on the one hand, they are not expected to be simply transferable from small molecules to long chains (72) since they contain in an averaged way the long-range interactions. On the other hand, some quantities such as the electronic gap are very sensitive to small variations in the parameters. Therefore it seems reasonable to adjust some of these parameters to experiments performed on $(CH)_x$.

As will be shown in the next section, Raman data and the electronic gap yield essentially the combinations α^2/t_0 and $\alpha^2/(K_0 t_0)$. Thus if one parameter is chosen from other considerations, the others are fixed. It appears to be most natural to choose t_0 which is expected to be determined essentially by the bond length (9). With a bond angle of 120°, the structural data (Section 7.2) would imply a mean bond length of about 1.4 Å for trans-$(CH)_x$. Using Kakitani's analysis (72), this would yield a parameter $t_0 \approx 3$ eV.

7.3.3 Bond Alternation in Even Chains

It will be shown now that a system described by our model Hamiltonian is unstable with respect to bond length alternation as described by the equation

$$u_n = (-1)^n u \tag{7.5}$$

Assuming cyclic boundary conditions, one observes that this formula is only applicable for even chains. The case of odd chains where a "misfit" is produced is treated in Section 7.3.4.

The question of whether bond length alternation does occur in long polyenes has already been studied in the very beginning of molecular orbital theory (52, 73), with negative results. The problem regained interest when it was observed that the optical absorption frequencies appeared to tend to a finite limit for long chains (74), in contrast to the theoretical predictions. The hypothesis of bond alternation which was advanced by Kuhn (8) to explain these experiments obtained support from detailed calculations of Platt (75), Labhart (76), and Ooshika (77) and was proved rigorously within molecular orbital theory by Longuet-Higgins and Salem (78). This finding is closely related to the "Peierls theorem" (54), which states that a one-dimensional metal is unstable with respect to a periodic lattice distortion, which in turn produces a gap in the electronic spectrum at the Fermi surface and therefore leads to an insulating ground state. There are minor formal differences originating in the different model Hamiltonians. Peierls uses the Fröhlich model (79) where the lattice displacements are coupled to the electronic density. Instead, in Hückel-type theories the relevant electronic operator coupled to the lattice is the bond order.

The instability of the ground state is derived as follows. Inserting Eq. (7.5) into the Hamiltonian $H = H_\sigma + H_\pi$, the electronic part is easily diagonalized (80), yielding the spectrum

$$\epsilon_k = -\cos ka \, [(2t_0)^2 + (4\alpha u \tan ka)^2]^{1/2} \tag{7.6}$$

where a is half the lattice constant (Fig. 7.7) and the wave vectors $k = 2\pi\nu/Na$ lie in the (extended) zone $-N/2 < \nu \leq N/2$ (N = total number of CH groups or π-electrons). The ground-state energy (per CH group) is obtained by summing over the occupied states:

$$E_0 = 2K_0 u^2 + \frac{2}{N} \sum_k n_k \, \epsilon_k \tag{7.7}$$

Consider first the limit of an infinite chain. The spectrum exhibits a gap $2\Delta_0 = 8\alpha u$ at the Fermi level, and the ground-state energy is obtained as

$$E_0 = 2K_0 u^2 - 4t_0 \frac{E(1 - \delta^2)}{\pi} \tag{7.8}$$

where $E(x)$ is the complete elliptic integral and $\delta \equiv 2\alpha u/t_0$ is the ratio between electronic gap and bandwidth. As a function of u, the ground-state energy has the form of a double-well potential as shown in Fig. 7.8, proving that the ground state is unstable with respect to bond alternation.

In order to fix the parameters K_0 and α, we note that, since experimentally the gap is 1.4 eV and by assumption the bandwidth is 12 eV, the parameter δ is 0.12, and we may expand E_0 in powers of δ. Then the minimalization can be performed analytically with the result

$$2\Delta_0 \approx \frac{16 t_0}{\exp(1 + 1/2\lambda)} \tag{7.9}$$

where $\lambda \equiv 2\alpha^2/\pi t_0 K_0$. With the measured value of the gap, we obtain $\lambda \approx 0.2$. The second derivative of E_0 with respect to u at the minimum yields the effective force constant for oscillations around the new equilibrium positions. One obtains (for $\delta \ll 1$)

$$K_{\text{eff}} \equiv \frac{1}{4} \frac{\partial^2}{\partial u^2} E_0 \approx 2\lambda K_0 \qquad (7.10)$$

This is the force constant for an optical phonon where the CH units move parallel to the chain axis 180° out of phase with a frequency (71)

$$\omega = \left(\frac{4K_{\text{eff}}}{M_{CH}}\right)^{1/2} \qquad (7.11)$$

Here it has been assumed that C and H atoms move rigidly together, whereas in reality the mode is split due to the relative C—H motions yielding the values $\omega_1 = 1474$ cm^{-1} and $\omega_2 = 1080$ cm^{-1} (29). Taking a value of 1300 cm^{-1} as representative, we obtain from Eq. (7.11) an effective force constant $K_{\text{eff}} \approx 20$ eV Å^{-2}. The parameters K_0 and α are now easily derived by using Eqs. (7.9) and (7.10) (Table 7.2). They agree essentially with those derived by Mele and Rice using a similar procedure (81) but disagree with those chosen by Su, Schrieffer, and Heeger (80) which take the elastic constant of ethane as representative for K_0. Although a "bare" force constant of 50 eV Å^{-2} may appear to be high, an adjustment of the parameters to experiments on polyacetylene seems reasonable. In doing so, however, one has to keep in mind that these numbers are renormalized quantities (including in an averaged way correlation, quantum effects, interchain coupling, and so on) with no precise microscopic significance.

The following simple argument indicates that the Hückel scheme must be considered as a very approximate theory for (CH)$_x$. Treating the dynamic displacements around the minimum u_0 as those of a harmonic oscillator, the zero-point motion associated with quantum fluctuations would produce a mean-square root displacement $\overline{\delta u} = (\hbar/2M\omega)^{1/2}$. With $\omega = 1300$ cm^{-1}, one finds $\overline{\delta u} = 0.032$ and 0.040 Å, respectively for the two parameter sets of Table 7.2.

Table 7.2 Elastic and Electronic Parameters

	K_0 (eV/Å^2)	t_0 (eV)	α (eV/Å)
(a)	50	3	6.9
(b)	21	2.5	4.1

[a] Values derived by authors; see also Refs. 81 and 71.
[b] Values derived by Su et al. (80).

THEORETICAL APPROACH

Therefore not only will the nonlinearity of the adiabatic potential (Fig. 7.8) be important, but also the Born–Oppenheimer approximation which may be considered as an expansion in $\delta u/u_0$ (82) has to be questioned.

As evidenced by Raman experiments, the chains in polyacetylene samples contain an average of, say, 50 CH groups. In order to compare theory with experiment it is therefore important to know to what extent the asymptotic results for infinite chains are modified for finite chains. Considering again even chains with cyclic boundary conditions, it is easy to see that the two cases $N = 4M + 2$ and $N = 4M$, where M is an integer, will behave differently. The reason is that for $N = 4M + 2$ the ground state of the undistorted chain is not degenerate and the electronic energy gain due to the lowering of occupied levels is of second order in the distortion amplitude, whereas for $N = 4M$ the (undistorted) ground state is degenerate leading to an energy gain proportional to the distortion (Fig. 7.9). In the latter case, the Jahn–Teller theorem is applicable, which states that a system with an orbitally degenerate ground state is unstable with respect to distortions which destroy those elements of symmetry responsible for the degeneracy. Therefore a distortion will occur, irrespective of chain length. In the former case, the decrease in electronic energy will not be sufficient to compensate the increase in elastic energy for short chains. This will only be possible for long enough chains where the electronic levels become quasidegenerate, leading to a "pseudo Jahn-Teller effect" (see Chapter 8 of Salem's book, Ref. 9).

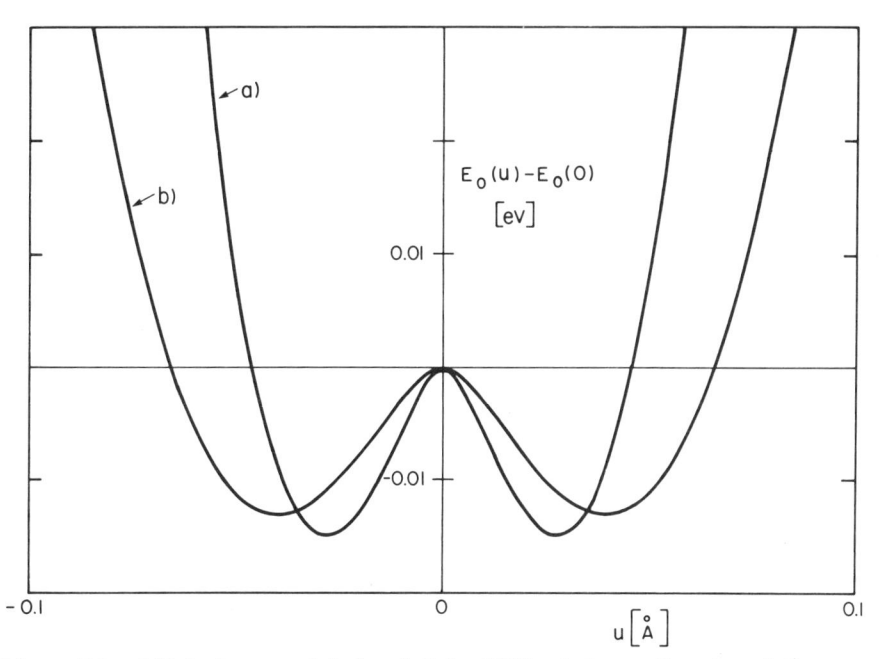

Figure 7.8 Adiabatic potential of an infinite $(CH)_x$ chain as a function of the mean displacement u for the two parameter sets of Table 7.2.

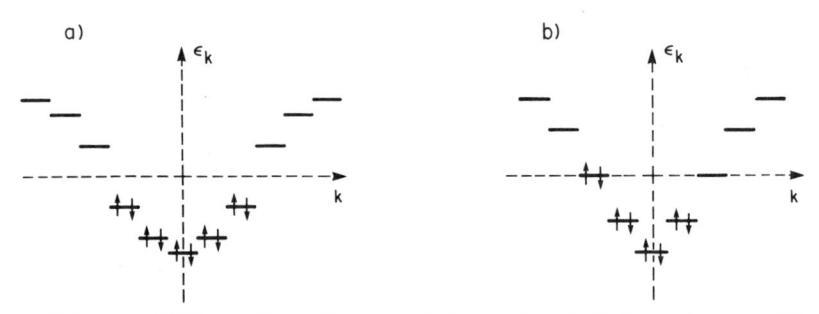

Figure 7.9 Level filling scheme for even chains and periodic boundary conditions: (a) $N = 4M + 2$, (b) $N = 4M$.

The numerical minimalization of the ground-state energy, Eq. (7.7), for finite chains and with a parameter $\lambda \approx 0.2$ shows that, in the case $N = 4M + 2$, chains containing less than 18 CH units are undistorted. Longer chains are distorted, and the distortion increases with N, approaching exponentially the asymptotic limit. For $N = 4M$, the chains are always distorted and the distortion decreases with N toward the asymptotic limit. Figure 7.10 shows how the effective elastic constant [Eq. (7.10)] and the electronic gap (taken as the difference between the lowest unoccupied and the highest occupied level) de-

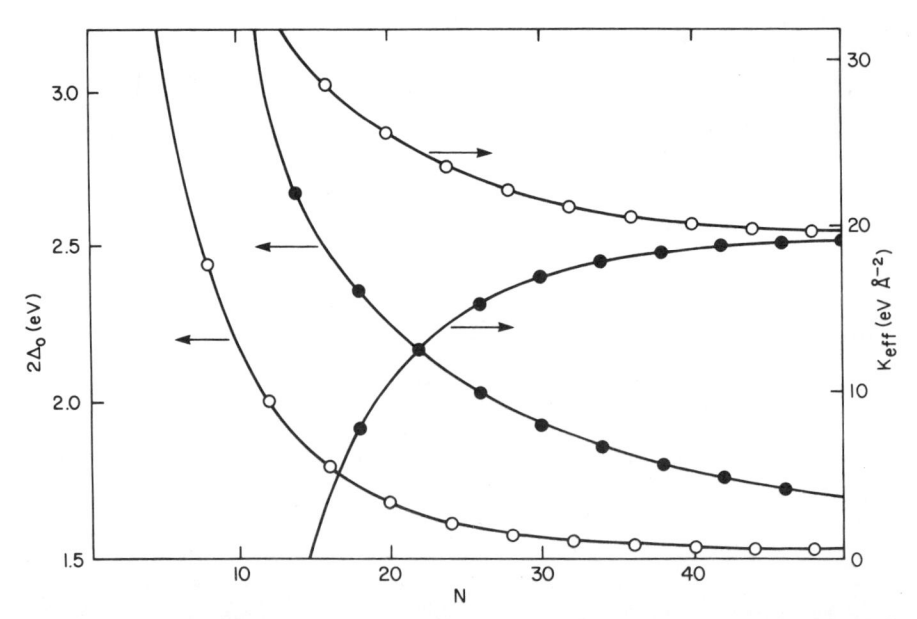

Figure 7.10 Electronic gap $2\Delta_0$ and effective elastic constant K_{eff} as function of chain length for even number N of CH units, $N = 4M + 2$ (●) and $N = 4M$ (○), and for periodic boundary conditions. The parameter set (a) of Table 7.2 has been used.

pend on N. The difference between the asymptotic limit of 1.53 eV for the gap and the value of 1.4 eV used to determine the parameters originates from the approximation made in deriving Eq. (7.9). The effective elastic constant varies with N like the distortion amplitude u, showing a "softening" for $N = 4M + 2$ with N approaching the "transition point," $N = 14$.

Experimental Raman frequencies measured on finite polyene segments decrease as a function of chain length (74) in a similar way as the sequence $N = 4M$. However, one should keep in mind that the present results have been derived with cyclic boundary conditions. The specific boundary conditions, for example, due to endgroups, may strongly influence the bond alternation structure as evidenced by numerical computation (71) and diffraction experiments (83).

7.3.4 Bond Alternation Defects in Odd Chains

Odd-numbered chains (with cyclic boundary conditions) do not fit into the simple scheme of bond alternation throughout the whole chain. This is commonly discussed in terms of a staggered displacement φ_n defined by the equation

$$u_n = (-1)^n \varphi_n \tag{7.12}$$

In even chains with simple bond alternation, φ_n is constant, (namely, either $+u_0$ or $-u_0$, reflecting the structural degeneracy), whereas in an odd-numbered chain, a transition from $\varphi_n = -u_0$ to $\varphi_n = +u_0$ has to occur somewhere. The odd electron is localized in a "dangling bond" around this lattice defect, as illustrated in Fig. 7.11. As discussed a long time ago (78, 84), the actual healing length of these bond alternation defects is expected to be much larger than in Fig. 7.11, namely, of the order of the critical length above which pseudo Jahn–Teller distortions become feasible, that is, of the order of 20 CH units. Explicit calculations for the electronic structure associated with such defects have been made by Pople and Walmsley (85) and later by Kventsel and Kruglyak (86), who show that the localized state corresponds to a level at midgap, namely, at energy zero [cf. Eq. 7.6]. It is in fact easy to show that the π-electron Hamiltonian, Eq. 7.2, has always a zero-energy eigenstate if the chain is odd-numbered. Seeking this state in the form $\Sigma_n \lambda_n c_{ns}$, one obtains a system of linear equations:

$$0 = t_{n-1,n} \lambda_{n-1} + t_{n,n+1} \lambda_{n+1} \qquad n = 1, \ldots, N \tag{7.13}$$

For $t_{0,1} = t_{N,N+1} = 0$, it has the solution

$$\lambda_{2m} = 0$$
$$\lambda_{2m+1} = (-1)^m \lambda_1 \prod_{l=1}^{m} \left(\frac{t_{2l-1,2l}}{t_{2l,2l+1}} \right) \tag{7.14}$$

Figure 7.11 Bond alternation defect in an odd-numbered chain.

where λ_1 has to be appropriately normalized. No such solution exists in an even chain. For a defect centered at an odd site as in Fig. 7.8, the ratio $t_{2l-1,2l}/t_{2l,2l+1}$ will exceed 1 at the left-hand side and become smaller than 1 at the right-hand side of the defect. Therefore the amplitudes λ_{2m+1} increase by approaching the impurity from either side and the wave-function associated with the midgap state is localized around the defect.

A more detailed theoretical treatment of bond alternation defects was stimulated by the suggestion that mobile bond alternation defects could be responsible for the narrow spin-resonance line which was observed in trans-$(CH)_x$ (87). Rice (88) and Su, Schrieffer, and Heeger (69, 80) proposed the ansatz

$$\varphi_n = u_0 \tanh\left(\frac{na}{\xi}\right) \tag{7.15}$$

for the staggered displacement field, where u_0 is the distortion amplitude of the homogeneous ground state (of the infinite chain) and ξ is a variational parameter. With Eqs. (7.4) and (7.14), the wavefunction of the localized state has the amplitudes

$$\lambda_{2m+1} \sim (-1)^m \left\{ \text{sech}\left[\frac{(2m+1)a}{\xi}\right] \right\}^{\xi\delta/a} \tag{7.16}$$

where δ is defined as before. The numerical minimalization of the defect energy yields a healing length $\xi \approx 7a$ and a defect energy $E_s = 0.42$ eV (69). Brazovskii (89) and Rice (88) propose the relation $\xi = a/\delta$, which is exact in the continuum limit (Section 7.3.5). In this limit, the defect energy is given as $E_s = 2\Delta_0/\pi$ (90). As already anticipated by Longuet-Higgins and Salem (78), such a defect would be able to move quite freely along the chain. From the kinetic energy involved in such a movement, one obtains an effective mass which is quite small, $M_s \approx 6m_e$, pointing to the importance of quantum effects (69). The lattice discreteness tends to pin the defects, but due to their large extent, the pinning energy is quite small, namely, about 2 meV (69). Therefore the defects are expected to be highly mobile above a temperature of, say, 30 K.

In infinite chains, the bond alternation defects may be considered as topological *excitations* due to specific boundary conditions at $\pm\infty$. In finite odd chains, *the ground state* is defected due to the odd numbers of electrons, its

THEORETICAL APPROACH

energy depending both on chain length and defect location within the chain. As detailed numerical calculations for the energetically favorable configurations (71) and for the nucleation of such defects (70, 91) show, the most stable position is in the middle of the chain.

7.3.5 Continuum Description and the Soliton Concept

Depending on which aspects are judged to be essential, different concepts may be invoked for the bond alternation defects. Brazovskii uses the term "polaron" to point out the combined effect of electron localization and lattice polarization (89). The name "kink" (88) and the notion "topological soliton" (69) stress the analogy to nonlinear lattice dynamical models, the latter pointing also to the origin of the stability of these defects. Since the word "soliton" has become fashionable in this context, we shall also use it. But it has to be understood in a physical sense as "stable, finite-energy, particle-like field pattern with some central physical nonlinear properties" (92) and not in the strict sense of mathematicians (as defined, e.g., in Ref. 93). Nevertheless, within the continuum model, the bond alternation defects represent a reflectionless potential for the band states, "a signature characteristic of solitonic states" (94).

Since the characteristic length for fluctuations, the soliton length, is much bigger than the lattice constant, a continuum description should represent a good approximation. In deriving the continuum limit from the original model Hamiltonian $H_\sigma + H_\pi$, Eqs. (7.1), (7.2), and (7.4), the staggered field φ_n defined by Eq. (7.12) becomes a continuous function $\varphi(x)$, and the electronic operators c_{ns} are replaced by field operators $\psi_{1s}(x)$ and $\psi_{2s}(x)$ representing particles close to $+k_F$ and $-k_F$, respectively. The resulting Hamiltonian is

$$\mathcal{H} = 2K_0 \int dx\, \varphi^2(x)$$

$$+ \sum_s \int dx\, \{-i\hbar v_F [\psi_{1s}^+(x)\frac{\partial}{\partial x}\psi_{1s}(x) - \psi_{2s}^+(x)\frac{\partial}{\partial x}\psi_{2s}(x)]$$

$$+ 4\alpha\sqrt{a}\, \varphi(x)[\psi_{1s}^+(x)\psi_{2s}(x) + \psi_{2s}^+(x)\psi_{1s}(x)]\} \quad (7.17)$$

where $v_F = 2t_0 a/\hbar$ is the Fermi velocity obtained from the linearization of the bare electronic spectrum around k_F. From Eq. (7.17) the following coupled system of eigenvalue equations is deduced for the eigenvalues ϵ_λ and the corresponding wave functions $v_\lambda(x)$ and $w_\lambda(x)$ associated with the field operators $\psi_{1s}(x)$ and $\psi_{2s}(x)$, respectively,

$$-i\hbar v_F \frac{\partial}{\partial x} v_\lambda + \Delta(x) w_\lambda = \epsilon_\lambda v_\lambda$$

$$i\hbar v_F \frac{\partial}{\partial x} w_\lambda + \Delta(x) v_\lambda = \epsilon_\lambda w_\lambda \quad (7.18)$$

These equations are known from the theory of inhomogeneous superconductors as the Bogoliubov–de Gennes equations (95). The wave functions $v_\lambda(x)$ and $w_\lambda(x)$ are appropriately normalized, and the gap parameter $\Delta(x)$ is related to the field $\varphi(x)$ by the equation

$$\Delta(x) = 4\alpha\sqrt{a}\,\varphi(x) \qquad (7.19)$$

By requiring the ground-state energy (or the thermodynamic potential at finite temperatures) to be a minimum, the gap parameter is determined self-consistently by the equation

$$\Delta(x) = -\frac{4\alpha^2 a}{K_0} \sum_{\lambda s} n_\lambda \left[v_\lambda^*(x) w_\lambda(x) + v_\lambda(x) w_\lambda^*(x)\right] \qquad (7.20)$$

where n_λ measures the occupation of level λ. In the homogeneous case where $\Delta(x) = \Delta_0 =$ constant, the wave functions are simple plane waves and the electronic spectrum has a gap $2\Delta_0$ at the Fermi level. From the self-consistency equation (7.20), the gap is obtained as

$$2\Delta_0 = \frac{W}{\cosh(1/2\lambda)} \qquad (7.21)$$

where W is the bandwidth and the parameter λ is defined as before (Section 7.3.3). For $\lambda \ll 1$, this value is bigger by a factor of $e/2 \approx 1.4$ than that obtained previously for the original model Hamiltonian, Eq. (7.9). This difference, which is due to the approximations made in deriving the continuum limit (in particular the linearization of the bare electronic spectrum), will also appear in other characteristic energies such as the soliton energy E_s. Alternatively, this effect may be compensated by slightly modifying the parameters as proposed by Cross and Fisher in the context of the spin-Peierls transition (96).

Inhomogeneous solutions have already been found by Bar-Sagi and Kuper (97) in the framework of superconductivity. Brazovskii (89) and Takayama, Lin-Liu, and Maki (90) have discussed the corresponding solutions for the inhomogeneous Peierls state. Differentiating Eq. (7.18) with respect to x and introducing new functions $f_\lambda^\pm = v_\lambda \pm w_\lambda$, one obtains a simple second-order differential equation:

$$\left[(\hbar v_F)^2 \frac{d^2}{dx^2} + \epsilon_\lambda^2 - \Delta^2(x) \pm \hbar v_F \frac{d\Delta}{dx}\right] f_\lambda^\pm = 0 \qquad (7.22)$$

With the ansatz $\Delta(x) = \Delta_0 \tanh(x/\xi)$, this equation is explicitly solved in terms of hypergeometric functions (98). Depending on the parameter $r = \xi\delta/a$ [Eq. (7.16)], there may be several bound states, but there is always at least one. The electronic spectrum of the continuum states is not changed by the inhomogeneity

THEORETICAL APPROACH

but, due to the phase shifts, the density of states is altered. It is quite remarkable that for $r = 1$ the self-consistency condition, Eq. (7.20), is identically satisfied (90), proving that the ansatz for $\Delta(x)$ represents an exact solution. There is one bound state at the middle of the gap, and the soliton energy defined as the difference between the energies of the inhomogeneous and homogeneous states is $E_s = 2\Delta_0/\pi$. The soliton state is triply degenerate since the midgap level can either be singly occupied (corresponding to a charge-neutral soliton carrying spin $\frac{1}{2}$) or unoccupied (spinless soliton with charge $+e$) or doubly occupied (spinless soliton with charge $-e$) (99).

7.3.6 The Soliton Lattice and the Commensurate–Incommensurate Transition

As discussed in Sections 7.3.3 and 7.3.4, the ground state of odd-numbered chains contains a single bond alternation defect (soliton), whereas that of even chains is defect free. Rice (88) and Su, Schrieffer and Heeger (69) recognized that additional defects can be created upon doping. The argument is simply that the energy required to generate a (charged) soliton ($2\Delta_0/\pi$ in the continuum limit) is smaller than the energy Δ_0 required to put an additional electron (hole) into the conduction (valence) band. The problem of finding the ground-state configuration of a one-dimensional electron–phonon system with $1 \pm \nu$ electrons per lattice site ($\nu \ll 1$) has been studied by Kotani (100), who finds that, due to Umklapp scattering, higher harmonics are involved in the lattice distortion pattern and that the electronic spectrum contains two gaps, one at $k_F = \pi(1 \pm \nu)/2a$ and the other at $\pi/a - k_F = \pi(1 \mp \nu)/2a$, as shown in Fig. 7.12. Starting from the single localized level at midgap, a symmetric band around this level is built up as $|\nu|$ increases. The associated distortion pattern has a period $2a/\nu$ and forms a "soliton lattice" (101), as illustrated in Fig. 7.13. Horovitz has been able to solve exactly the self-consistent set of equations, Eqs. (7.18) and (7.20), for the case of a soliton lattice by exploiting the equivalence of the fermion model to a "massive" boson model (102). For $\delta \ll 1$ and $\nu \ll 1$, the bandwidth W_1 and gap Δ_1 (see Fig. 7.12) are given by

$$W_1 \approx 8\Delta_0 \exp\left(-\frac{\pi\delta}{2\nu}\right)$$

$$\Delta_1 \approx \Delta_0 \left(1 - \frac{2\nu\delta}{\pi} + \frac{\delta^2}{4}\right) \tag{7.23}$$

The energy of the soliton lattice per kink is found to be

$$\frac{E_{SL}}{\nu} = E_s\left[1 + 4\exp\left(-\frac{\pi\delta}{\nu}\right) - \frac{\nu\delta}{\pi} + \frac{\delta^2}{6}\right] \tag{7.24}$$

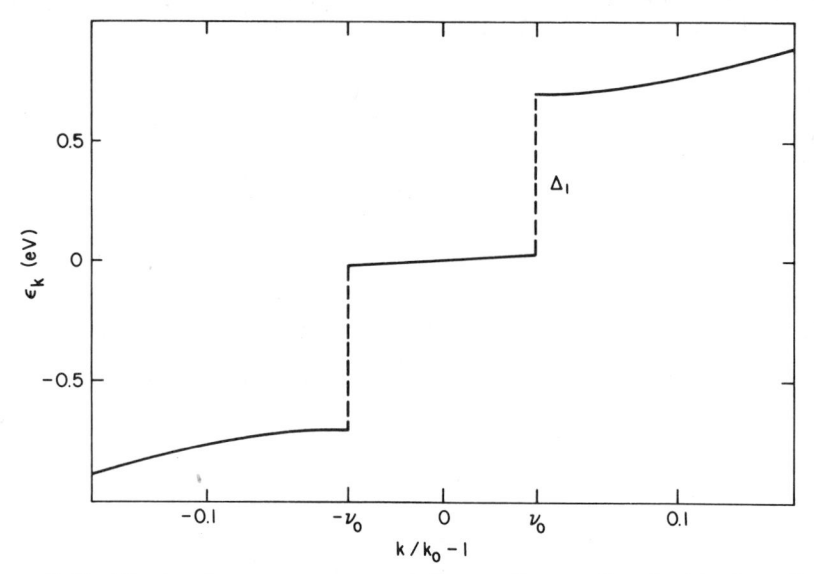

Figure 7.12 Electronic spectrum around $k_0 = \pi/2a$ associated with the soliton lattice of Fig. 7.13.

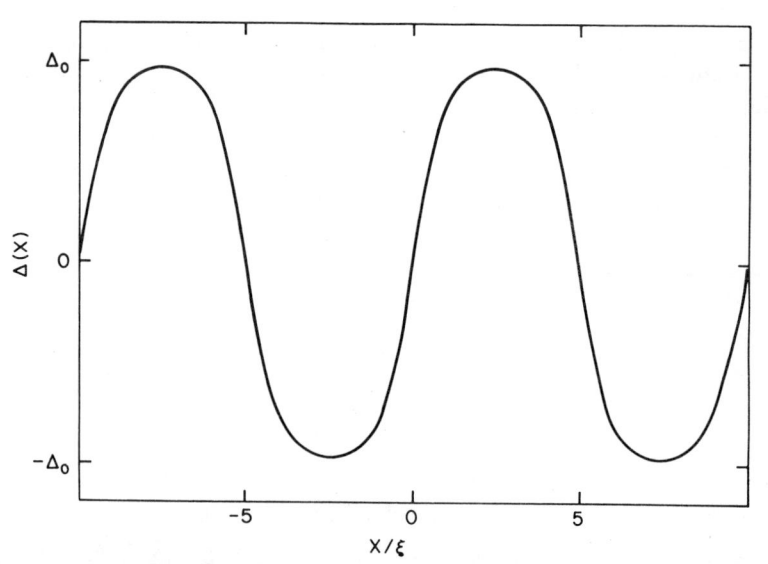

Figure 7.13 Distortion pattern of a soliton lattice in its stable configuration. The parameter ζ denotes the soliton length $\zeta \approx v_F/\Delta_o \approx 2a/\pi\delta$ and $\delta \equiv 2\Delta_o/W$ has been taken as 0.14 yielding $\nu_o = 4.4\%$.

THEORETICAL APPROACH

The interesting point is that, as a function of ν, this energy has a minimum determined from the equation

$$\exp\left(\frac{\pi\delta}{2\nu_0}\right) = \frac{2\pi}{\nu_0} \tag{7.25}$$

suggesting that, for $\nu < \nu_0$, a phase separation will occur into a region with a soliton lattice of period ν_0 and one without solitonic defects. The characteristic values of δ for polyacetylene imply $\nu_0 \approx 3$–5%. Increasing ν beyond ν_0 requires more and more energy due to the compression of the soliton lattice, and finally a transition to an incommensurate phase is expected to occur. A detailed analysis of this transition cannot be given by the theory of Horovitz since it assumes the order parameter $\Delta(x)$ to be real, whereas, at least for $\nu \approx \nu_0$, a more complete treatment based on a complex order parameter including phase solitons is required.

7.3.7 Correlation Effects

For an interacting electron system the self-consistent Hartree-Fock approximation determines the best single-particle states. As is well known from band theory, correlation, defined as that effect of the interaction which is not taken into account by the Hartree-Fock approximation, can be quite important (103). The ab initio calculations reviewed in Section 7.3.1 indicate that this is also true in the case of polyacetylene. The Hückel scheme which is by definition a one-particle theory does not take into account correlation provided that "bare" parameters are used. If the parameters are fitted to experimental data correlation is built in, though in an averaged manner.

Let us then assume that our model Hamiltonian $H_0 = H_\sigma + H_\pi$, Eqs. (7.1), (7.2), and (7.4), is defined in terms of bare constants, and add the interaction between π-electrons in the form of a Pariser–Parr–Pople Hamiltonian (104),

$$\mathcal{H}_{int} = U \sum_m n_{m\uparrow} n_{m\downarrow} + \frac{1}{2} \sum_{\substack{mm' \\ ss'}}' V_{mm'} n_{ms} n_{m's'} - \sum_{\substack{mm' \\ s}} V_{mm'} n_{ms} \tag{7.26}$$

where $n_{ms} = c_{ms}^+ c_{ms}$ is the π-electron density at site m and the prime indicates that the term $m = m'$ is excluded from the sum.

Correlation effects will be strong if $U/t_0 \gg 1$. In fact, neglecting the coupling to the lattice ($\alpha = 0$) and the long-range interaction ($V_{mm'} = 0$), we are left with the one-dimensional Hubbard model the ground state of which is known to be antiferromagnetic (105). Estimates indicate that $U/t_0 \geq 1$ for polyacetylene (106). On this basis and in an attempt to resolve an apparent discrepancy between theoretical and experimental values for the optical gap, Ovchinnikov, Ukrainskii, and Kventsel (10) argued that the ground state of a $(CH)_x$ chain should be antiferromagnetic and the gap mainly a result of correlation. Recently, Kondo (107) and Nakano and Fukuyama (108) extended this argument to include the dependence of the effective exchange integral $J_{n,n+1} = 4t_{n,n+1}^2/U$ on bond length which leads to a "spin-Peierls instability" (109). In this model, not only a lattice dimerization is produced, but also soliton-like excitations are predicted to occur, in a similar way as discussed previously (108). Therefore it will not be easy to discriminate between this limiting case and the simple Hückel picture on the basis of experimental data.

In a recent analysis dealing with the full Hamiltonian $H = H_0 + H_{int}$, Horsch finds that the on-site Coulomb interaction enhances both the lattice distortion and the condensation energy by such an amount that one is tempted to consider electronic correlation as "principal driving mechanism for the dimerization" (110). The ground state, although not truly antiferromagnetic, is found to have antiferromagnetic short-range order. The long-range part of the interaction counteracts the lattice dimerization, but this effect depends strongly on how the matrix elements $V_{m'}$ are screened by the σ-electrons.

Just the opposite behavior is expected if one starts from a Fröhlich model where the phonons are coupled to the electronic density instead of the bond order. The lattice distortion associated with an electronic charge-density wave decreases with increasing U, whereas the long-range part tends to stabilize the lattice dimerization (111, 112). These findings suggest that the electronic ground state cannot be described by a simple order parameter like bond order or spin density or charge density but does exhibit a more complicated structure.

Quite generally, it seems difficult at present to decide about the relevance of correlations for polyacetylene, mainly since several parameters are not known with sufficient accuracy. Further *ab initio* calculations including correlations should be useful in this respect. On the other hand, Hückel-type theories with effective parameters do represent a very practical ordering scheme for actual experiments.

7.3.8 Predictions for Polyacetylene

Given the morphology and structure depicted in Section 7.2 and the present theoretical framework, it is natural to ask which kind of predictions can be made for specific experiments. Since most of the theory has been developed with measured properties in mind, these predictions may not be unbiased. Nevertheless, they should represent a useful frame of reference for the experimental data to be presented in the following chapters.

Hückel theory predicts that $(CH)_x$ is a Peierls insulator, but due to a distribution of finite chain lengths several characteristic features will be smeared out. In particular, the square-root singularity due to the gap in the optical absorption spectrum will be rounded off and the diffraction spots originating from bond alternation will be broadened. Both effects are expected to be enhanced by quantum fluctuations. Interchain correlation will further smear out the band gap.

Since the effective force constant for vibrations parallel to the chain axis also depends on chain length, the corresponding Raman lines will be broadened.

Bond alternation defects can occur in both isomers; but while they are predicted to be highly mobile in the trans form, they will be pushed toward a chain end in the cis form for energetic reasons. Neutral defects will carry spin and produce characteristic spin resonance lines which should strongly depend on the degree of isomerization (motional narrowing in the trans form). The localized electronic state associated with these defects will produce an additional optical absorption peak at midgap and, as explicit calculations show (113, 94), with considerable oscillator strength.

Removing or adding charge by introducing acceptor or donor molecules will first tend to saturate the neutral bond alternation defects. Due to this spin compensation, the ESR signal produced by spin-carrying defects will decrease. Additional doping will then proceed via creation of charged defect pairs which are expected to be tightly bound in the cis isomer (Fig. 7.14), whereas in the trans form they tend to form a soliton lattice with a large lattice constant (about $25a$).

The dopant molecules may be of two different kinds. Either they are completely fixed once they are introduced or they are able to adjust themselves according to fluctuations in their surrounding. We may use the term "quenched impurities" for the former and "annealed impurities" for the latter case, respectively (114). Since the dopants are charged, they will strongly be bound to the oppositely charged defects, namely, with a binding energy of about 0.3 eV (69). Furthermore, the impurities tend to screen the charge of the bond alternation defects such that the Coulomb repulsion between charged defects which otherwise would strongly influence the defect structure (115) will not be important.

Disregarding impurities, heavily doped $(CH)_x$ would exhibit an incommensurate Peierls distortion. By pinning the phase of the electronic order parameter (116, 117), quenched impurities will have a strong phase-disordering effect at high concentration, lowering the stability of the Peierls phase and eventually leading to a disordered metallic phase where the Peierls gap is closed (118). With annealed impurities, the Peierls distortion will possibly survive and an impurity band is expected to be formed within the gap at high concentration. It should be mentioned, however, that alternative mechanisms have been proposed for the semiconductor–metal transition, namely, specific changes of the Fermi surface (67) and essential modifications of the band structure (119).

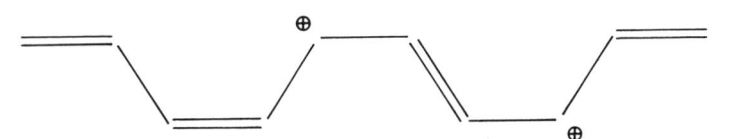

Figure 7.14 Charged defect pair in cis-$(CH)_x$.

Finally we point out that, due to the fibrillar morphology, polyacetylene will always appear as an inhomogeneous material, at least on large enough scales. This will be of importance for electrical transport properties and for optical spectra.

7.3.9 Recent Developments

A recent review on the electronic properties of polymers (including polyacetylene) as obtained from *ab initio* calculations has been given by Ladik and Suhai (120). Brédas et al. propose an interesting method for reducing the computational effort involved in such calculations by deriving an effective one-electron Hamiltonian for polymers from *ab initio* results obtained for small molecules (121).

In contrast to *ab initio* calculations, the semiempirical method uses experimental information to determine the basic parameters, and, therefore, different parameter sets are usually obtained if different experimental data are used. Pietronero et al. propose the following set of parameters for the SSH Hamiltonian, Eqs. (7.1), (7.2), and (7.4), including weak on-site correlation: $K_0 = 46$ eV Å^{-2}, $t_0 = 2.5$ eV, and $\alpha = 6.5$ eV Å^{-1} (122). These values are quite close to those derived in Sections 7.3.2 and 7.3.3 [parameter set (*a*) in Table 7.2]. Note that this parameter set is consistent with recent x-ray data giving a bond alternation amplitude $u \approx 0.03$ Å (46).

The soliton concept has been further refined to include quantum fluctuations and electron–electron correlation. Quantum fluctuations yield a quite important reduction in energy, $\delta E_s \approx -E_s/4$, but a negligible mass renormalization (123). On-site Coulomb correlation (Hubbard parameter U) modifies both the soliton extent and its energy (124). In particular, as has been already noted by Rice and Mele using a simple model (125), the energy of the charged soliton is greatly enhanced due to the double occupation of the midgap level such that, depending on the value of U, the charged soliton may be unstable. This would preclude charged solitons from contributing to dc transport.

The diffusion of solitons has been studied by Maki by including interactions with both acoustic and optical phonons. He finds a temperature-dependent diffusion constant that is predominantly determined by the coupling to acoustic phonons, at least in the accessible temperature range ($T \leq 400$ K) (126).

An important question is how the solitonic states are affected by the Coulomb potential of dopant ions. Conventional donor and acceptor states with electronic levels in the gap (127, 128) appear to be energetically favored for certain choices of the screening (127). Thus the existence of charged solitons in doped polyacetylene is at least uncertain. Quite unexpectedly, even the neutral solitons would be rather strongly pinned to the impurities, namely, by about 10 meV (129). Therefore the spin-carrying defects become immobile at low temperatures ($T \leq 100$ K).

In addition to the soliton and the soliton lattice (102, 130) a new intrinsic defect which had shown up previously in numerical simulations (70) has been found recently as an exact solution of the continuum model (131, 132). It exhibits a local indentation of the dimerization amplitude and two localized electronic states with levels in the gap. It carries charge and spin and is closely related to the conventional polaron (133).

Although the applicability of the soliton model for $(CH)_x$ may be rather restricted, it offers a very interesting theoretical framework for other materials (134) and even for field theory (135, 136).

7.4 PRISTINE AND LIGHTLY DOPED POLYACETYLENE: A ONE-DIMENSIONAL SEMICONDUCTOR

In this section, we want to concentrate on a few basic experiments which have been performed on undoped and lightly doped $(CH)_x$. Their interpretation should help to answer the question as to what extent the expectations and predictions developed in the last section have been verified in polyacetylene. Details of the doping procedure and dopant-specific properties, particularly at high doping concentration, are discussed in the next section.

7.4.1 Magnetic Properties

It is generally found (87, 137–140) that pristine cis material has a spin susceptibility of 2.8×10^{-8} emu/mol carbon, which is one to two orders of magnitude smaller than that of the trans material. It has been therefore concluded that the isomerization of the samples introduces additional spin-carrying defects located on the chain. The ESR spectrum of trans polyacetylene consists of a single line at $g = 2.003$, and the concentration of spins is 10^{-4}–10^{-3} per carbon atom equivalent to 100–1000 ppm. ESR as well as static susceptibility measurements show that the susceptibility obeys a Curie law between 300 and 1.5 K (141, 142), that is, an activation energy for a transition to the paramagnetic state from a nonparamagnetic state would be extremely small if not negligible. The ESR line of cis has a Gaussian-type shape, whereas the trans lines have nearly Lorentzian character (140). The ESR linewidth of cis is of the order of 10 G and almost temperature independent, whereas the linewidth of trans decreases strongly with increasing temperature to values below 1 G at room temperature (Fig. 7.15). These results have been interpreted as motional narrowing in $trans$-$(CH)_x$. An alternative explanation of exchange narrowing has been excluded since no correlation between spin density and linewidth for trans samples could be found and because this effect is not expected to generate the observed strong temperature dependence of the linewidth. Hence the ESR data provide evidence of mobile paramagnetic defects in trans and of immobile defects of far lower concentration in cis or cis-rich material. Note the interesting different magnetic properties of $(CH)_x$ described here and the doped polymer matrices discussed in Chapter 6.

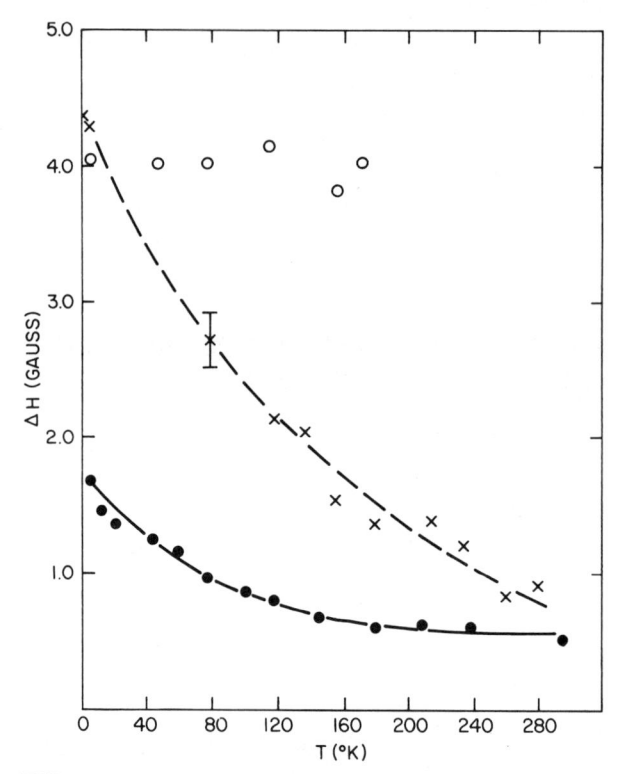

Figure 7.15 ESR linewidths as function of temperature; (x) *trans*-$(CH)_x$; (●) *trans*-$(CD)_x$; (o) *cis*-$(CD)_x$. (From Weinberger et al. [140].)

This conclusion has been confirmed by dynamic nuclear polarization measurements by Nechtschein et al. (143). Figure 7.16 shows the enhancement of the NMR amplitude (at the nuclear Larmor frequency ν_N) in *trans*-$(CH)_x$ by pumping at and near the electronic Larmor frequency ν_e. This enhancement, the Overhauser effect, gives definite evidence for the mobility of the unpaired electronic spin. It contrasts with the observation of the "solid state" effect in cis in the same experiment, where forbidden transitions at $\nu_e \pm \nu_N$ are induced due to static electronic spins. Measurements of the proton relaxation rate T_1^{-1} show that T_1^{-1} is proportional to $\nu_N^{-1/2}$ for undoped *trans*-$(CH)_x$ over the frequency range 10–340 MHz. This gives support for a model of one-dimensional spin diffusion with an intrachain diffusion rate $D = 6 \times 10^{13}$ s^{-1} and an anisotropy in spin diffusion of 10^6. The value of D, however, is two to three orders of magnitude higher than the D value derived from the observed ESR linewidth ($\Delta H \sim 1G$) (140) or that obtained from spin-echo measurements (144).

In order to resolve this discrepancy, Holczer et al. (145) interpreted their ESR data in terms of diffusive and localized spins assuming an exchange between the two spin species so that the linewidth could be approximated by $\Delta w = (1 - c)$

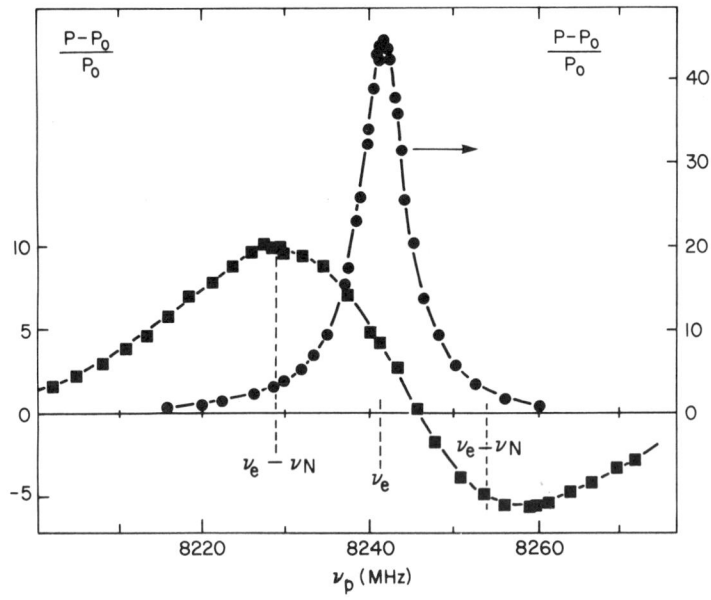

Figure 7.16 Enhancement of proton NMR amplitude P as function of the microwave pumping frequency ν_p near $\nu_e = 8240$ MHz for cis-rich (■) and trans (●) undoped $(CH)_x$. P_0 is the NMR signal amplitude without pumping. (From Nechtschein [143].)

$\Delta w_D + c\Delta w_L$, where c is the concentration of the localized spins, Δw_D and Δw_L are the linewidths of the diffusive and the localized spins, respectively. The $\Delta w_L/\Delta w_D$ ratio was estimated to be 30 at room temperature, so that for $c \geq 0.1$ the ESR linewidth would be dominated by localized spins. In this picture, spin localization would be caused by cooling to low temperatures or oxidation. The low-temperature localization is consistent with the pinning energy of 2 meV, which was estimated in the soliton theory. In the nuclear relaxation process, on the other hand, the diffusing spins are predominantly sensed by the protons and thus determine the observed high D value.

The spin susceptibility of $cis\text{-}[CH(AsF_5)_y]_x$ was found to be independent of temperature above 100 K (Pauli susceptibility) and to increase linearly with doping concentration for $0.003 < y < 0.12$, as shown in Fig. 7.17 (144). It has therefore been argued that the doping is inhomogeneous for all concentrations of AsF_5 and that the doped parts of the polymer would form metallic clusters, giving rise to a temperature-independent Pauli susceptibility.

The variation of the spin susceptibility of $trans\text{-}(CH)_x$ (146) as a function of AsF_5 concentration y differs distinctly from that of $cis\text{-}(CH)_x$. The concentration of Curie law spins, N_c, decreases from 400 ppm for $y = 0$ to about 120 ppm for $y = 0.002$ and continues to fall rapidly to practically immeasurably small densities for $y = 0.05$ (Fig. 7.18), whereas the onset of Pauli susceptibility is not

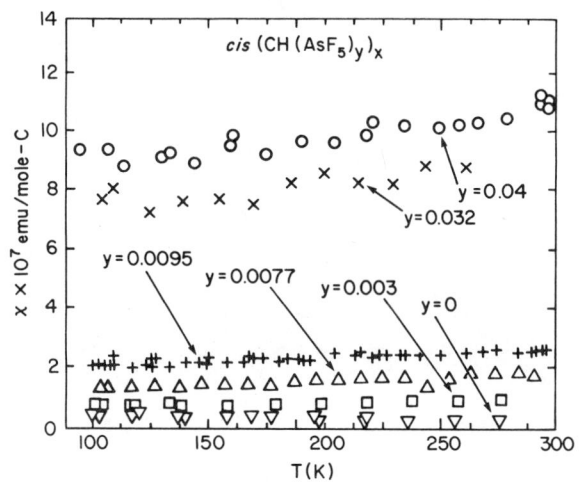

Figure 7.17 Spin susceptibility as function of temperature for cis-$(CH)_x$ doped with various levels of AsF_5. (From Tomkiewicz [144].)

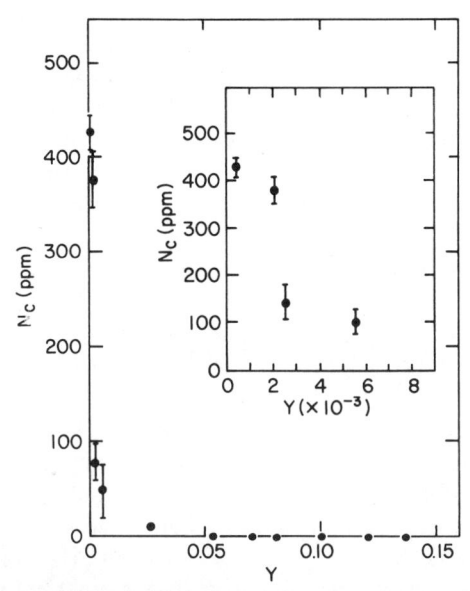

Figure 7.18 Concentration of Curie-law spins (N_c) for $trans$-$(CH)_x$ doped with various levels of AsF_5. (From Ikehata [146].)

observed before $y = 0.07$ is reached. It is not clear at the present time whether these drastic differences in the magnetic behavior are isomer specific or due to differences in the doping procedure or the doping mechanism in the two isomer forms (see also Section 7.5.1). This question is worth further consideration and work, since the results obtained on *trans*-(CH)$_x$ are in good agreement with the expectations of the soliton model. It shows that the initially uncharged spin-carrying defects lose their spin upon doping. For y around 0.01, where the conductivity is of the order of 1 to 10 Ω^{-1} cm^{-1} (146), the results suggest that transport occurs via spinless charge carriers.

The question of experimental evidence for soliton formation has been addressed in recent review papers (147, 148). Attempts have been made to determine the temperature dependence of the diffusion rate of the neutral soliton by evaluating results on the temperature dependence of the proton spin relaxation time T_1 (149) and of the ESR linewidth (150,151). The analysis of the temperature dependence of the proton relaxation time T_1 gives a diffusion constant that increases with decreasing temperature, whereas the analysis of the ESR linewidth gives the contrary behavior. Proton and ^{13}C spin–lattice relaxation measurements in iodine-doped (CH)$_x$ seem to be explicable in terms of one-dimensional spin diffusion (152). A motional narrowing transition of the proton NMR linewidth in undoped and AsF$_5$-doped *trans*-(CH)$_x$ at 155 K may find its explanation in terms of solitonlike excitations or translational fluctuations of the chains (153).

7.4.2 Optical Properties

The absorption spectrum of pristine (CH)$_x$ of either isomer form is typical for a semiconductor. The absorption edge of *cis*-(CH)$_x$ is at about 2 eV (154). Its low-energy tail structure may be due to partial isomerization, but an intrinsic optical gap of about 1.8 eV may be derived from the data. The absorption coefficient reaches a structured peak of about 4×10^5 cm^{-1}, with maxima at 2.1, 2.3, and 2.4 eV. The maximum value is similar to what is commonly observed for direct transitions in inorganic semiconductors, although it is nearly one order of magnitude below the maximum absorption coefficient due to valence–conduction band transitions, for example in Si.

The intrinsic absorption edge of trans material (154) is shifted to smaller energies with respect to that of cis, and the optical gap is located at about 1.4 eV (Fig. 7.19). The peak absorption of *trans*–(CH)$_x$ is structureless, and the maximum of the absorption coefficient of about 3×10^5 cm^{-1} is located at 1.9 eV. The shape of the absorption spectrum expected to show a square-root singularity for a perfect linear chain is probably smeared out due to the chain length distribution, interchain coupling, disorder, and fluctuations. Figure 7.20 shows the reflectance spectra measured by Fincher et al. (40) with linearly polarized light on stretch-aligned trans films. The reflectivity is significantly

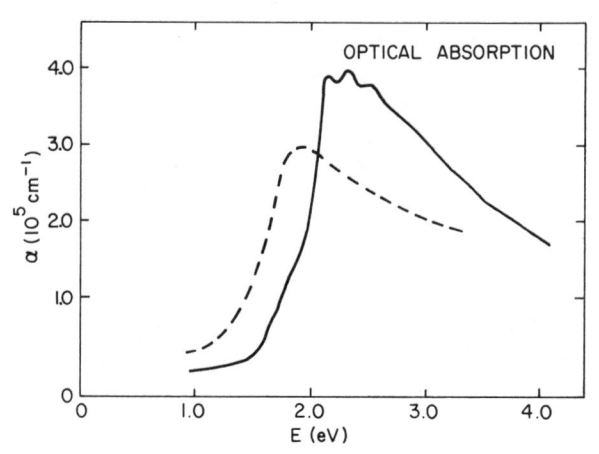

Figure 7.19 Absorption spectrum of *cis*- (full line) and *trans*- (broken line) $(CH)_x$. (From Fincher et al. [154].)

greater for light polarized parallel than for light polarized perpendicular to the fiber orientation. Kramers–Kronig analysis was used to analyze the reflectance data and to obtain the dielectric constant $\epsilon = \epsilon_1 + i\epsilon_2$; ϵ_1 is negative at high photon energies crossing zero at 1.9 eV, reaches a maximum of 19 at 1.45 eV, and approaches a value of approximately 4 in agreement with the dielectric measurements in the GHz range (155). The effective number of valence electrons per (CH) unit contributing to the optical constants can be calculated from the optical conductivity $\sigma = \epsilon_0 \omega \epsilon_2$ using the appropriate sum rule (156). This number was found to be about 0.5 to unity, showing that the interband transitions observed in this energy range pretty well exhaust the oscillator strength of the π-band system. The implication made by Fincher et al. (40) that this numerical

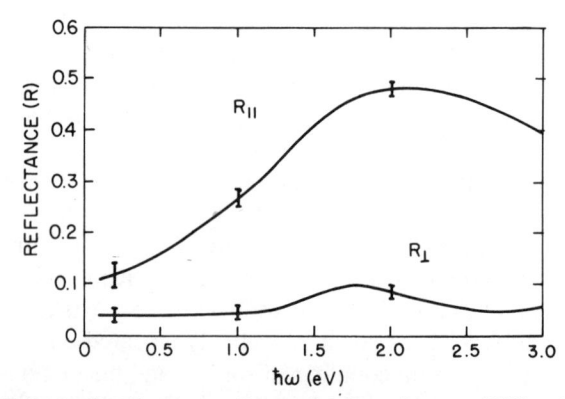

Figure 7.20 Polarized reflectance of stretch-aligned *trans*-$(CH)_x$. $R_\parallel$ and $R_\perp$ refer to light polarization parallel and perpendicular to the alignment. (From Fincher et al. [154].)

result gives evidence for interband transitions rather than exciton transitions is not generally valid since it is possible that nearly all the oscillator strength goes into the excitonic transition, in particular for large binding energies (157).

A better argument for the preponderance of interband transitions comes from photoconductivity studies (158, 159), which show that free carriers are excited by the light in the pertinent wavelength range. On the other hand, it has been concluded from energy loss experiments (160) that undoped $(CH)_x$ has an excitonic or indirect absorption edge. The Kramers–Kronig analysis of the polarized reflectance data obtained on stretch-aligned films yields a maximum of $K^{\|}_{max} = 4.8 \times 10^5$ cm^{-1}, which is nearly twice $K_{max} = 3 \times 10^5$ cm^{-1} for unoriented trans-$(CH)_x$. A factor of 2 would be expected for a random orientation under the assumption that $K^{\perp}$ is negligible. This qualitative agreement is satisfactory in view of the difficulties in determining K quantitatively from thin-film transmission measurements due to uncertainties in fill factor and exact thickness as well as the problem of light scattering from the irregular fiber structure.

The absorption spectra in the visible and near IR show also strong changes upon doping (113, 161, 162). Usually at low doping levels, the interband transition is suppressed, and concurrently a band builds up centered in the region between 0.5 and 0.8 eV, that is, at about half the band gap energy. There is already a weak absorption in undoped $(CH)_x$ centered near 0.9 eV. Upon dilute doping with AsF_5, the intensity of this band increases roughly proportional to the doping and shifts toward lower energies (Fig. 7.21). The total oscillator strength is conserved through the decrease of the interband transition. It is generally accepted that the p-type doping-induced absorption is due to transitions from the valence band to unoccupied localized states at midgap. Calculations within the soliton model both for the induced absorption and the suppression of the interband absorption have given approximate quantitative agreement with experiment (113, 94). It should be noted, however, that optical absorption crosssections of the order of 10^{-16}–10^{-15} cm^2 as derived from the experiment are also found for localized deep level impurities in inorganic semiconductors (163). Furthermore, conservation of oscillator strength is, in principle, required by the sum rules for optical transitions. So it seems that the optical data in the region from 0.5 to 4 eV may be interpreted with the soliton model but by themselves do not provide a definite proof for the existence of solitons in lightly doped $(CH)_x$.

Doping of polyacetylene affects the optical spectra in the IR profoundly, the changes being practically independent of the dopant (28, 164). This suggests that the new features are characteristic of changes of the polyacetylene chain itself. Experimental results obtained with iodine shown in Fig. 7.22 are thus representative for other dopants.

Upon doping, the overall transmission drops, and new broad lines build up at 1400 and 900 cm^{-1} with increasing dopant concentration. In $(CD)_x$, the strong lines are found at 1120 and 780 cm^{-1} (165). The internal modes of a soliton-type defect have been calculated by Mele and Rice (81) in agreement with the

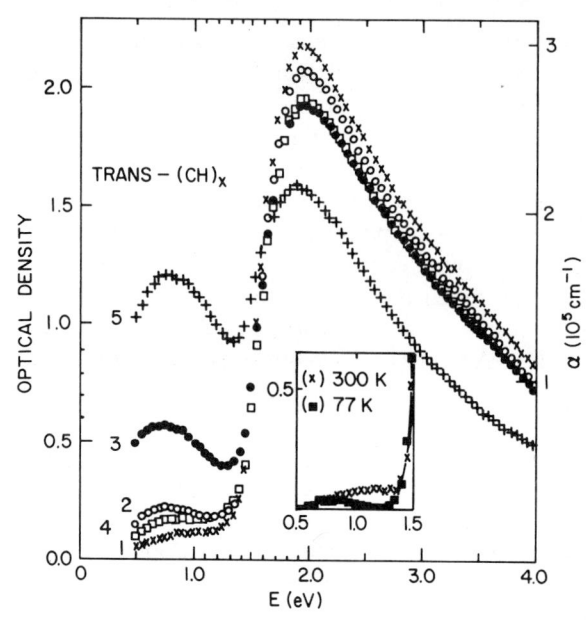

Figure 7.21 Absorption spectra of *trans*-$(CH)_x$: (1) undoped; (2) $y = 0.0001$ (AsF_5); (3) $y = 0.001$ (AsF_5); (4) compensated with NH_3; (5) $y = 0.005$ (AsF_5). Part of the spectrum for undoped $(CH)_x$ at two different temperatures is shown in the inset. (From Suzuki [113].)

experimental results. The strong intensities of the two new absorption bands and their independence on the doping agent give additional evidence for the existence of solitons at low doping concentrations.

Dopant-induced absorption bands at 0.17 and 0.11 eV attributed to local IR active modes of the charged soliton and a band centered at 0.5–0.8 eV explained as a transition to the soliton state at midgap have been considered to be hallmarks of solitons. However, as shown by Horovitz for a simple model, both the relative intensity and the frequency of the dopant-induced IR lines are independent of the charge configuration and thus not exclusively connected with solitons (166). Part of the model assumptions have been criticized by Mele and Rice (167) but the general conclusion of Horovitz should remain valid. By comparing theory and experiment, Maki and Nakahara conclude that the intensity of the observed midgap absorption is about twice that expected for one soliton per dopant molecule (168). This may indicate that bound pairs of a charged and neutral soliton (i.e., polarons) are created upon doping (169). Transitions between deep impurity levels and band states would represent an alternative mechanism for the midgap absorption. Experimentally, both the IR modes and the "midgap" absorption have been observed even in the metallic state (170, 171). While the persistence of the IR absorption peaks agrees with the theory of Horovitz (166), the maximum in the optical absorption around 0.5 eV could be related to

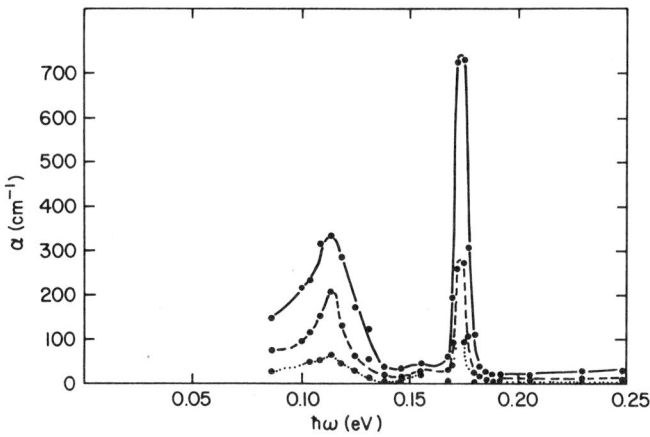

Figure 7.22 Additional absorption (over the undoped sample) for cis-$(CH)_x$ doped with various levels of iodine: $y = 0.0003$ (I_3^-) for upper curve; $y = 0.0016$ and $y = 0.00005$ for two lower curves, respectively. (From Fincher et al. [164].)

Anderson localization. Indeed, theoretical investigations show that $\sigma(\omega)$ goes through a maximum even in a regime where both the dc conductivity and the magnetic susceptibility are metal-like (172, 173).

A series of photoconductivity experiments have been performed to prove the formation of charged solitons in view of the theoretical prediction (70, 174) that photoexcited carriers should relax into solitonic states. Based on the existence of photoconductivity in $trans$-$(CH)_x$ and its absence in cis-$(CH)_x$, it has been conjectured that solitons are formed in $trans$-$(CH)_x$ by light (175). A simultaneous investigation of photoconduction and of IR absorption did not show the expected increase in absorption due to a soliton-induced state at midgap (176). These latter results have been taken as evidence against formation of charged solitons by photoexcited carriers.

7.4.3 Electrical Transport Properties

Probably the most prominent phenomenon observed with $(CH)_x$ films is the dramatic increase of the dc conductivity upon doping (12). The critical concentration for the transition from the semiconducting to the highly conducting state is not well defined but generally found to be of the order of 0.01 (177). In the following, we discuss experimental results below $y \approx 0.01$.

The temperature dependence of the conductivity of undoped and doped samples has been measured in the range from 40 mK to room temperature (178), the range for an individual sample being a function of sample resistance. The log σ versus T^{-1} curves for undoped and weakly doped samples show usually a curvature which may arise from disorder or/and a distribution of activation energies. Sometimes plots of log σ versus $T^{-1/4}$ give straight lines for $trans$-$(CH)_x$ (179), at low dopant concentrations.

In order to investigate whether the carrier transport occurs preferentially parallel or perpendicular to the fiber axes, the electrical conductivity has been investigated as a function of fiber orientation. Measurements on stretch-aligned films show that for elongation ratios of ~3, conductivity ratios $\sigma_\parallel/\sigma_\perp$ of about 10 are obtained for undoped films and of about 12 and 16 after doping with iodine and AsF$_5$, respectively (39). The anisotropy of the conductivity of shear flow-oriented material is small compared to that of stretch-aligned films, though scanning electron microscope pictures give the impression of perfect alignment of the fibers.

cis-(CH)$_x$ lightly doped with AsF$_5$ ($y \leq 0.01$) has been investigated by Tomkiewicz et al. (180) at high electric fields ($E \geq 10^3$ V/cm) and at temperatures below 77 K. The resistance was found to be proportional to exp (E_0/E) and to reach an asymptotic value at sufficiently high fields independent of temperature. At low fields, Ohm's law was valid and the temperature dependence of the resistance could thus be fitted by a relationship $R = R_0 \exp(T_0/T)^{1/2}$. The field dependence as well as the temperature dependence of the resistance are suggestive of "cermets," that is, conducting (metallic) particles embedded in an insulating matrix (181). If the data are evaluated applying the theory for granular metals, then one finds that the assumed particles must be separated by approximately 60 Å.

The thermoelectric power is positive for undoped (CH)$_x$ and for material doped with iodine or AsF$_5$, indicating p-type conduction and consistence with the charge transfer and the formation of negative ions. In the case of iodine-doped samples, the thermoelectric power is independent of temperature and of dopant concentration below about $y = 0.001$, that is, below the semiconductor-metal (S-M) transition (177). These data have been explained by hopping of the carriers whose concentration is about 2×10^{18} cm^{-3}. If these results are combined with the conductivity data, the mobility of the holes is found to be activated, to have a room temperature value of 5×10^{-5} cm^2/Vs, and to increase steeply with increasing iodine doping.

Kwak et al. (182) also have investigated the thermoelectric power of AsF$_5$-doped (CH)$_x$. At low concentration below $y = 0.01$, they find in contrast to Park et al. (177) for iodine doping, that the thermoelectric power is temperature dependent and can be fitted to a semiconductor model. They obtain the position of the Fermi level at 0.4 ± 0.1 eV in agreement with the activation energy for the conductivity. Evaluation of these data on the basis of semiconductor statistics gives a nonactivated mobility which decreases from 3–6 cm^2/Vs for undoped (CH)$_x$ to somewhat smaller values with increasing AsF$_5$ concentration. Except that different dopants have been used, no obvious reason for the discrepancy between the two papers can be given. Unfortunately, quantitative data of the carrier mobility are not available since the sample morphology is not amenable to meaningful Hall measurements (183).

Besides the above quoted interpretations in terms of a simple semiconductor picture (182), the metallic droplet model (180) or hopping between localized states (177), the interesting possibility of transport by charged solitons has been proposed. In this case, the conductivity would be activated with an activation energy equal to the binding energy E_b between the charged soliton and the nearby impurity, namely, of the order of 0.3 eV (cf. Section 7.3.8). A specific model for phonon-assisted hopping between the electronic midgap states associated with the solitons has been proposed by Kivelson (184). This mechanism requires both neutral and charged solitons to be present.

Electrical transport measurements by Epstein et al. (185) in the frequency range between 10 and 10^5 Hz seem to confirm Kivelson's theory (184, 186) on intersoliton hopping of charge carriers whereas similar measurements by other authors (187, 188) indicate that the material behaves like an inhomogeneous conductor.

Despite the fact that the optical experiments provide a quite consistent picture of a semiconductor with localized levels at midgap for weakly doped $(CH)_x$, the question as to which is the dominant transport mechanism cannot be answered unambiguously at present, and further experiments are required to clarify the situation. Of course, inhomogeneity due to the partial crystallinity of the material will always complicate the interpretation of optical and electrical measurements.

7.5 THE SEMICONDUCTOR-METAL TRANSITION AND HEAVILY DOPED POLYACETYLENE

The physical properties of strongly doped $(CH)_x$ will be discussed in this section. At first sight, there is a bewildering maze of experimental data. Sometimes there are even contradictory results and consequently also contradictory interpretations with seemingly equivalent procedures. The experimental results are discussed within the context of the nature of the highly conducting phase and the characterization of the transition region.

7.5.1 Dopants and Doping Procedures

The most common method of doping is exposure of the films to the vapor of electron attracting substances such as I_2 and AsF_5 (Table 7.3), giving rise to p-type conduction. The dopant can also be introduced by treating the film with appropriate solutions. Films have also been doped electrochemically (190) giving a readily controllable means of doping. The highest doping levels which can be achieved lie between $y = 0.1$ and $y = 0.3$ referring to molecules incorporated per (CH) unit.

The n-type conductivity of $(CH)_x$ is obtained by doping with electron-donating species (Table 7.3). However, n-type doped films have been barely investigated, though doping can be achieved simply by immersing the film into a tetrahydrofuran solution of sodium naphthalide, for example. The reason why

Table 7.3 Conductivities of Polyacetylene with Various Dopants After Reference 189.[a]

Electron-Attracting Dopants	p-Type	Electron-Donating Dopants	n-Type
Br	10^{-3}	Li	2×10^2
Cl	10^{-4}	Na	10^2
I	10^2	K	50
AsF$_5$	10^3		
SbF$_5$	5×10^2		
ClO$_4$	10^3		
PF$_5$	10^2–10^3		

[a]Values are given in Ω^{-1} cm^{-1}.

there have been only few investigations on n-type doped films lies in the increased sensitivity of the doped films to air and moisture.

The chemical nature of the dopant has been investigated by various experimental techniques. Raman studies (31, 191, 192) indicate that I_3^- and I_5^- ions are present in (CHI$_y$)$_x$, the most significant portion of the halogen being built in as I_3^-. On the other hand, it appears from photoemission spectra that I_5^- is the relevant charged dopant species (193). In the case of AsF$_5$, x-ray absorption indicates a disproportionation reaction leading to the incorporation of one AsF$_6^-$ ion per two AsF$_5$ molecules (194).

It has often been claimed that after doping to the highly conducting state, the isomeric composition is trans, irrespective of whether the original film was cis or trans. The persistence of intrinsic absorption features characteristic for cis material (154, 162) indicates, however, that the cis–trans isomerization is not catalyzed at low concentrations of, for example, BF$_3$ or AsF$_5$. It has also been reported (195) that, on exposing fully AsF$_5$ or AgClO$_4$ doped cis–trans films to the compensating agent ammonia, the cis and trans Raman lines recuperate to approximately the same relative and absolute intensities as in the undoped starting material. Even stronger evidence against isomerization induced by doping with AsF$_5$ was found in Knight shift measurements by Peo et al. (196), which indicate that at doping levels up to $y = 0.05$ the cis/trans ratio is hardly affected at all (Fig. 7.23). From these reports, one must conclude that the physical properties of doped films may well depend on the isomer form of the starting material. Differences in the magnetic properties, for example, as discussed in Section 7.4.1 are not surprising in the light of these findings.

The question of how the dopant molecule diffuses into the (CH)$_x$ films and whether the dopant is homogeneously distributed (even in pure trans starting material) is of crucial importance for any interpretation. Investigations of x-ray diffraction on doped (CH)$_x$ by various groups (43, 197) show that the dopants I_2, Br$_2$, and AsF$_5$ are intercalated in the crystalline regions.

HEAVILY DOPED POLYACETYLENE

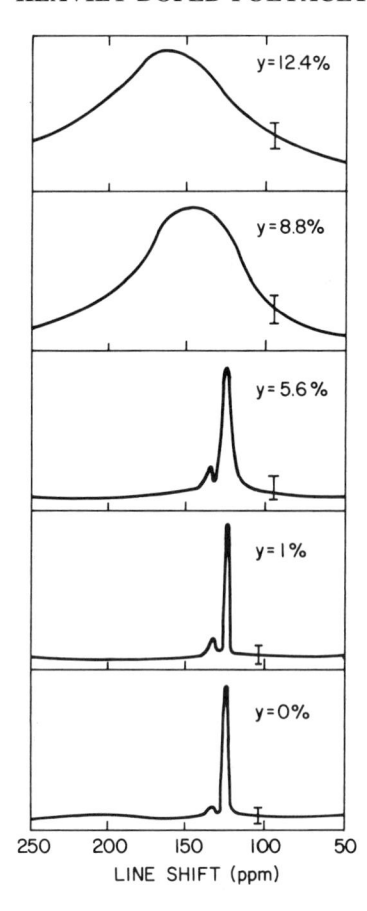

Figure 7.23 C^{13} NMR absorption spectrum for cis-$(CH)_x$ doped with various levels of AsF_5. In undoped $(CH)_x$ the line at 127 ppm is attributed to cis-$(CH)_x$ and the small satellite at 135 ppm to traces of trans-$(CH)_x$. The cis/trans ratio is hardly affected by doping up to 5%. (From Peo et al. [196].)

The diffusion constant of iodine in $(CH)_x$ was determined to be about 10^{-13} cm^2/s by careful measurements of iodine uptake as a function of time (198). The diffusing species are most likely iodine atoms, a notion which is supported by the fact that the iodine accepted by $(CH)_x$ is proportional to the square root of the iodine vapor pressure.

A remarkable feature of iodinated $(CH)_x$ films is that the iodine content as well as the conductivity of thin films ($d \leq 0.2$ μm) drops continuously with time in a dynamic vacuum. This indicates that the iodine is relatively loosely bound to the polyacetylene. These observations may be obscured in the case of thick layers ($d \approx 10$–100 μm) since readsorption processes within the porous material may give rise to a substantially increased time of desorption. Other dopants are probably incorporated differently and are more strongly bound to the $(CH)_x$.

The original mixed amorphous–crystalline state and the very nature of the diffusion into the fibers imply some nonuniformity of the doping for any dopant. Reliable measurements of the degree of homogeneity probably come from techniques such as Auger photoemission spectroscopy which can probe over a length scale smaller than the fiber dimensions (193).

Probably, due to uncertainties in sample definition, there are often contradictory results with respect to the physical properties. We mention here the nonlinear dc conductivity in strongly iodine-doped $(CH)_x$ where two groups (179, 199) report qualitatively different field dependences and consequently arrive at opposing interpretations.

^{13}C NMR studies of AsF_5-doped polyacetylene (200) confirm that no extensive conversion of cis to unoxidized trans occurs upon doping. The down-field shift of the NMR line in the metallic state in this work is interpreted as a chemical shift rather than as a Knight shift (196). Mössbauer studies of iodine-doped $(CH)_x$ confirm the existence of I_3^- and I_5^- linear molecules (201, 202).

7.5.2 The Nature of the Highly Conducting State and the Semiconductor–Metal Transition

The electrical conductivity of $(CH)_x$, as already described in Section 7.4.3, increases above a critical dopant concentration very rapidly to high values which are close to those of metals. In the highly conducting state, ESR measurements give evidence for Pauli susceptibility (137, 146); and in NMR experiments a Knight shift (196) is observed. All these results are reminiscent of the insulator–metal transition of heavily doped semiconductors (203), which is often described in terms of an Anderson transition (204). However, there is a significant difference if one extends this rather superficial comparison to the optical properties. Usually, one observes in the case of highly conducting inorganic semiconductors, such as InSb (205) and CdO (206), a Burstein shift (i.e., a shift of the absorption edge to higher energies with increasing conductivity), and the optical transition from the valence to the conduction band remains intact even in the metallic state. For the highly conducting polyacetylene, no Burstein shift is observed and the π–π^* transition disappears completely in all cases measured so far except for iodine doping. Therefore, the analogy to inorganic semiconductors is not straightforward, and only careful and accurate experiments will allow a conclusion to be drawn as to which kind of insulator–metal transition is operative in polyacetylene.

Unfortunately, unknown variables in material preparation and doping procedure often seem to give rise to considerable spread in data. Thus the transition from the semiconducting to the highly conducting state is often qualitatively different for various dopants and sometimes even for the same dopant. For iodine, the increase in conductivity as a function of dopant concentration is often rather gradual (155) and can be approximated by a power law, $\sigma \sim y^n$, over the concentration range of $10^{-4} \leq y \leq 10^{-1}$ with $n \approx 3$–4. Sometimes, the concentration range between 10^{-4} and 10^{-2} is characterized by a law $\sigma \sim y^{1/2}$ to be expected in the weak doping regime of a semiconductor, and only above a concentration of $y = 10^{-2}$ does the conductivity increase steeply. For other dopants, a sharp rise in conductivity is observed in a small concentration range (177). In Fig. 7.24, results obtained with AsF_5 and I_2 illustrate the differences of the $\sigma(y)$ characteristics. The highest conductivities achieved at nominally the same doping levels vary also, for iodine doping between 10 and 1.5×10^3 Ω^{-1} cm^{-1} and for AsF_5 doping between 10^2 and 2.5×10^3 Ω^{-1} cm^{-1} at room temperature. Therefore, any differences in physical quantities such as mobility which have been derived from conductivity data and which have been attributed to the various dopants have to be treated with caution.

For all dopants, the constants ΔE and σ_0 derived from $\sigma = \sigma_0 \, e^{-\Delta E/kT}$ in a limited temperature range close to room temperature change rapidly in going through the transition (177, 198), although the values of σ_0 change from sample to sample. Thus, the preexponential factor σ_0 increases from 0.8 to 20 Ω^{-1} cm^{-1} for an iodine-doped sample with $\sigma_{max} = 10$ Ω^{-1} cm^{-1}, whereas σ_0 increases by

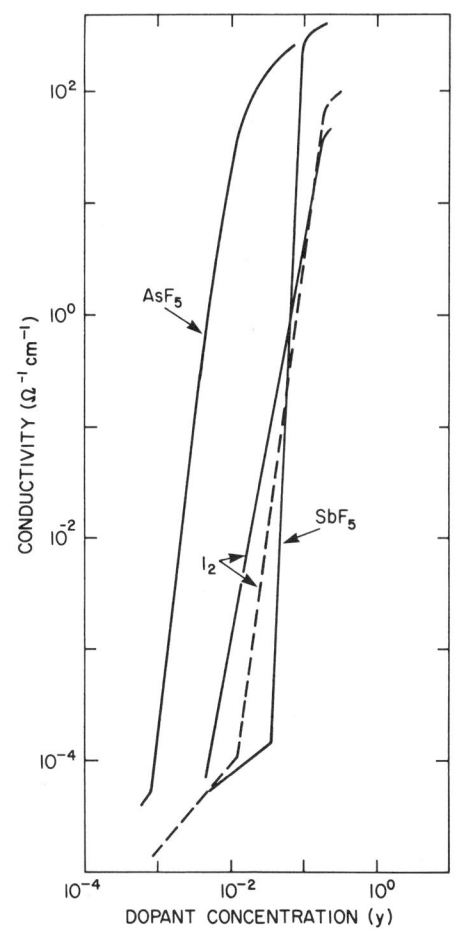

Figure 7.24 Direct-current conductivity of *trans*-$(CH)_x$ as function of concentration for various dopants. In the case of iodine, the concentration refers to the number of I atoms per CH group. The curve with AsF_5 has been taken from Park et al. [177].

the same factor of 20 to 30 but from 8 to 200 Ω^{-1} cm^{-1} in another sample with σ_{max} = 200 Ω^{-1} cm^{-1}. The drop in activation energy from 0.3 to 0.02 eV always seems to occur in the concentration range between 10^{-3} and about 2×10^{-2} and to be practically independent of the dopant used. Differences between I_2- and AsF$_5$-doped samples exist at the highest concentration levels of about $y \approx 10^{-1}$: The conductivity of [CH(AsF$_5$)$_{0.1}$]$_x$ shows a maximum at 250 K, being activated only below this temperature, whereas σ of I_2-doped samples remains activated over the whole temperature range.

The thermoelectric power shows for both dopants, I_2 (177, 207) as well as AsF$_5$ (182), a strong drop in the concentration range $10^{-3} \leq y \leq 10^{-2}$ in which ΔE and σ_0 were also found to change rapidly. The thermopower drops in this concentration range by about two orders of magnitude from about 10^3 μV/K. The data thus indicate a change in transport mechanism at $y \approx 10^{-2}$. Above this concentration, only a small decrease in the thermopower with increasing y is observed. At the highest doping level, the temperature dependence of the thermopower is markedly different for AsF$_5$-doped samples: that of [CH(AsF$_5$)$_{0.1}$]$_x$ is strictly linear with temperature, whereas that for [CH(I$_3$)$_{0.7}$]$_x$ is curved. However, in both cases the thermopower decreases smoothly to zero with $T \rightarrow 0$. Between $10^{-2} \leq y < 10^{-1}$, the thermopower of AsF$_5$-doped (CH)$_x$ as a function of temperature is analogous to that of the iodine-doped samples, and the conductivity is also slightly activated. The electrical measurements seem to indicate, therefore, that the differences between I_2- and AsF$_5$-doped samples are rather minor and might be associated with a different charge transfer between (CH)$_x$ and the dopant.

At first sight, the optical spectra show a pronounced species dependence in the visible. In the case of I_2 doping, the main point to note is that even at the highest doping level (and independent of the dc conductivities ($\sigma \leq 500$ Ω^{-1} cm^{-1}, obtained at the highest doping), the fundamental gap is still observable (208). Figure 7.25 shows the optical conductivity $\sigma = c\epsilon_0 nK$ of I_2-doped samples over a wide energy range where the intrinsic absorption is still visible as a separate feature. In (CH)$_x$ doped with AsF$_5$ or BF$_3$ (162), however, the bandgap is no longer visible in the highly conductive state. The optical properties in this energy range are those of a metal. The persistence or disappearance of the optical gap shows a certain parallelism with the observations made in the electrical measurements. AsF$_5$-doped samples show, at least above 250 K, metallic behavior consistently in the optical and electrical properties, whereas I_2-doped (CH)$_x$ does not.

The IR spectra are characterized by some features which are common to all dopants. We see in Fig. 7.25 that the optical conductivity of I_2-doped (CH)$_x$ has a maximum at about 0.5 eV. We see further that σ decreases with decreasing photon energy down to 0.005 eV equivalent to 40 cm^{-1}. The value of 0.005 eV is still about one order of magnitude higher than the dc value. The continuing decrease of σ contrasts with the expected Drude behavior of a free carrier system in the metallic state. A maximum in absorbance at about 0.5 eV is common to

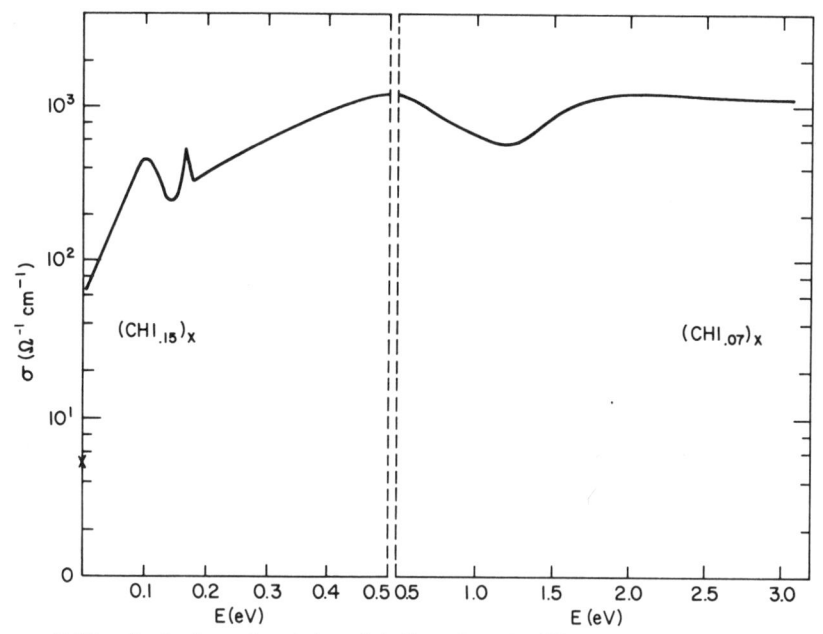

Figure 7.25 Optical conductivity of iodine-doped $(CH)_x$. Note the two different energy scales.

all dopants investigated so far, the position of which varies with concentration by 0.2 to 0.3 eV. Also, a decrease in absorbance with decreasing photon energy has been observed quite generally. Thus, it seems that the Drude behavior is always absent in the IR, though a definite answer concerning this question can only be given if in addition to the absorption data the reflectivity in the pertinent wavelength range is also known.

The dopant-induced IR bands at 1400 and 900 cm^{-1} (see Section 7.4.2) clearly persist up to the highest doping levels, independent of the species. This can be seen for iodine in Fig. 7.25, but also for SbF$_5$ (209) and AsF$_5$ (28) they stay unshifted in frequency and, if at all, are only moderately broadened. Inhomogeneity of the samples due to highly and weakly doped regions cannot play a dominant role since otherwise the gap would also remain visible in the case of AsF$_5$ or SbF$_5$ doped samples.

The frequency range between the IR and dc has not yet been covered completely, and measurements of the ac conductivity exist only up to 10 GHz. It has been reported (179, 210) that the conductivity of highly doped samples is independent of frequency up to 10 GHz and thus equal to the dc value.

The magnetic susceptibility of heavily doped polyacetylene ($y \geq 0.1$) is generally found to be temperature independent, with a value of about 4×10^{-6} emu/mol (137, 211, 212). This Pauli-type susceptibility is indicative of metallic behavior and a density of states of 0.1–0.2 (eV)$^{-1}$. Below $y = 0.1$, the magnetic susceptibility is characterized by two sudden changes as a function of AsF_5 concentration according to Ref. 146. In the concentration range between 10^{-4} and 10^{-3}, the Curie-law spins drop from 4×10^{-4} spins/C-atom to unmeasurably small values remaining low at higher concentrations. For $5 \times 10^{-2} < y < 10^{-1}$, the Pauli susceptibility increases sharply from very small values of $\chi_p \leq 10^{-7}$ emu/mol to 3×10^{-6} emu/mol. Above $y \approx 10^{-1}$, χ_p is independent of concentration. Thus the drop in Curie-law spins occurs before the transition from the semiconducting to the highly conducting state, whereas a Pauli susceptibility indicating a metallic state is only observed above this transition. Similarly, no Knight shift (196) is detected below $y \approx 7 \times 10^{-2}$, and only above this critical concentration is a shift of about 30 ppm and strong line broadening observed. In contrast to these results are those of Ref. 144, where linear increase of the Pauli susceptibility is presented for samples doped with AsF_5 in the concentration range $3 \times 10^{-3} \leq y \leq 4 \times 10^{-2}$. The linear extrapolation of these data to $y = 10^{-1}$, however, yields a value for χ_p which agrees with that given by other authors.

Two models have been discussed mainly to interpret the semiconductor–metal transition in polyacetylene, the metallic droplet model, and the soliton model. In the metallic droplet model, it is assumed that highly conducting regions of a size of about 60 Å are already formed at the lowest doping levels, without specifying the mechanism that leads to metallic behavior within these droplets (137, 144). Doping proceeds by increasing the size or number of the metallic regions and the transition takes place at the percolation threshold. This model, which is heavily based on those magnetic measurements that yield a Pauli susceptibility already at low doping levels and a linear increase in the susceptibility with doping, explains rather naturally the nonlinear electric transport at high electric fields (180). It disagrees, however, with magnetic susceptibility measurements by other groups (137) and the NMR results (196). These experiments suggest that an intermediate phase exists between $y_1 \approx 0.01$ and $y_2 \approx 0.05$ where the conductivity is metallic, but no Pauli susceptibility is observed.

Within the soliton model, this intermediate phase is described as a "soliton liquid." For $y < y_1$, charged solitons created upon doping are strongly bound to the impurities and form a "soliton glass" which melts at y_1 due to "cooperative screening and quantum effects" (99). The depinning of the charged but spinless solitons leads to the increase in conductivity at y_1 without changing the magnetic susceptibility. Finally, at y_2 the electronic gaps are filled due to a disorder-induced quenching of the Peierls distortion (118), leading to a finite density of states at the Fermi level and consequently to a Pauli susceptibility. Additional support for the soliton model comes from the observed maximum in the optical absorption around 0.5 eV and the characteristic modes in the IR. However, the fact that these same features are seen up to the highest doping level where it is hard to imagine that solitons should still exist is difficult to explain within this model.

Whereas it seems impossible to decide at present whether the metallic droplet model or the soliton model is more appropriate to describe the semiconductor–metal transition, mainly since contradictory results have been reported for the magnetic susceptibility, the situation seems to be more clear with regard to the highly conducting state, $y \geqslant y_2$. Not only the values for the electrical conductivity, but also the linear temperature dependence of the thermopower, the Pauli susceptibility, and the Drude behavior of the optical absorption spectrum above approximately 0.2 eV are characteristic for a metallic state. Because of the amorphous structure and the heavy doping, disorder will play a major role in the charge transport. Fluctuation-induced tunneling between well-conducting regions as proposed by Sheng (213) reproduces quite well the temperature dependence of the conductivity. On the other hand, localization, enhanced by the one-dimensional nature of $(CH)_x$ chains, together with strong electron–phonon coupling, could lead to polaronic effects. Polarons might in fact be the origin of the maximum in the optical absorption as discussed many years ago in a different context (214, 215).

7.6 OTHER CONDUCTING POLYMERS

7.6.1 Undoped Polymers

Considering the question of which properties an organic polymeric material must have in order to be highly conductive, it is generally agreed that a polymer chain should have an extended π-electron system requiring a sufficiently high degree of polymerization and planarity. Furthermore, the interchain distance should be small enough to allow the transfer of the charges from one chain to a neighboring one.

Systematic investigations have been made of the conductivity as a function of the number of monomer repeat units showing an increase in conductivity with the number of, for example, phenylene rings in the case of polyphenylenes. However, it was also found that the preparation conditions could shift the conductivity by several orders of magnitude (216) so that the question as to the influence of the number of repeat units on conductivity is not clearly answered. The influence of the variation of the chemical structure on the electrical conductivity has also been investigated (217). It was found that exchanging phenylene units by thiophene diyl units in polyarylene vinylenes gave a substantial increase in conductivity from 10^{-14} to 10^{-7} Ω^{-1} cm^{-1}. A decrease of the conductivity from 10^{-14} to 10^{-20} Ω^{-1} cm^{-1} was observed if the phenylene units were replaced by thiadiazole diyl units (Table 7.4). These results suggest that heteroaromatic electron excess (deficient) units in the chain may increase (lower) the conductivity. However, since one has also to expect that in these conductivity ranges σ is very sensitive to traces of accidentally introduced donors or acceptors, the observed change in conductivity may not be directly related to the changes in the

Table 7.4 Influence of Chemical Structure of Various Poly(arylene vinylenes) on Conductivity[a]. (From Kossmehl [217].)

Structure	Conductivity
$[-C_6H_4-CH=CH-C_6H_4-CH=CH-]_n$ (phenylene–vinylene)	$\sigma \approx 2 \cdot 10^{-14}\ (\Omega^{-1}\text{cm}^{-1})$
$[-C_6H_4-CH=CH-C_4H_2S-CH=CH-]_n$ (phenylene–vinylene–thiophene–vinylene)	$\sigma \approx 3 \cdot 10^{-7}\ (\Omega^{-1}\text{cm}^{-1})$
$[-C_4H_2S-CH=CH-C_4H_2S-CH=CH-]_n$ (thiophene–vinylene)	$\sigma \approx 4 \cdot 10^{-8}\ (\Omega^{-1}\text{cm}^{-1})$
$[-C_6H_4-CH=CH-C_3HN_2S-CH=CH-]_n$ (phenylene–vinylene–thiadiazole–vinylene)	$\sigma \approx 10 \cdot 10^{-19}\ (\Omega^{-1}\text{cm}^{-1})$

π-electron system of the polymeric molecule. Therefore it seems to us that in addition to this pioneering work, additional careful studies are required to allow clear-cut conclusions to be made.

7.6.2 Doped Aromatic Polymers

Polymers which have been made highly conducting by doping (pyro-polymers excluded) are poly(p-phenylene) (PPP) (218), poly(p-phenylene vinylene) (PPV) (219), poly(p-phenylene sulfide) (PPS) (220–222), polypyrrole (PYR) (223), and polythiophene (224). The essential molecular building units are either the six-membered phenylene rings or the five-membered hetero rings which are linked together directly or indirectly by a C_2H_2 group and an S atom in the case of PPV and PPS, respectively.

As discussed above, it may be concluded that due to the existence of π-electrons in the molecules forming the polymer building unit, the conductivity of the infinite chain could be high. However, if the resonance integrals between the atoms linking the rings are small, then the π-electrons will tend to become localized on their molecular sites. Correspondingly, the optical absorption spectrum does not differ significantly from that of the monomeric unit and the optical gap should not shift if the number of molecular units is increased. In the case of the oligomers of the phenylenes, a red shift is observable which is however small compared to those of polyenes and polyacenes (Fig. 7.26). Thus the delocalization of the electrons is expected to be smaller for PPP than for $(CH)_x$. The bond connecting the phenyls is extremely long (~ 1.5 Å), which suggests a low bond order and little electron delocalization in agreement with the above conclusion.

The optical absorption edge of PPS, also shown in Fig. 7.26, indicates that

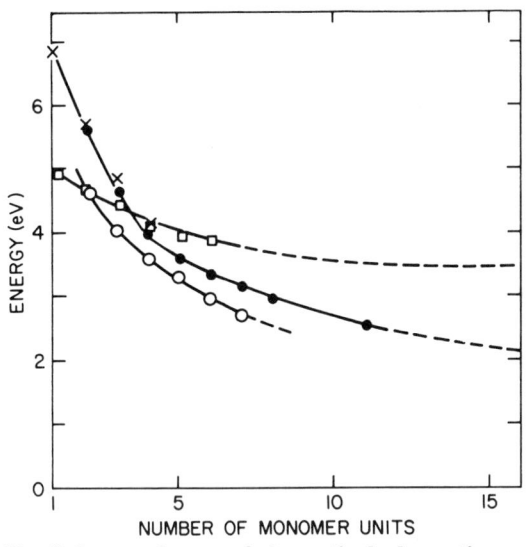

Figure 7.26 Shift of the maximum of the optical absorption as function of the number of monomer units for different types of polymers: poly-p-phenylene (□); polyacene (O) polyenes (•).

there is practically no delocalization of the electrons along the chain for PPS. This is not unexpected, since it is also well known that the PPS forms a "zigzag" chain and in addition that the phenylene rings are not coplanar (225). Similar optical and structural information is missing for polypyrrole and polythiophene.

The electrical conductivity of these polymers in the undoped state is low and in general correlates with their optical gaps. The conductivities range between 10^{-10} and 10^{-16} Ω^{-1} cm^{-1}. Upon doping with electron-accepting or donating species, the conductivity of these polymers can be raised by many orders of magnitude to values between 1 and 10^3 Ω^{-1} cm^{-1}. The conductivity of the "metallic" state decreases with decreasing temperature typically by a factor of 2 from room temperature to 20 K, except for PPS for which σ decreases much stronger by a factor of 10^3–10^4 between 300 and 100 K. If the conductivity is plotted against $T^{-1/4}$, then straight lines are often obtained. It is likely that this behavior is indicative of some localization effects arising from disorder. The thermoelectric power measured for PYR and PPS gave positive values in correspondence with the electron-accepting property of the dopants used. The magnitude of the thermoelectric power was two to three orders of magnitude smaller than that typically observed for semiconductors.

The optical properties in the visible and infrared change dramatically upon doping. For those polymers investigated, a general pattern seems to exist for the change of their optical transmission. The dopant induces an absorption peak below the intrinsic absorption edge and the optical gap remains visible even in the highly conducting state, similar to what is observed in (CH)$_x$ doped with iodine (Fig. 7.27). A behaviour typical for a Drude electron gas is not observed.

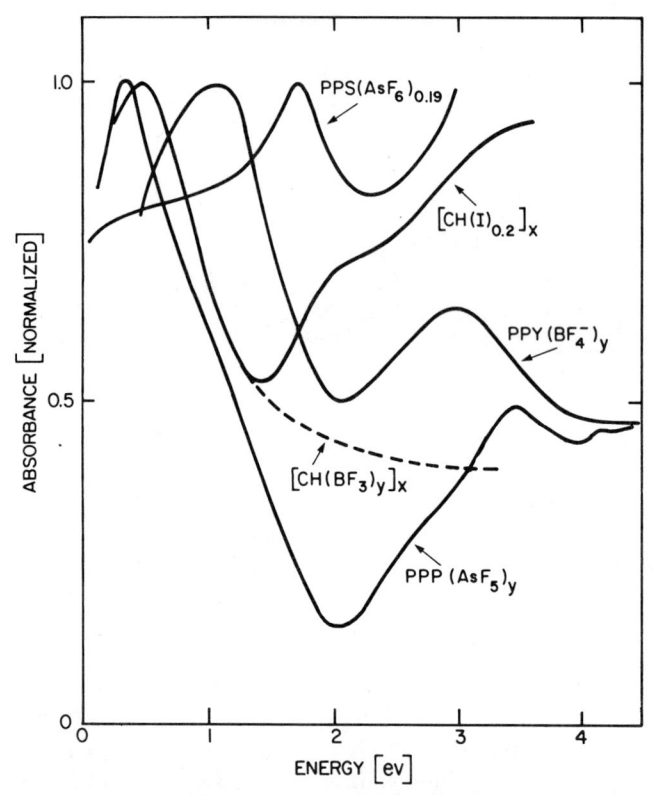

Figure 7.27 Optical absorbance of various doped polymers: poly-p-phenylene (PPP) (From Shacklette et al. [218]); polypyrrole (PPY) (From Kanazawa et al. [223]); poly(p-phenylene sulfide) (PPS) (From Clarke et al. [220].)

Upon doping, the infrared spectrum of PPS shows new lines and a change in the relative intensity of lines characteristic for undoped PPS. The changes in intensity of the IR lines correlate well with x-ray results which indicate a strong disorder introduced by the doping. The IR spectra of heavily doped PPP and PPV are featureless and the lines characteristic of the polymer are no longer visible.

Both PPP and PPV lose their high conductivity upon exposure to NH_3. In the case of PPP, the original IR spectrum of the undoped material is recovered after exposure to NH_3. It has therefore been concluded that the enhanced conductivity is due to the formation of a charge-transfer complex between dopant and polymer.

7.7 CONCLUDING REMARKS

The great effort to explore the electrical and related properties of polyacetylene has been stimulated by both its simple chemical structure and the discovery that metallic conductivity can be achieved by doping this material. Thus, it is not surprising that many more data are available for $(CH)_x$ than for other conducting polymers. These latter materials will certainly receive more attention in the future since they behave similarly to $(CH)_x$ in many respects, suggesting that concepts developed for polyacetylene may be generalized to include a much wider class of polymers. Furthermore the comparison with other types of materials like doped semiconductors, amorphous semiconductors and amorphous metals could yield new insight into the basic phenomena of charge transport in disordered materials.

In addition to the use of $(CH)_x$ in experimental solar cells (226), it has been shown that photoelectrochemical cells (227) and lightweight rechargeable batteries (228) can be fabricated. Organic batteries based on $(CH)_x$ and related materials, in particular, appear to hold considerable technological promise capitalizing as they do on the combined electronic and mechanical properties of polymeric systems.

Despite the enormous amount of experimental data available for $(CH)_x$ and a growing number of theoretical studies, there are still many open questions, especially those concerning the semiconductor–metal transition, details of electrical transport and magnetic susceptibility, or the morphology and structure of polyacetylene. These problems are partly due to experimental difficulties leading often to conflicting results and partly due to the lack of an established theoretical framework to interpret the data. It appears therefore that an improvement in the material and in the doping procedures and carefully designed experiments are required. In addition, thorough theoretical studies are needed in order to judge the relative importance of one-dimensional fluctuations versus interchain coupling, electron–phonon coupling versus correlation, and structural disorder versus dopant-induced defects.

ACKNOWLEDGMENTS

The authors have benefited from numerous discussions with many people, in particular with A. R. Bishop, A. J. Heeger, G. Lieser, M. Nechtschein, M. Peo, W. Rehwald, S. Roth, S. Strässler, Y. Tomkiewicz, and G. Wegner.

REFERENCES

1. H. Inokuchi and H. Akamatu, in *Solid State Physics,* F. Seitz and D. Turnbull, Eds., Academic, New York, 1955, Vol. 12, p. 93.
2. A. A. Berlin, *J. Polym. Sci.,* **55,** 621 (1961).
3. H. A. Pohl, in *Modern Aspects of the Vitreous State,* J. D. Mackenzie, Ed., Butterworth, London, 1962, p. 72; A. Rembaum, J. Moacanin, and H. A. Pohl, in *Progress in Dielectrics,* J. B. Birks, Ed., Temple, London, 1965, Vol. 6, p. 41.
4. H. A. Pohl and R. P. Chartoff, *J. Polym. Sci. A,* **2,** 2787 (1964).

5   W. A. Little, *Phys. Rev.*, **134,** A1416 (1964); see also the paper by Gutfreund and Little in Ref. 6.
6   J. T. Devreese, R. P. Evrard, and V. E. van Doren, Eds., *Highly Conducting One-Dimensional Solids,* Plenum, New York, 1979.
7   R. L. Greene, G. B. Street, and L. J. Suter, *Phys. Rev. Lett.*, **34,** 577 (1975); R. L. Greene and G. B. Street, in *Chemistry and Physics of One-Dimensional Metals,* H. J. Keller, Ed., Plenum, New York, 1977, p. 167.
8   H. Kuhn, *J. Chem. Phys.*, **16,** 840 (1948); **17,** 1198 (1949).
9   L. Salem, *The Molecular Orbital Theory of Conjugated Systems,* Benjamin, London, 1966.
10 A. A. Ovchinnikov, I. I. Ukrainskii, and G. V. Kventsel, *Usp. Fiz. Nauk,* **108,** 81 (1972); [*Sov. Phys. Usp.*, **15,** 575 (1973)].
11 T. Ito, H. Shirakawa and S. Ikeda, *J. Polym. Sci. Polym. Chem. Ed.*, **12,** 11 (1974).
12 C. K. Chiang, C. R. Fincher, Y. W. Park, A. J. Heeger, H. Shirakawa, E. J. Louis, S. C. Gau, and A. G. MacDiarmid, *Phys. Rev. Lett.*, **39,** 1098 (1977).
13 G. Natta, G. Mazzanti, and P. Corradini, *Accad. Nazl. Lincei, Sci. Fis. Mat. Nat.*, **25,** 3 (1958).
14 W. H. Watson, W. C. McMordie, and L. G. Lands, *J. Polym. Sci.*, **55,** 137 (1961).
15 L. S. Meriwether, E. C. Coltrup, G. W. Kennerly, and R. N. Reusch, *J. Org. Chem.*, **26,** 5155 (1961).
16 L. B. Luttinger, *J. Org. Chem.*, **27,** 1591 (1962).
17 H. Noguchi and S. Kambara, *J. Polym. Sci. B,* **1,** 553 (1963).
18 W. E. Daniels, *J. Org. Chem.*, **29,** 2963 (1964).
19 F. Ciardelli, E. Benedetti, and O. Pieroni, *Makromol. Chem.*, **103,** 1 (1967).
20 P. S. Woon and M. F. Farona, *J. Polym. Sci. Polym. Chem. Ed.*, **12,** 1749 (1974).
21 D. Braun and W. Quarg, *Angew. Makromol. Chem.*, **29/30,** 163 (1973).
22 T. Ogawa, R. Cedeno, and E. T. Herrera, *Makromol. Chem.*, **180,** 785 (1979).
23 E. H. Land, *J. Opt. Soc. Am.*, **12,** 957 (1951).
24 J. H. Edwards and W. J. Feast, *Polym. Comm.*, **21,** 595 (1980).
25 T. Ito, H. Shirakawa, and S. Ikeda, *J. Polym. Sci. Polym. Chem. Ed.*, **13,** 1943 (1975).
26 H. Shirakawa and S. Ikeda, *Polym. J.*, **2,** 231 (1971).
27 H. Kuzmany, *Phys. Status Solidi B,* **97,** 521 (1980).
28 J. F. Rabolt, T. C. Clarke, and G. B. Street, *J. Chem. Phys.*, **71,** 4614 (1979).
29 F. Inagaki, M. Tasumi, and T. Miyazawa, *J. Raman Spectrosc.*, **3,** 335 (1975).
30 J. Harada, Y. Furukawa, and M. Tasumi, *J. Chem. Phys.*, **73,** 4746 (1980).
31 S. Lefrant, L. S. Lichtmann, H. Temkin, and D. B. Fitchen, *Solid State Comm.*, **29,** 191 (1979); L. S. Lichtmann and D. B. Fitchen, *Synth. Met.*, **1,** 139 (1980).
32 G. Lieser, G. Wegner, W. Müller, and V. Enkelmann, *Makromol. Chem. Rapid Commun.*, **1,** 621 (1980).

33 U. Soga, S. Kawakami, H. Shirakawa, and S. Ikeda, *Makromol. Chem. Rapid Commun.*, **1**, 523 (1980).
34 T. Ito, H. Shirakawa, and S. Ikeda, *Kobunshi Ronbunshu*, **33**, 339 (1976).
35 H. Shirakawa and S. Ikeda, *Synth. Met.*, **1**, 175 (1980).
36 G. E. Wnek, J. C. W. Chien, F. E. Karasz, M. A. Druy, Y. W. Park, A. G. MacDiarmid, and A. J. Heeger, *J. Polym. Sci. Polym. Lett. Ed.*, **17**, 779 (1979).
37 F. E. Karasz, J. C. W. Chien, R. Galkiewicz, G. E. Wnek, A. J. Heeger, and A. G. MacDiarmid, *Nature*, **282**, 286 (1979).
38 S. Ikeda and H. Shirakawa, *Preprints of Short Communications, 26th International Symposium on Macromolecules*, I. Lüderwald and R. Weis, Eds., Mainz, 1979, p. 1479.
39 Y. W. Park, M. A. Druy, C. K. Chiang, A. G. MacDiarmid, A. J. Heeger, H. Shirakawa, and S. Ikeda, *J. Polym. Sci. Polym. Lett. Ed.*, **17**, 195 (1979).
40 C. R. Fincher, Jr., D. L. Peebles, A. J. Heeger, M. A. Druy, Y. Matsumura, A. G. MacDiarmid, H. Shirakawa, and S. Ikeda, *Solid State Comm.*, **27**, 489 (1978).
41 M. A. Druy, C. H. Tsang, N. Brown, A. J. Heeger, and A. G. MacDiarmid, *J. Polym. Sci. Polym. Phys. Ed.*, **18**, 429 (1980).
42 W. H. Meyer, *Synth. Met.*, **4**, 81 (1981).
43 R. H. Baughman, S. L. Hsu, G. P. Pez, and A. J. Signorelli, *J. Chem. Phys.*, **68**, 5405 (1978).
44 R. H. Baughman, S. L. Hsu, L. R. Anderson, G. P. Pez, and A. J. Signorelli, in *Molecular Metals*, W. E. Hatfield, Ed., Plenum, New York, 1979, p. 187.
45 G. Lieser, G. Wegner, W. Müller, V. Enkelmann, and W. H. Meyer, *Makromol. Chem. Rapid Commun.*, **1**, 627 (1980).
46 C. R. Fincher, C. E. Chen, A. J. Heeger, A. G. MacDiarmid, and J. B. Hastings, *Phys. Rev. Lett.*, **48**, 100 (1982).
47 K. Shimamura, F. E. Karasz, J. A. Hirsch, and J. C. W. Chien, *Makromol. Chem. Rapid Commun.* to appear (1981).
48 N. Mermilliod, L. Zuppiroli, and B. François, *J. Phys. (Paris)*, **41**, 1453 (1980).
49 P. R. Newman, M. D. Ewbank, C. D. Mauthe, M. R. Winkle, and W. D. Smolyncki, *Solid State Commun.*, **40**, 975 (1981).
50 K. Guckelsberger, P. Rödhammer, E. Gmelin, M. Peo, K. Menke, J. Hocker, S. Roth, and K. Dransfeld, *Z. Phys. B*, **43**, 189 (1981).
51 D. Moses, A. Denenstein, A. Pron, A. J. Heeger, and A. G. MacDiarmid, *Solid State Commun.*, **36**, 216 (1980).
52 J. E. Lennard-Jones, *Proc. Roy. Soc. Lond. A*, **158**, 280 (1937).
53 C. A. Coulson, *Proc. Roy. Soc. Lond. A*, **169**, 413 (1939).
54 R. F. Peierls, *Quantum Theory of Solids*, Clarendon, Oxford, 1955, p. 108.
55 E. Wigner, *Trans. Farad. Soc. Lond.*, **34**, 678 (1938).
56 E. Hückel, *Z. Phys.*, **70**, 204 (1931); **76**, 628 (1932).
57 R. Pariser and R. G. Parr, *J. Chem. Phys.*, **21**, 466, 767 (1953).
58 J. A. Pople, *Trans. Farad. Soc. Lond.*, **49**, 1375 (1953).
59 W. J. Hehre, R. F. Steward, and A. J. Pople, *J. Chem. Phys.*, **51**, 2657 (1969).

60 C.C.J. Roothaan, *Rev. Mod. Phys.*, **23**, 61 (1951).
61 G. Del Re, J. Ladik, and G. Biczó, *Phys. Rev.*, **155**, 997 (1967).
62 J.-M. André and G. Leroy, *Int. J. Quantum Chem.*, **5**, 557 (1971).
63 M. Kertész, J. Koller, and A. Azman, *J. Chem. Phys.*, **67**, 1180 (1977).
64 A. Karpfen and J. Petkov, *Solid State Comm.*, **29**, 251 (1979); *Theor. Chim. Acta*, **53**, 75 (1979).
65 S. Suhai, *J. Chem. Phys.*, **73**, 3843 (1980).
66 W. B. Fowler, *Phys. Rev.*, **151**, 657 (1966).
67 P. M. Grant and I. P. Batra, *Solid State Comm.*, **29**, 225 (1979).
68 A. Karpfen and R. Höller, *Solid State Comm.*, **37**, 179 (1981).
69 W. P. Su, J. R. Schrieffer, and A. J. Heeger, *Phys. Rev. Lett.*, **42**, 1698 (1979).
70 W. P. Su and J. R. Schrieffer, *Proc. Natl. Acad. Sci. USA*, **77**, 5626 (1980).
71 D. Vanderbilt and E. J. Mele, *Phys. Rev. B*, **22**, 3939 (1980).
72 T. Kakitani, *Prog. Theor. Phys.*, **51**, 656 (1974).
73 C. A. Coulson, *Proc. Roy. Soc. Lond. A*, **164**, 383 (1938).
74 B. Hudson and B. Kohler, *Ann. Rev. Phys. Chem.*, **25**, 437 (1974).
75 J. R. Platt, *J. Chem. Phys.*, **25**, 80 (1956).
76 H. Labhart, *J. Chem. Phys.*, **27**, 957 (1957).
77 Y. Ooshika, *J. Phys. Soc. Japan*, **12**, 1238 (1957).
78 H. C. Longuet-Higgins and L. Salem, *Proc. Roy. Soc. Lond. A*, **251**, 172 (1959).
79 H. Fröhlich, *Proc. Roy. Soc. Lond. A*, **215**, 291 (1952).
80 W. P. Su, J. R. Schrieffer, and A. J. Heeger, *Phys. Rev. B*, **22**, 2099 (1980).
81 E. J. Mele and M. J. Rice, *Phys. Rev. Lett.*, **45**, 926 (1980).
82 S. A. Brazovskii and I. E. Dzyaloshinskii, *Zh. Eksp. Teor. Fiz.*, **71**, 2338 (1976) [*Sov. Phys. JETP*, **44**, 1233 (1977)].
83 J.C.J. Bart and C.H. MacGillavry, *Acta Crystallogr. B*, **24**, 1569 (1968).
84 Y. Ooshika, *J. Phys. Soc. Japan*, **14**, 747 (1959).
85 J. A. Pople and S. H. Walmsley, *Mol. Phys.*, **5**, 15 (1962).
86 G. V. Kventsel and Y. A. Kruglyak, *Theor. Chim. Acta*, **12**, 1 (1968).
87 I. B. Goldberg, H. R. Crowe, P. R. Newman, A. J. Heeger, and A. G. MacDiarmid, *J. Chem. Phys.*, **70**, 1132 (1979).
88 M. J. Rice, *Phys. Lett.*, **71A**, 152 (1979).
89 S. A. Brazovskii, *Pis'ma Zh. Eksp. Teor. Fiz.*, **28**, 656 (1978) [*Sov. Phys. JETP Lett.*, **28**, 606 (1979)]; *Zh. Eksp. Teor. Fiz.*, **78**, 677 (1980) [*Sov. Phys. JETP*, **51**, 342 (1980)].
90 H. Takayama, Y. R. Lin-Liu, and K. Maki, *Phys. Rev. B*, **21**, 2388 (1980).
91 W. P. Su, *Solid State Comm.*, **35**, 899 (1980).
92 A. R. Bishop, in *Solitons and Condensed Matter Physics*, A. R. Bishop and T. Schneider, Eds., Springer Series in Solid-State Sciences, Berlin, 1978, Vol. 8, p. 85.
93 A. C. Scott, F. Y. F. Chu, and D. W. McLaughlin, *Proc. IEEE*, **61**, 1443 (1973).
94 J. T. Gammel and J. A. Krumhansl, *Phys. Rev. B*, **24**, 1035 (1981).

REFERENCES

95 P. G. de Gennes, *Superconductivity of Metals and Alloys*, Benjamin, New York, 1966, Chap. 5.
96 M. C. Cross and D. S. Fisher, *Phys. Rev. B*, **19**, 402 (1979).
97 J. Bar-Sagi and C. G. Kuper, *Phys. Rev. Lett.*, **28**, 1556 (1972).
98 L. D. Landau and E. M. Lifshitz, *Quantum Mechanics*, Pergamon, New York, 1958, p. 72.
99 W. P. Su, S. Kivelson, and J. R. Schrieffer, in *Physics in One Dimension*, J. Bernasconi and T. Schneider, Eds., Springer Series in Solid-State Sciences, Berlin 1981, Vol. 23, p. 201.
100 A. Kotani, *J. Phys. Soc. Japan*, **42**, 416 (1977).
101 M. J. Rice and J. Timonen, *Phys. Lett.*, **73A**, 368 (1979).
102 B. Horovitz, *Phys. Rev. Lett.*, **46**, 742 (1981).
103 S. T. Pantelides, D. J. Mikish, and A. B. Kunz, *Phys. Rev. B*, **10**, 2602 (1974).
104 I. Ohmine, M. Karplus, and K. Schulten, *J. Chem. Phys.*, **68**, 2298 (1978).
105 E. H. Lieb and F. Y. Wu, *Phys. Rev. Lett.*, **20**, 1445 (1968).
106 R. A. Harris and L. M. Falicov, *J. Chem. Phys.*, **51**, 5034 (1969).
107 J. Kondo, *Physica*, **98B**, 176 (1980).
108 T. Nakano and H. Fukuyama, *J. Phys. Soc. Japan*, **49**, 1679 (1980).
109 E. Pytte, *Phys. Rev. B*, **10**, 4637 (1974).
110 P. Horsch, *Phys. Rev. B*, **24**, 7351 (1981).
111 H. Barentzen, O. E. Polansky, N. Tyutyulkov, and C. I. Ivanov, *Z. Phys. B*, **38**, 211 (1980).
112 D. Baeriswyl and J. J. Forney, *J. Phys. C*, **13**, 3203 (1980).
113 N. Suzuki, M. Ozaki, S. Etemad, A. J. Heeger, and A. G. MacDiarmid, *Phys. Rev. Lett.*, **45**, 1209 (1980); Erratum in *Phys. Rev. Lett.*, **45**, 1463 (1980).
114 Shang-keng Ma, *Modern Theory of Critical Phenomena*, Benjamin, London, 1976, Chap. 10.
115 Y. R. Lin-Liu and K. Maki, *Phys. Rev. B*, **22**, 5754 (1980).
116 P. A. Lee, T. M. Rice, and P. W. Anderson, *Solid State Comm.*, **14**, 703 (1974).
117 H. Fukuyama and P. A. Lee, *Phys. Rev. B*, **17**, 535 (1978).
118 M. J. Rice and E. J. Mele, *Proceedings of the International Conference* on *Low Dimensional Synthetic Metals*, Helsingor, 1980, *Chem. Scripta 17*, 121 (1981); E. J. Mele and M. J. Rice, *Phys. Rev. B*, **23**, 5397 (1981).
119 R. V. Kasowski, E. Caruthers, and W. Y. Hsu, *Phys. Rev. Lett.*, **44**, 676 (1980).
120 J. Ladik and S. Suhai, in *Theoretical Chemistry*, Vol. 4, C. Thomson, Ed., Royal Society of Chemistry, London, 1981, p. 49.
121 J. L. Brédas, R. R. Chance, R. Silbey, G. Nicolas, and Ph. Durand, *J. Chem. Phys.*, **75**, 255 (1981).
122 L. Pietronero, S. Strässler, and P. Horsch, in Proceedings of the International Conference on Low-Dimensional Conductors, A. J. Epstein and E. M. Conwell, Eds., *Mol. Cryst. Liq. Cryst.*, in press.
123 M. Nakahara and K. Maki, preprint.
124 K. R. Subbaswamy and M. Grabowski, *Phys. Rev. B*, **24**, 2168 (1981).

125 M. J. Rice and A. J. Mele, *Solid State Commun.*, **35**, 487 (1980).
126 K. Maki, preprint.
127 G. W. Bryant and A. J. Glick, preprint.
128 D. Baeriswyl, preprint.
129 M. Kertész and P. R. Surján, *Solid State Commun.*, **39**, 611 (1981).
130 M. Nakahara and K. Maki, *Phys. Rev. B*, **24**, 1045 (1981).
131 S. A. Brazovskii and N. N. Kirova, *Pis'ma Zh. Eksp. Teor. Fiz.*, **33**, 4 (1981) [*Sov. Phys.-JETP Lett.*, **33**, 4 (1981)].
132 D. K. Campbell and A. R. Bishop, *Phys. Rev. B*, **24**, 4859 (1981).
133 For a short review on the various defect structures see D. Baeriswyl, in *Proceedings of the International Conference on the Physics of Intercalation Compounds*, Springer Series in Solid-State Sciences, L. Pietronero and E. Tosatti, Eds., Springer-Verlag, Berlin, 1981.
134 W. P. Su and J. R. Schrieffer, *Phys. Rev. Lett.*, **46**, 738 (1981).
135 J. Goldstone and F. Wilczek, *Phys. Rev. Lett.*, **47**, 986 (1981).
136 R. Jackiew and J. R. Schrieffer, preprint.
137 Y. Tomkiewicz, T. D. Schultz, H. B. Brom, T. C. Clarke, and G. B. Street, *Phys. Rev. Lett.*, **43**, 1532 (1979).
138 B. François, M. Bernard, and J. J. André, *J. Chem. Phys.*, **75**, 4142 (1981).
139 A. Snow, P. Brandt, D. Weber, and N. L. Yang, *J. Polym. Sci. Polym. Lett. Ed.*, **17**, 263 (1979).
140 B. R. Weinberger, E. Ehrenfreund, A. Pron, A. J. Heeger, and A. G. MacDiarmid, *J. Chem. Phys.*, **72**, 4749 (1980).
141 H. Shirakawa, T. Ito, and S. Ikeda, *Makromol. Chem.*, **179**, 1565 (1978).
142 M. Schwoerer, U. Lauterbach, W. Müller, and G. Wegner, *Chem. Phys. Lett.*, **69**, 359 (1980).
143 M. Nechtschein, F. Devreux, R. L. Greene, T. C. Clarke, and G. B. Street, *Phys. Rev. Lett.*, **44**, 356 (1980); F. Devreux, K. Holczer, M. Nechtschein, J. C. Clarke, and R. L. Greene, in *Physics in One Dimension*, J. Bernasconi and J. Schneider, Eds., Springer Series in Solid-State Sciences, Berlin, 1981, Vol. 23, p. 194.
144 Y. Tomkiewicz, N. S. Shiren, T. D. Schultz, K. Mortensen, M. L. Thewalt, J. D. Kuptsis, R. G. Schad, B. H. Robinson, T. C. Clarke, G. B. Street, H. Thomann, L. R. Dalton, and H. B. Brom, in *Physics in One Dimension*, J. Bernasconi and T. Schneider, Eds., Springer Series in Solid-State Sciences, Berlin, 1981, Vol. 23, p. 214; Y. Tomkiewicz, T. D. Schultz, H. B. Brom, A. R. Taranko, T. C. Clarke, and G. B. Street, *Phys. Rev. B*, **24**, 4348 (1981).
145 K. Holczer, J. P. Boucher, F. Devreux, and M. Nechtschein, *Phys. Rev. B*, **23**, 1051 (1981).
146 S. Ikehata, J. Kaufer, T. Woerner, A. Pron, M. A. Druy, A. Sivak, A. J. Heeger, and A. G. MacDiarmid, *Phys. Rev. Lett.*, **45**, 1123 (1980).
147 A. J. Heeger and A. G. MacDiarmid, in Proceedings of the International Conference on Low-Dimensional Conductors, A. J. Epstein and E. M. Conwell, Eds., *Mol. Cryst. Liq. Cryst.*, **77**, 1 (1981).
148 S. Etemad, A. J. Heeger, L. Lauchlan, T.-C. Chung, and A. G. MacDiarmid, in

REFERENCES

Proceedings of the International Conference on Low-Dimensional Conductors, A. J. Epstein and E. M. Conwell, Eds., *Mol. Cryst. Liq. Cryst.*, **77**, 43 (1981).

149 K. Kume, K. Mizumo, K. Mizoguchi, K. Nomura, J. Tanaka, M. Tanaka, and H. Fujimoto, in Proceedings of the International Conference on Low-Dimensional Conductors, A. J. Epstein and E. M. Conwell, Eds., *Mol. Cryst. Liq. Cryst.*, in press.

150 N. Kinoshita and M. Tokumoto, *J. Phys. Soc. Japan*, **50**, 2779 (1981).

151 K. Holczer, F. Devreux, M. Nechtschein, and J. P. Travers, *Solid State Commun.*, **39**, 881 (1981).

152 F. Masin, G. Gusman, and R. Deltour, *Solid State Commun.*, **40**, 415, 513 (1981).

153 S. Ikeda, M. Druy, T. Woerner, A. J. Heeger, and A. G. MacDiarmid, *Solid State Commun.*, **39**, 1239 (1981).

154 R. Fincher Jr., M. Ozaki, M. Tanaka, D. Peebles, L. Lauchlan, A. J. Heeger, and A. G. MacDiarmid, *Phys. Rev. B*, **20**, 1589 (1979).

155 G. Mihaly, G. Vancso, S. Pekker, and A. Janossy, *Synth. Met.*, **1**, 357 (1980).

156 D. L. Greenaway and G. Harbeke, *Optical Properties and Band Structure of Semiconductors,* Pergamon, Oxford, 1968.

157 J. C. Phillips, *Solid State Phys.*, **18**, 55 (1966).

158 T. Tani, W. D. Gill, P. M. Grant, T. C. Clarke, and G. B. Street, *Synth. Met.*, **1**, 301 (1980); T. Tani, P. M. Grant, W. D. Gill, G. B. Street, and T. C. Clarke, *Solid State Comm.*, **33**, 499 (1980).

159 L. Lauchlan, S. Etemad, T.-C. Chung, A. J. Heeger, and A. G. MacDiarmid, *Phys. Rev. B.*, **24**, 3701 (1981).

160 J. J. Ritsko, E. J. Mele, A. J. Heeger, A. G. MacDiarmid, and M. Ozaki, *Phys. Rev. Lett.*, **44**, 1351 (1980).

161 H. Shirakawa, T. Sasaki, and S. Ikeda, *Chem. Lett.*, 1113 (1978).

162 M. Tanaka, A. Watanabe, and J. Tanaka, *Bull. Chem. Soc. Japan*, **53**, 645 (1980).

163 H. G. Grimmeis, *Ann. Rev. Mat. Sci.*, **7**, 341 (1977).

164 C. R. Fincher, M. Ozaki, A. J. Heeger, and A. G. MacDiarmid, *Phys. Rev. B*, **19**, 4140 (1979).

165 S. Etemad, A. Pron, A. J. Heeger, A. G. MacDiarmid, E. J. Mele, and M. J. Rice, *Phys. Rev. B*, **23**, 5137 (1981).

166 B. Horovitz, in Proceedings of the International Conference on Low-Dimensional Conductors, A. J. Epstein and E. M. Conwell, Eds., *Mol. Cryst. Liq. Cryst.*, **77**, 285 (1981); *Phys. Rev. Lett.*, **47**, 1491 (1981).

167 E. J. Mele and M. J. Rice, *Phys. Rev. Lett.*, **47**, 1492 (1981).

168 K. Maki and M. Nakahara, *Phys. Rev. B*, **23**, 5005 (1981).

169 B. Horovitz, preprint.

170 M. Tanaka, A. Watanabe, and J. Tanaka, *Chem Scrip.*, **17**, 131 (1981).

171 H. Kiess, D. Baeriswyl, and G. Harbeke, in Proceedings of the International Conference on Low-Dimensional Conductors, A. J. Epstein and E. M. Conwell, Eds., *Mol. Cryst. Liq. Cryst.*, **77**, 147 (1981).

172. W. Götze, *Phil. Mag. B,* **43,** 219 (1981).
173. P. Prelovšek, *Phys. Rev. B,* **23,** 1304 (1981).
174. W. P. Su, in Proceedings of the International Conference on Low-Dimensional Conductors, A. J. Epstein and E. M. Conwell, Eds., *Mol. Cryst. Liq. Cryst.,* **77,** 265 (1981).
175. S. Etemad, T. Mitani, M. Ozaki, T.-C. Chung, A. J. Heeger, and A. G. MacDiarmid, *Phys. Rev. B,* **24,** 1 (1981); L. Lauchlan, S. Etemad, T.-C. Chung, A. J. Heeger, and A. G. MacDiarmid, *Phys. Rev. B,* **24,** 3701 (1981).
176. H. Kiess, R. Keller, D. Baeriswyl, and G. Harbeke, preprint.
177. Y. W. Park, A. J. Heeger, M. A. Druy, and A. G. MacDiarmid, *J. Chem. Phys.,* **73,** 946 (1980).
178. J. F. Kwak, T. C. Clarke, R. L. Greene, and G. B. Street, *Solid State Comm.,* **31,** 355 (1979).
179. A. J. Epstein, H. W. Gibson, P. M. Chaikin, W. G. Clark, and G. Grüner, *Phys. Rev. Lett.,* **45,** 1730 (1980).
180. Y. Tomkiewicz, K. Mortensen, M. C. Thewaldt, T. C. Clarke, and G. B. Street, *Phys. Rev. Lett.,* **45,** 490 (1980).
181. B. Abeles, *Appl. Solid State Sci.,* **6,** 1, (1976).
182. J. F. Kwak, W. D. Gill, R. L. Greene, K. Seeger, T. C. Clarke, and G. B. Street, *Synth. Met.,* **1,** 213 (1980).
183. K. Seeger, W. D. Gill, T. C. Clarke, and G. B. Street, *Solid State Comm.,* **281,** 873 (1978).
184. S. Kivelson, *Phys. Rev. Lett.,* **46,** 1344 (1981).
185. A. J. Epstein, H. Rommelmann, M. Abkowitz, and H. W. Gibson, in Proceedings of the International Conference on Low-Dimensional Conductors, A. J. Epstein and E. M. Conwell, Eds., *Mol. Cryst. Liq. Cryst.,* **77,** 81 (1981); *Phys. Rev. Lett.,* **47,** 1549 (1981).
186. S. Kivelson, in Proceedings of the International Conference on Low-Dimensional Conductors, A. J. Epstein and E. M. Conwell, Eds., *Mol. Cryst. Liq. Cryst.,* **77,** 65 (1981).
187. C. K. Chiang and A. D. Franklin, *Solid State Commun.,* **40,** 775 (1981).
188. M. Gamoudi, J. J. André, B. François, and M. Maitrot, to be published in *J. Phys. (Paris),* June 1982.
189. A. G. MacDiarmid and A. J. Heeger, *Synth. Met.,* **1,** 101 (1980).
190. R. J. Nigrey, A. G. MacDiarmid, and A. J. Heeger, *J. Chem. Soc. Chem. Comm.,* 594 (1979).
191. I. Harada, M. Tasumi, H. Shirakawa, and S. Ikeda, *Chem. Lett.,* 1411 (1978).
192. S. L. Hsu, A. J. Signorelli, G. P. Pez, and R. H. Baughman, *J. Chem. Phys.,* **69,** 106 (1978).
193. H. R. Thomas, W. R. Salaneck, C. B. Duke, E. W. Plummer, A. J. Heeger, and A. G. MacDiarmid, *Polymer,* **21,** 1238 (1980).
194. T. C. Clarke, R. H. Geiss, W. D. Gill, P. M. Grant, J. W. Macklin, H. Morawitz, J. F. Rabolt, G. B. Street, and D. J. Sayers, *Synth. Met.,* **1,** 21 (1979).
195. G. B. Street and T. C. Clarke, *ACS Adv. Chem.,* **186,** 177 (1980).

REFERENCES

196 M. Peo, H. Foerster, K. Menke, J. Hocker, J. A. Gardner, S. Roth, and K. Dransfeld, *Solid State Comm.*, **38**, 467 (1981).
197 G. B. Street and T. C. Clarke, *IBM J. Res. Develop.*, **25**, 51 (1981).
198 H. Kiess, W. Meyer, D. Baeriswyl, and G. Harbeke, *J. Elec. Mat.*, **9**, 763 (1980).
199 K. Seeger, W. Mayer, A. Philipp, and W. Röss, Proceedings of the International Conference on Low-Dimensional Synthetic Metals, Helsingor, 1980, Chem. Scripta, **17**, 129 (1981).
200 T. C. Clarke and J. C. Scott, *Solid State Commun.*, **41**, 389 (1982).
201 G. Kaindl, G. Wortmann, S. Roth, and K. Menke, *Solid State Commun.*, **41**, 75 (1982).
202 T. Matsuyama, H. Sakai, H. Yamaoka, Y. Maeda, and H. Shirakawa, *Solid State Commun.*, **40**, 563 (1981).
203 M. N. Alexander and D. F. Holcomb, *Rev. Mod. Phys.*, **40**, 815 (1968).
204 N. F. Mott, *Metal Insulator Transitions*, Taylor and Francis, London, 1974.
205 M. Tanenbaum and H. B. Briggs, *Phys. Rev.*, **91**, 1561 (1953).
206 H. Finkenrath, *Z. Phys.*, **159**, 112 (1960).
207 Y. W. Park, A. Denenstein, C. K. Chiang, A. J. Heeger, and A. G. MacDiarmid, *Solid State Comm.*, **31**, 355 (1979).
208 G. Harbeke, H. Kiess, W. Meyer, and D. Baeriswyl, *Proceedings of 15th International Conference* on *Physics* of *Semiconductors*, Kyoto, 1980, *J. Phys. Soc. Japan*, **49**, Suppl. 17, 856 (1980).
209 H. Kiess and G. Harbeke, unpublished results.
210 R. M. Grant and M. Krounbi, *Solid State Comm.*, **36**, 291 (1980).
211 R. B. Weinberger, J. Kaufer, A. J. Heeger, A. Pron, and A. G. MacDiarmid, *Phys. Rev. B*, **20**, 223 (1979).
212 P. Bernier, M. Rolland, M. Galtier, A. Montaner, M. Regis, M. Candille, C. Benoit, M. Aldissi, C. Linaya, F. Schué, J. Sledz, J. M. Fabre, and L. Giral, *J. Phys.* (Paris), **40**, L-297 (1979).
213 P. Sheng, *Phys. Rev. B*, **21**, 2180 (1979).
214 P. Gerthsen, E. Kauer, and H. G. Reik, *Festkörperprobleme*, **5**, 1 (1966).
215 I. G. Austin and N. F. Mott, *Adv. Phys.*, **18**, 41 (1969).
216 H. Naarmann, *Ber. Bunsenges. Phys. Chem.*, **83**, 427 (1979).
217 G. Kossmehl, *Ber. Bunsenges. Phys. Chem.*, **83**, 417 (1979).
218 L. W. Shacklette, R. C. Chance, D. M. Ivory, G. G. Miller, and R. H. Baughman, *Synth. Met.*, **1**, 307 (1980).
219 G. E. Wnek, J.C.W. Chien, F. E. Karasz, and C. P. Lillya, *Polym. Comm.*, **20**, 1441 (1979).
220 T. C. Clarke, K. K. Kanazawa, V. Y. Lee, J. F. Rabolt, J. R. Reynolds, and G. B. Street, *J. Polym. Sci. Polym. Phys. Ed.*, to appear.
221 J. F. Rabolt, T. C. Clarke, K. K. Kanazawa, J. R. Reynolds, and G. B. Street, *J. Chem. Soc. Chem. Comm.*, 347 (1980).
222 R. R. Chance, L. W. Shacklette, G. G. Miller, D. M. Ivory, J. M. Sowa, M. L. Elsenbaumer, and R. H. Baughman, *J. Chem. Soc. Chem. Comm.*, 348 (1980).

223 K. K. Kanazawa, A. F. Diaz, W. D. Gill, P. M. Grant, G. B. Street, G. P. Gardine, and J. F. Kwak, *Synth. Met.*, **1,** 329 (1980).
224 G. Kossmehl, private communication.
225 B. J. Tabor, E. P. Magré, and J. Boon, *Eur. Polym. J.*, **7,** 1127 (1971).
226 M. Ozaki, D. L. Peebles, B. R. Weinberger, C. K. Chiang, S. C. Gau, A. J. Heeger, and A. G. MacDiarmid, *Appl. Phys. Lett.*, **35,** 83 (1979).
227 S. N. Chen, A. J. Heeger, Z. Kiss, A. G. MacDiarmid, S. C. Gau, and D. L. Peebles, *Appl. Phys. Lett.*, **36,** 98 (1980).
228 D. MacInnes, Jr., M. A. Druy, P. J. Nigrey, D. R. Nairns, A. G. MacDiarmid, and A. J. Heeger, *Chem. Commun.*, in press; A. G. MacDiarmid and A. J. Heeger, in Proceedings of the International Conference on Low-Dimensional Conductors, A. J. Epstein and E. M. Conwell, Eds., *Mol. Cryst. Liq. Cryst.*, in press.

INDEX

α-phase, *see* Form II
α-relaxation, 71, 91, 92
ab initio calculations, 18, 21, 22, 277, 294
Absorption, *see* Optical properties
Acetylene polymerization, 270
Acronyms:
 polymers, 5, 314
 small molecules, 11
Activation energy:
 concentration dependence, 223
 electrical transport, 222, 226, 227, 236, 304, 309
 field dependence, 223
 hole transport, 222, 226, 227, 236, 261
 polarization, 95
 polaron, 223
 β-relaxation, 95
 trapping levels, 79
Affinities, electron, 165, 177
Amorphous:
 phase, 125, 241
 solids, 235, 239
Annihilation:
 biexcitonic, 168
 energy transfer, 172
 singlet-triplet, 193
 triplet-exciton, 196
 triplet-triplet, 193, 197
Antimony pentachloride, doped polycarbonate, 256
Applications of polymers, 2, 96, 148, 205, 217, 317
Aromatic polymers:
 electron affinities, 177
 ionization energies, 177
Arrhenius law, 73

β-phase, *see* Form I

β-relaxation, 71, 78, 91, 94, 95
Bandgap, 19, 20, 165, 268, 278, 281, 284, 288, 291, 311
Band structure, 15, 19, 20, 23, 35, 278, 281
Band theory limits, 23
Biexcitonic annihilation, 168
Blocking layer, 126
Bond:
 alternation, 19, 268, 271, 278, 280
 defects, 285, 289, 293
 length, 18, 19
 order, 280, 292
Born-Oppenheimer approximation, 277, 282
Bulk states, 49, 80

Carbazole polymers:
 charge generation, 246, 249
 electronic transport, 220, 236
 energy transfer, 187, 195
 see also individual listings
Carbyne:
 bandgap, width, 20
 bond length, 20
 tight binding model, 20
Carrier, *see individual charge topics*
Catalyst, 270, 275
 Luttinger, 275
 Ziegler, 270, 275
Cermet, 304
Chain, theory:
 even chains, 280
 finite length, 283
 odd chains, 285
Chain length, 271
Characteristic temperature, 226
Charge decay, 82, 83, 86
Charge density:
 box distribution, 74

327

depth, 69
distribution, 69, 80
polarization, 65
in polymers, 80
real, 64
spatial distribution, 81, 82
surface, 65
Charge generation:
efficiency, 247
field dependence, 250
Onsager mechanism, 247
photosensitization, 251
Charge injection, 47, 126
metal-polymer contact, 47
Charge measuring techniques, 64
capacitive probe, 67
dynamic capacitive probe, 67
e-beam sampling, 69, 82
isothermal discharge, 70
pressure pulse, 69, 82
thermal pulse, 67, 81
thermally stimulated current, 70
Charge storage, 3, 59
applications, 96
dipole charge, 60, 90
real charge, 60, 77
Charge transport, *see* Electronic transport
Charging techniques, 61, 121
charge deposition, 61
corona, 62, 121
electric beam, 63
liquid contact, 62
thermal methods, 61, 121
Clausius-Mossotti equation, 130
CNDO calculations, 18, 38
Concentration dependence:
energy transfer, 167, 171, 179, 189, 195
hopping activation energy, 223
hopping transport, 224, 227, 229, 257
photosensitization, 255
Conductivity, electrical, *see* Electronic transport
Conformational trap, 228, 246
Continuous-time random walk, 237
Continuum description, 287
Core level spectra, 31
Correlation effects, 20, 269, 278, 291, 294
Coulomb-trap, 233
energy, 49, 247
Critical concentration, 171, 174
Crystal orbital theory, 277
CTRW, *see* Continuous-time random walk

Defects, bond, 285, 289, 293
Dielectric response, 25, 37, 45, 300
Diffraction:
electron, 271
x-ray, 271, 274, 306
Dipole-dipole resonance, 170
Dipole polarization, 62, 90, 123
depolarization, 70
orientation model, 123
polarization techniques, *see* Charging techniques
relaxation distribution, 94
relaxation frequency, 72, 90
rigid dipole model, 130
thermal vibration, 130
Discharge:
emission-limited, 217
fractional, 94
space-charge limited, 217
thermally stimulated, 70
Disorder:
diagonal, 24, 223
hopping time distribution, 233
hopping transport, 233
off-diagonal, 24, 223
Dispersion:
hopping transport, 233
Monte-Carlo simulation, 238, 243
multiple-trapping transport, 243
parameter, 233, 240, 244
Distribution:
charge, 69, 74, 80, 81
hopping time, 233, 238
relaxation, 94
trap, 80, 243
Dopant, 214, 305
acceptor, 268, 305
donor, 268, 305
molecular dispersion, 214
DRIC, *see* Radiation induced condivitity, delayed
Drift mobility, 85, 219
measurement, 217
Dye sensitization, 251

Effective mass, 17, 19, 22
Effective temperature, 226
EFS, *see* Excimer forming sites
Electrets, 3, 60, 96, 177, 121, 139
Electrocaloric effect, 113
Electromechanical coupling, 113
Electron:
affinities, 165, 177
energy loss spectroscopy, 32, 33, 39, 44

INDEX

hopping transport, 224
Electronic states, 3, 13, 14
  ab initio calculations, 18, 21, 278, 294
  bandgap, 19, 20, 165, 268, 278, 281, 284, 288, 291, 311
  band structure, 15, 19, 20, 23, 35, 278, 281
  excited states, 161
  ground state, 281, 284
  localized states, 24, 162, 286
  mid gap level, 285, 289, 302
  relaxation energy, 26
Electronic transport, 4, 17, 268, 269, 303, 308, 311, 315
  chemical control, 256
  conductivity, definition, 17
  hopping, 223, 227, 256
  photoconductivity, 36, 42, 215, 300, 303
  radiation induced, 64
  thermally stimulated, 70
  thermoelectric power, 304, 310
Electrostrictive coupling, 128
Emission-limited discharge, 217
Energy gap, *see* Bandgap
Energy-transfer, 3
  applications, 198, 205
  processes, 161
EXAFS, 46
Exchange narrowing, 260
Excimer, 174, 194, 195
  fluorescence, 175, 188, 191, 195
  forming sites, 174, 189, 192, 193, 198, 201
  phosphorescence, 191, 194, 195, 196
  transport traps, 246
Exciplex, 174, 189
Excitations, 25, 161
  deactivation, 178, 179, 180
Exciton, 164
  charge transfer, 166
  diffusion, 167
  Frenkel, 162, 164
  hopping, 163, 166, 167, 188, 192, 195, 197, 199, 204
  singlet, 166, 169
  triplet, 166, 172
  Wannier, 162, 164
Extended tight-binding theory, 19

FEP, *see* Fluoroethylenepropylene
Fermi level, 278

Ferroelectricity, 124
Fluctuation formalism, 121
Fluorescence, *see* Optical properties
Fluoroethylenepropylene:
  charge decay, 83, 84, 87
  charge density, 80
  charge storage, 60, 64
  electron mobility, 86
  electron range, 63
  hole mobility, 86
  hole transit, 85
  melting point, 62
  polarization, 91
  radiation induced current, 88
  α-relaxation, 78, 91
  β-relaxation, 91
  p-relaxation, 78
  spatial trap distribution, 80
  surface traps, 79
  thermally stimulated current, 78, 79, 83, 88
  trap level, 79, 89
  volume traps, 79
  see also Poly(terafluorethylene)
Form I, II, III crystal, 96, 122, 131
Förster theory, transfer, 170, 197, 193, 199. 200
Frenkel excitons, 164
  self-trapped, 169

γ-relaxation, 71, 91
Gaussian transport, 233
Generation efficiency, *see* Charge generation
Grain boundary relaxation, 125
Graphite, interband transitions, 34
Ground state energy, 281, 284

Hartree-Fock theory, 20, 21, 278
Heterocharge, 121
Hole hopping transport, 221, 227
  concentration dependence, 223, 229
  coulomb traps, 233
  disorder effects, 233
  field dependence, 225
  morphology effect, 233
  temperature dependence, 223
  thickness dependence, 238
  trap-free, 233
  trap-limited, 230
Homocharge, 121
Hopping time distribution, 238
Hopping transport, 163, 227
  electron, 224

exciton, 163, 188, 192
hole, 221, 227
polaron, 223
Hückel theory, 22, 277, 279, 282
  continuum description, 287
  effective elastic constant, 284
  force constant, 282
  ground state energy, 281, 284
  parameters, 280, 294
  SSH Hamiltonian, 279
Hysteresis curves, 124

Impurities, 229, 233, 293, 294
Inelastic electron scattering, see Electron energy loss spectroscopy
Inelastic neutron scattering, 32
Intermediate crystalline phase, 126
Intermolecular interaction, 25, 216
Intersystem crossing, 180
Intramolecular interaction, 25
Inverse piezoelectric effect, 145
Ionization energies, electron, 165, 177
Isomers, 271
Isomerization, 271, 272, 295, 306
N-Isopropylcarbazole:
  absorption, 221
  chemical structure, 11
  dispersed in Lexan, 216, 220
  energy gap, 165
  Frenkel exciton, 165
  hopping transport, 223, 226
  masterplot, 239
  trapping, 231

Jablonski diagram, 178
Jahn-Teller theorem, 283

Kramers-Kronig relations, 33, 39, 45, 142
KYNAR, see Polycarbonate

Lambert-Beer's law, 179
LCAO theory, 20, 278
LEXAN, see Polycarbonate
Light scattering, 181
Localization parameters:
  clusters, 244
  excited states, 256
  hopping transport, 223
  temperature dependence, 229
Localized states, 24, 162, 216, 220

Magnetic properties, 260, 295
  ESR, 260, 295, 299, 308
  exchange, 260, 295
  motional narrowing, 26, 259, 295
  NMR, 296, 299, 308
  Overhauser effect, 296
  solid-state effect, 296
  spin diffusion, 261
Masterplot, 239
Measurement techniques:
  charge density, 66
  charge depth, 69
  charge distribution, 69
  charge generation efficiency, 217, 246
  charge storage phenomena, 61, 64
  charge transport, 217
  chemical techniques, 186
  drift mobility, 217
  energy transfer, 180
  fluorescence, 181
  inelastic electron scattering, 32
  optical spectroscopy, 32
  piezoelectricity, 145
  photoemission, 29
  phosphorescence, 182
  pulsed spectroscopy, 183
  pyroelectricity, 146
  single-photon counting, 183
  spin resonance, 182
  thermally stimulated current, 70
  transit time, 217
  uv absorption, 181
  see also individual listings
Metallic conductivity, 4
  state, 269, 313
Metal-polymer contact, 47, 126
Mobility, 17, 19, 22, 86, 224, 227, 233
  edge, 78
  hopping, 86, 224, 225, 227, 233
  minimum band mobility, 23
  molecular crystals, 228
  polaron hopping, 233
Molecular doping:
  chemical control of conductivity, 256
  concept, 216
  hopping transport, 220, 227
  magnetic properties, 258
Molecular-ion model, 24, 28
Molecular-orbital theory, 276, 280
Monte-Carlo simulations, 88, 238, 243
Morphology, 233, 273

INDEX

hopping transport, 233
Multiple trapping transport, 243
Mylar, see Poly(ethylene terephthalate)

Naphthalene polymers, energy transfer, 197
NIPCA, see N-Isopropylcarbazole
Non-Gaussian transport, 233, 242
Norrish splitting, 203
Nylon, piezoelectricity, 139

Off-diagonal disorder, 24, 233
Onsager mechanism, 247
On-site energy, diagonal disorder, 25, 233, 245
Optical properties:
  absorption, 42, 44, 47, 178, 181, 221, 253, 268, 281, 299, 300, 302, 303, 315, 316
  dielectric constant, 27, 37, 45, 46, 300
  delayed fluorescence, 191, 193, 195, 196
  energy transfer, 161, 180
  excited states, 164
  fluorescence, 175, 181, 195, 197, 199
  infrared, 271, 302, 310
  intra-band excitation, 34
  photoemission, 30, 31, 36, 38, 41
  properties, 299
  Raman, 271, 272, 306
  reflectance, 46, 299
  spectra, 310
  spectroscopy, 32
  sum rule, 300
  see also Measurement techniques
Oriented amorphous phase, 126
Oxygen adsorption, 49

PACN, see Polyacenapthalene
Pariser-Parr-Pople model, 277, 291
PBLG, see Poly(γ-benzyl-L-glutamate)
PC, see Polycarbonate
PCEVE, see Poly(2-N carbazole ethyl vinyl ether)
PE, see Polyethylene
Peierls theorem, 19, 277, 281, 293
Perrins model, 175, 202
PET, see Poly(ethylene terephthalate)
Photoconductivity, 36, 42, 215. See also Charge generation; Electronic transport

Photoelectronic properties, 215
Photoemission, 29-31, 36, 38, 41
Photogeneration, 246. See also Charge generation
Photosensitization, 251
Piezoelectricity, 3, 111
  applications, 96, 148
  constant, 118, 124, 131-134, 138, 140, 144, 148
  constant of various polymers, 138, 140
  definition, 111
  dimensional effect, 135
  electrostriction effects, 135
  formation, 121. See also Charging techniques
  fluctuation formalism, 121
  general equations, 114
  α-helical backbone, 118
  intrinsic crystal effect, 135
  inverse, 112, 145
  polar crystals, 127
  polymer-ceramic composites, 140
  polypeptides, 117
  polyvinylidene fluoride, 132
  relaxation, 142
  space-charge effects, 126
  spontaneous polarization, 115
  strain gradient, 113
  thermodynamic theory, 128
  uniform strain, 113
Plasmon, 15, 35
Plastic scintillators, 198
PMLG, see Poly(γ-methyl-L-glutamate)
PMMA, see Poly(methyl methacrylate)
Polarization:
  charge, 60, 65, 77, 126
  decay, 72, 95
  dipole, 60, 65, 72-75, 90-95, 123
  distribution, 94
  energy, 26, 28, 30, 52
  formation, 121. See also Charging techniques
  permanent, 65, 112
  polar crystals, 128, 130
  poling field dependence, 132
  residual, 65. 112
  spontaneous, 115, 122, 125, 128, 130-133
  strain dependence of spontaneous polarization, 135
Polaron, 233, 295

Poling parameters, 90, 121, 125
Poly(acenaphthalene):
  energy transfer, 198
  repeat unit, 6
Polyacenes, optical absorption, 315, 316
Polyacetylene:
  adiabatic potential, 283
  applications, 317
  bond alternation, 19, 285
  bond length, 19
  correlation effects, 20, 291
  elastic parameters, 282
  electrical properties, 303
    activation energy, 304, 309
    charged defect pair, 293
    conductivities, dopants, 306
      309, 311
    effective mass, 19
    electronic parameters, 282
    mobility, 19, 304
    soliton, 287, 289
    thermoelectric power, 272
    $T^{-1/4}$ law, 303, 315
    $T^{-1/2}$ law, 304
  isomers, 272
  magnetic properties, 295
  morphology, 274
  one-dimensional model, 16
  preparation, 270
  repeat unit, 5
  soliton concept, 287
  spectroscopic properties, 299
    absorption, 268, 281, 299,
      300-303
    electronic gap, 284
    energy band, 17, 19
    ESR, 296, 298
    inelastic electron scattering, 39
    NMR, 297, 307
    photoemission, 38
    reflectance, 300
  structure, 270, 274, 276, 279
  synthesis, 270
  theory, 276
  transition, 289, 305, 308, 312
Poly(N-acryloylcarbazole), energy transfer, 195
Poly(γ-benzyl-L-glutamate):
  Cole-Cole plot, 144
  piezoelectric constant, 140, 144
  piezoelectricity, 118
Poly(2-N carbazolylethyl vinyl ether):
  energy transfer, 194
  excimer forming sites, 194
  repeat unit, 9, 194
Polycarbonate:
  doped:
    charge generation, 246
    charge transport, 220
    magnetic properties, 258
    photosensitization, 251
  dipole polarization, 91, 93
  molecularly doped, 215, 220
  piezoelectric constant, 140
  β-relaxation, 91, 93
  repeat unit, 9
Polydiacetylene:
  band structure, 20, 40
  electrical conductivity, 269
  mobility, 23
  optical absorption, 42
  photoconductivity, 42
  photoemission, 41
  structure, 21
Polyenes, optical absorption, 315, 316
Polyethylene:
  band structure, 22, 35
  charge injection, 62
  dielectric response, 37
  effective mass, 22
  mobility, 22, 23
  photoconductivity, 36
  photoemission, 36
  radiation induced current, 88
  β-relaxation, 91
  γ-relaxation, 91
  repeat unit, 5
  thermally stimulated current, 91
  trap depth, 79
Poly(ethylene terephthalate):
  bulk states, 84
  charge decay, 83
  charge density, 80
  charge distribution, 82
  charge storage, 60
  dipole relaxation, 95
  melting point, 62
  polarization, 91
  radiation induced current, 88
  α-relaxation, 91
  β-relaxation, 91
  repeat unit, 8
  surface states, 84
  trap level, 79
Polymer-ceramic composites, 140
  permittivity, 141
  piezoelectric constant, 141

INDEX

Poly(γ-methyl-L-glutamate):
  Cole-Cole plot, 145
  piezoelectric constant, 118, 140
  piezoelectricity, 118
  repeat unit, 10
Poly(methyl methacrylate):
  bulk state density, 51
  contact charge exchange, 51, 62
  isothermal decay of polarization, 96
  metal-polymer contact, 51
  optical absorption, 43
  poling temperature, 93
  α-relaxation, 91
  β-relaxation, 91, 95
  repeat unit, 7
  thermally stimulated current, 91
Poly(methyl vinyl ketone), energy transfer, 204
Polypeptides:
  fluctuation formalism, 121
  α-helical backbone, 120
  piezoelectricity, 119
Poly(p-phenylene), optical absorption, 315, 316
Poly(p-phenylene sulfide):
  electrical conductivity, 315
  optical absorption, 315
  thermoelectric power, 315
Poly(p-phenylene vinylene), optical absorption, 315, 316
Polypropylene:
  charge decay, 83
  repeat unit, 5
Polypropyleneoxide, piezoelectricity, 118
Polypyrole:
  optical absorption, 315, 316
  thermoelectric power, 315
Polystyrene:
  band width, 23
  contact charge exchange, 50, 62
  electron affinity, 177
  electron energy loss, 44, 47
  energy transfer, 198, 200
  EXAFS, 46
  fluorescence, 199
  ionization energy, 177
  metal-polymer contact, 50
  polarization energy, 25
  repeat unit, 5, 198
Poly(tetrafluoroethylene):
  charge decay, 83
  charge storage, 60
  electron range, 63
  repeat unit, 5
  trapping level, activation energy, 79
  see also Fluoroethylenepropylene
Poly(vinylbenzophenone):
  energy transfer, 204
  repeat unit, 7, 204
Poly(N-vinylcarbazole):
  absorption, 221
  charge generation, 249
  charge transfer complex, 248
    excitation energy dependence, 249
    field dependence, 250
    generation process, 247, 251
    Onsager mechanism, 247
    sensitization, 248, 249
    xerographic gain, 247
  charge transport, 220, 224, 227, 236, 242, 244
    activation energy, 226, 233
    brominated carbazole, 236, 241
    dispersion, 236, 245
    hopping, 223
    localization, 236, 245
    polaron, 233
    trap-free mobility, 233
  energy gap, 165
  energy transfer, 187, 195
    excimer decay, 189
    excimer forming sites, 189 190-192
    liquid nitrogen temperature, 190
    related polymers, 195
    room temperature, 188
    singlet-triplet annihilation, 193
  fluorescence, 188, 191, 192, 195
  morphology, 233, 241
  phosphorescence, 195
  repeat unit, 8
Poly(N-vinyl-3, 6 dibromocarbazole), energy transfer, 195
Poly(vinyl chloride):
  dipole polarization, 91
  piezoelectric constant, 140
  piezoelectricity, 139
  repeat unit, 5
  β-relaxation, 95
Poly(vinylidene fluoride):
  electrostriction constant, 134
  ferroelectricity, 124
  form I, II, III, 122, 123, 128
  glass transition, 125

INDEX

grain boundary relaxation, 125
hysteresis curve, 124
melting point, 123
oriented amorphous phase,
  125
α-peak, 91
γ-peak, 91
α-phase, *see* Form II
β-phase, *see* Form I
permittivity, 141
piezoelectricity, 90, 111, 121, 128,
  132, 139
  constants, 118, 131-134, 138,
    140, 148
  hysteresis, 124
  polar crystal theory, 127
  rigid dipole model, 130
  theoretical predictions, 131, 133
polarization, 112
  decay, 96
  dipole, 90, 123
  isothermal decay, 96
  non-uniformity, 127
  space-charge, 126
  spontaneous, 122, 132
  switching model, 125
poling parameter, 90, 121, 125
pseudo-crystalline phase, 126
pyroelectricity, 90, 111, 132, 139
  constants, 92, 96, 131-134, 137,
    138, 140, 148
  polar crystal theory, 127
  rigid dipole model, 130
  theoretical predictions, 131, 133
rigid-dipole model, 130
semicrystalline polymer, 134
switching model, 125
Young's modulus, 134
Poly(1-vinylnaphthalene):
  energy transfer, 197
  exciton hopping model, 197
  fluorescence, 197
  repeat unit, 6
Poly(2-vinylnaphthalene):
  energy transfer, 196
  excimer forming sites, 196
  fluorescence, 196
  repeat unit, 6, 196
Poly(vinyl phenylketone):
  copolymerized, 203
  energy transfer, 202
  repeat unit, 6, 202
Poly(N-vinylphthalamide):
  energy transfer, 205

repeat unit, 7, 205
Poly(2-vinylpyridene):
  bulk state density, 51
  contact charge exchange, 51
  dielectric response, 27, 37, 45
  metal-polymer contact, 51
  photoemission, 29, 30
  polarization energy, 26
  reflecativity, 46
  repeat unit, 26, 51
Poole-Frenkel effect, 232
PP, *see* Polypropylene
PPP, *see* Poly(*p*-phenylene)
PPS, *see* Poly(*p*-phenylene sulfide)
PPV, *see* Poly(*p*-phenylene vinylene)
PS, *see* Polystyrene
PTFE, *see* Poly(tetrafluoroethylene)
PVB, *see* Poly(vinyl benzophenone)
PVC, *see* Poly(vinyl chloride)
PVCA, *see* Poly(*N*-vinylcarbazole)
2Br-PVCA, *see* Brominated poly
  (*N*-vinylcarbazole)
3Br-PVCA, *see* Brominated poly
  (*N*-vinylcarbazole)
PVDF, *see* Poly(vinylidene fluoride)
PVF_2, *see* Poly(vinylidene fluoride)
PVK, *see* Poly(*N*-vinylcarbazole)
P1VN, *see* Poly(1-vinylnaphthalene)
P2VN, *see* Poly(2-vinylnaphthalene)
PVPI, *see* Poly(*N*-vinylphthalamide)
PVPK, *see* Poly(vinyl phenylketone)
PYR, *see* Polypyrrole
Pyroelectricity, 3, 111
  applications, 96, 148
  constant, 118, 133, 138-141,
    147, 148
  definition, 111
  general equations, 114
  heterogeneity, 117
  intrinsic crystal effect, 137
  polar crystals, 127
  polymer-ceramic composites, 140
  poly(vinylidene fluoride), 92, 96,
    132
  relation to piezoelectricity, 133
  space-charge effect, 126

Quantum fluctuations, 282, 294
Quenching factor, 167, 171, 181

Radiation induced conductivity,
  64, 87
  delayed, 87
Raman chain length, 271

INDEX

Random walk, 167, 174
 continuous time, 237
Real charge storage, 77
Relaxation:
 α, 71, 91, 92
 β, 71, 78, 91, 94, 95
 γ, 71, 91
 ρ, 71, 78
 dipole, 71, 72, 94
 energy, 26
 piezoelectric, 142
Resonance integral, 16
Rhodamine B:
 chemical structure, 12
 photosensitization, 252
 trapping, 252
 xerographic discharge, 254
RIC, see Radiation induced conductivity
Rigid dipole model, 130

Scher-Montroll theory, 233
Semiconductor-metal transition, 308
Selenium, hole transport, 239
Shake up satellites, 31
Single-photon counting, 183
Single step:
 singlet transfer, 168
 triplet transfer, 172
Singlet-singlet annihilation, 166
Singlet-triplet annihilation, 193
60° switching model, 125
Soliton:
 concept, 287
 lattice, 289
 liquid, glass, 312
 properties, 294, 299, 302, 312
Space-charge-limited discharge, 220
Spectroscopy, techniques:
 absorption, 32
 contact charge exchange, 50
 delayed fluorescence, 182
 fluorescence, 181
 inelastic electron scattering, 32
 photoemission, 29
 phosphorescence, 182
 picosecond, 184, 185
 uv absorption, 180
Spontaneous polarization, 122
Stern-Vollmer theory, 167
Surface states, 47, 78, 80
Susceptibility, 295, 297, 311

TCNQ, see Tetracyanoquinodimethane
Teflon, see Poly(tetrafluoroethylene)
Tetracyanoquinodimethane:
 chemical structure, 11
 polymer salts, conductivity, 257
Thermalization radius, 247, 249, 250
Thermally stimulated current, 70
 for box-shaped change distributions, 75, 77
 dipole polarization, 90, 91, 94
 distribution of relaxation frequences, 84
 fractional, 94
 open circuit, 75, 79, 83
 short circuit, 72, 88, 91, 93
 theory, 72, 76
Thiapyrylium dye:
 chemical structure, 11
 doped polycarbonate, 252
 photosensitization, 255
Tight binding theory, 15
 energy bands, 17, 19, 21, 22
 extended tight-binding, 19
 limits, 23
 one-dimensional, 16
 resonance integral, 16
TPA, see Triphenylamine
TPM, see Triphenylmethane
TSC, see Thermally stimulated current
TTA, see Tri-p-tolylamine
Transition:
 commensurate-incommensurate, 289
 semiconductor-metal, 305, 308, 312
Transport:
 dispersion, 234, 236, 237, 245
 effects of morphology, structures, 236, 242
 electron hopping, 85, 224
 hole hopping, 85, 224, 225, 227
 hopping mechanism, 220, 227
 hopping properties, 87, 222, 226, 229, 230, 238
 non-dispersive, 234
 time-of-flight technique, 217
 transit time, 86, 218
 trap-limited, 77, 230
 see also Electrical properties
Trap:
 densities, 77, 80
 distribution, 80, 243
 free hopping, 233
 levels, 77

limited hopping, 230
  surface, 47, 78, 80
  volume, 47, 78, 80
  see also individual charge topics
Triboelectricity, 46
  bulk states, 49
  contact charge exchange, 50
  extrinsic surface state, 47
  metal-polymer contact, 48
  molecular-ion model, 49
  polymer-polymer contact, 48
Triphenylamine:
  chemical structure, 11
  doped polycarbonate, 221
    dispersive transport, 233
    dye sensitization, 251
    hopping transport, 220
    localization raius, 223, 256
    Onsager mechanism, 250
    optical absorption, 221
    photogeneration, 250
    trap-limited hopping, 230
Triphenylmethane, hopping
    transport, 226
Tri-p-tolylamine, chemical structure,
    11
  DC conductivity, 258
  doped polycarbonate, 258
  magnetic properties, 260

Xerographic discharge, 219, 248,
    254